HAESE MATHEMATICS

Specialists in mathematics publishing

Mathematics

for the international student
Mathematics HL (Option): Sets, Relations and Groups

HL Topic 8

FM Topic 4

Catherine Quinn

Chris Sangwin

Robert Haese

Michael Haese

for use with IB Diploma Programme

MATHEMATICS FOR THE INTERNATIONAL STUDENT
Mathematics HL (Option): Sets, Relations and Groups

Catherine Quinn	B.Sc.(Hons), Grad.Dip.Ed., Ph.D.
Chris Sangwin	M.A., M.Sc., Ph.D.
Robert Haese	B.Sc.
Michael Haese	B.Sc.(Hons.), Ph.D.

Haese Mathematics
152 Richmond Road, Marleston, SA 5033, AUSTRALIA
Telephone: +61 8 8210 4666, Fax: +61 8 8354 1238
Email: info@haesemathematics.com.au
Web: www.haesemathematics.com.au

National Library of Australia Card Number & ISBN 978-1-921972-32-4

Published by Haese Mathematics.
152 Richmond Road, Marleston, SA 5033, AUSTRALIA

First Edition 2013
Reprinted 2014

Artwork by Brian Houston.

Cover design by Piotr Poturaj.

Typeset in Australia by Deanne Gallasch. Typeset in Times Roman $10\frac{1}{2}$.

Printed in Singapore by Opus Group.

Acknowledgements: While every attempt has been made to trace and acknowledge copyright, the authors and publishers apologise for any accidental infringement where copyright has proved untraceable. They would be pleased to come to a suitable agreement with the rightful owner.

FOREWORD

Mathematics HL (Option): Sets, Relations and Groups has been written as a companion book to the Mathematics HL (Core) textbook. Together, they aim to provide students and teachers with appropriate coverage of the two-year Mathematics HL Course, to be first examined in 2014.

This book covers all sub-topics set out in Mathematics HL Option Topic 8 and Further Mathematics HL Topic 4, Sets, Relations and Groups.

The aim of this topic is to introduce students to the basic concepts, techniques and main results in abstract algebra, specifically for sets, relations and group theory.

Detailed explanations and key facts are highlighted throughout the text. Each sub-topic contains numerous Worked Examples, highlighting each step necessary to reach the answer for that example.

Theory of Knowledge is a core requirement in the International Baccalaureate Diploma Programme, whereby students are encouraged to think critically and challenge the assumptions of knowledge. Discussion topics for Theory of Knowledge have been included on pages 114 and 124. These aim to help students discover and express their views on knowledge issues.

The accompanying student CD includes a PDF of the full text and access to specially designed software.

Graphics calculator instructions for Casio fx-9860G Plus, Casio fx-CG20, TI-84 Plus and TI-*n*spire are available from icons located throughout the book.

Fully worked solutions are provided at the back of the text, however students are encouraged to attempt each question before referring to the solution.

It is not our intention to define the course. Teachers are encouraged to use other resources. We have developed this book independently of the International Baccalaureate Organization (IBO) in consultation with experienced teachers of IB Mathematics. The Text is not endorsed by the IBO.

In this changing world of mathematics education, we believe that the contextual approach shown in this book, with associated use of technology, will enhance the students understanding, knowledge and appreciation of mathematics and its universal applications.

We welcome your feedback.

Email: *info@haesemathematics.com.au*

Web: *www.haesemathematics.com.au*

CTQ CS
RCH PMH

ACKNOWLEDGEMENTS

The authors and publishers would like to thank all those teachers who offered advice and encouragement on this book.

USING THE INTERACTIVE STUDENT CD

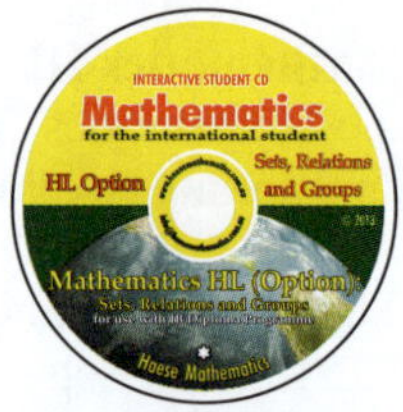

The interactive CD is ideal for independent study.

Students can revisit concepts taught in class and undertake their own revision and practice. The CD also has the text of the book, allowing students to leave the textbook at school and keep the CD at home.

By clicking on the relevant icon, a range of interactive features can be accessed:

- Graphics calculator instructions for the **Casio fx-9860G Plus**, **Casio fx-CG20**, **TI-84 Plus** and the **TI-*n*spire**
- Interactive links to software

INTERACTIVE LINK

GRAPHICS CALCULATOR INSTRUCTIONS

TABLE OF CONTENTS

SYMBOLS AND NOTATION USED IN THIS BOOK		6
A	Sets	9
B	Ordered pairs	20
C	Functions	34
D	Binary Operations	44
E	Groups	55
F	Permutation Groups	68
G	Subgroups	80
H	Cyclic Groups	86
I	Homomorphism	92
J	Isomorphism	96
K	Cosets and Lagrange's Theorem	103
	Review set A	107
	Review set B	108
	Review set C	110
	Review set D	111
THEORY OF KNOWLEDGE (Mathematical Paradox)		114
APPENDIX (Methods of proof)		115
THEORY OF KNOWLEDGE (Axioms and Occam's razor)		124
WORKED SOLUTIONS		125
INDEX		174

SYMBOLS AND NOTATION USED IN THIS BOOK

$\approx$	is approximately equal to
$>$	is greater than
$\geqslant$	is greater than or equal to
$<$	is less than
$\leqslant$	is less than or equal to
$\{......\}$	the set of all elements ......
$\{x_1, x_2, ...\}$	the set with elements $x_1, x_2, ...$
$\{x \mid \ \}$	the set of all x such that
$n(A)$	the number of elements in the finite set A
A'	the complement of the set A
$\in$	is an element of
$\notin$	is not an element of
$\varnothing$	the empty (null) set
$\mathbb{N}$	the set of all natural numbers $\{0, 1, 2, 3, ...\}$
$\mathbb{Z}$	the set of integers $\{0, \pm1, \pm2, \pm3, ...\}$
$\mathbb{Z}^+$	the set of positive integers $\{1, 2, 3, ...\}$
$\mathbb{Q}$	the set of rational numbers
$\mathbb{Q}^+$	the set of positive rational numbers $\{x \mid x \in \mathbb{Q}, \ x > 0\}$
$\mathbb{Q}'$	the set of irrational numbers
$\mathbb{R}$	the set of real numbers
$\mathbb{R}^+$	the set of positive real numbers $\{x \mid x \in \mathbb{R}, \ x > 0\}$
$\mathbb{C}$	the set of complex numbers $\{a + ib \mid a, b \in \mathbb{R}\}$
$\mathbb{U}$	the universal set
$\cup$	union
$\cap$	intersection
$A \backslash B$	the difference of the sets A and B, $A \backslash B = \{x \mid x \in A \text{ and } x \notin B\}$
$A \Delta B$	the symmetric difference of the sets A and B, $A \Delta B = (A \backslash B) \cup (B \backslash A)$
$A \times B$	The Cartesian product of sets A and B, $A \times B = \{(a, b) \mid a \in A, \ b \in B\}$
$\mathbb{Z}_p$	the set of equivalence classes $\{0, 1, 2, ..., p-1\}$ of integers modulo p
$\subseteq$	is a subset of
$\subset$	is a proper subset of
$P(A)$	the power set of the set A
$\Rightarrow$	implies that
$\not\Rightarrow$	does not imply that
$\Leftrightarrow$	if and only if

xRy	x is related to y
$f:\ A \to B$	f is a function under which each element of set A has an image in set B
$f:\ x \mapsto y$	f is a function under which x is mapped to y
$f(x)$	the image of x under the function f
f^{-1}	the inverse function of the function f
$f \circ g$	or $f(g(x))$ the composite function of f and g
$\|x\|$	the modulus or absolute value of x
$[a, b]$	the closed interval $a \leqslant x \leqslant b$
$]a, b[$	the open interval $a < x < b$
$\sum_{i=1}^{n} u_i$	$u_1 + u_2 + u_3 + ... + u_n$
$\prod_{i=1}^{n} u_i$	$u_1 \times u_2 \times u_3 \times .... \times u_n$
$\max\{a, b\}$	the maximum value of a or b
$a \mid b$	a divides b
$\gcd(a, b)$	the greatest common divisor of a and b
$\text{lcm}(a, b)$	the lowest common multiple of a and b
$a \equiv b (\text{mod}\, n)$	a is congruent to b modulo n
$\frac{dy}{dx}$	the derivative of y with respect to x
$f'(x)$	the derivative of $f(x)$ with respect to x
$\frac{d^2y}{dx^2}$	the second derivative of y with respect to x
$f''(x)$	the second derivative of $f(x)$ with respect to x
$\frac{d^ny}{dx^n}$	the nth derivative of y with respect to x
$f^{(n)}(x)$	the nth derivative of $f(x)$ with respect to x
$\int y\, dx$	the indefinite integral of y with respect to x
$\ln x$	the natural logarithm of x
sin, cos, tan	the circular functions
csc, sec, cot	the reciprocal circular functions
arcsin, arccos, arctan	the inverse circular functions
$\text{cis}\ \theta$	$\cos\theta + i \sin\theta$
$n!$	$n \times (n-1) \times (n-2) \times .... \times 3 \times 2 \times 1$
$\binom{n}{r}$	$\frac{n!}{r!(n-r)!}$

$\{G, *\}$	the group with non-empty set G and binary operation $*$
a^{-1}	the inverse of the group element $a \in G$
e	the identity element of a group
$\|G\|$	the order of the group G
$\|g\|$	the order of the group element $g \in G$
$(a_1 a_2 .... a_r)$	the permutation which maps a_1 to a_2, a_2 to a_3, ..., a_r to a_1
S_n	the set of all permutations of 1, 2, 3, ..., n
D_n	the dihedral group of degree n
$H < G$	H is a subgroup of G
$\langle g \rangle$	the cyclic group with generator g
$G \cong H$	G is isomorphic to H
gH	$\{g * h \mid h \in H\}$

A SETS

Many ideas relating to set theory were an essential part of the growth of mathematics. However, it was not until **Georg Cantor** (1845 - 1918) that the study of sets was developed as a formal theory.

A **set** S is a collection of objects, called the **elements** or **members** of the set.

If x is an element of S, we write $x \in S$. If x is not an element of S, we write $x \notin S$.

A set must be **well-defined**, which means that if x is some element or object, then either $x \in S$ or $x \notin S$, and the definition of S will determine which of these two possibilities holds.

For example, if S is the collection of vowels in the English alphabet, then we write $S = \{\text{a, e, i, o, u}\}$. There is no ambiguity about what determines membership of S, so S is well-defined and hence S is a set. We see that $\text{a} \in S$ but $\text{d} \notin S$.

By contrast, consider a collection of the 10 best actors in the world. This collection is not well-defined as 'best' is too subjective and does not clearly define membership. This collection is therefore not a set.

A set is called a **finite set** if it contains a finite number of elements.
A set is called an **infinite set** if it contains an infinite number of elements.

The number of distinct elements in a set S is denoted $n(S)$ or $|S|$. For finite sets, Cantor called $n(S)$ the **cardinality** or **cardinal number** of the set S.

For example, $S = \{\text{a, e, i, o, u}\}$ has cardinality $n(S) = 5$.

For infinite sets, the definition of cardinality is more complex, and is not discussed here.

Where $n(S)$ is small, it is usually easy to list all the elements in the set individually. For larger sets, we do not wish to list every element, so we instead use the 'set-builder' notation $\{x \mid x$ has some specified property$\}$. This notation is read as 'the set containing all elements x, such that x has that property'.

For example, $\{x \mid x$ is an IB student enrolled in Mathematics HL$\}$ describes all IB students studying HL mathematics.

NUMBER SETS

You should already be familiar with the following infinite sets of numbers:

- $\mathbb{N}$ is the set of natural numbers $\{0, 1, 2, ...\}$ (Note that 0 is omitted in some other texts.)
- $\mathbb{Z}$ is the set of integers $\{0, \pm 1, \pm 2, ...\}$
- $\mathbb{Q}$ is the set of rational numbers $\{x \mid x = \frac{p}{q},\ p, q \in \mathbb{Z},\ q \neq 0\}$
- $\mathbb{Q}'$ is the set of irrational numbers, which are numbers that are real but not rational
- $\mathbb{R}$ is the set of real numbers
- $\mathbb{C}$ is the set of complex numbers $\{z \mid z = a + ib,\ a, b \in \mathbb{R}\}$
- $\mathbb{Z}^+$, $\mathbb{Q}^+$, and $\mathbb{R}^+$ denote the positive elements of $\mathbb{Z}$, $\mathbb{Q}$, and $\mathbb{R}$ respectively.
  For example, $\mathbb{Z}^+ = \{1, 2, 3, ...\}$.

Note that the set of real numbers is difficult to describe, but is considered to be well-defined nevertheless. We know a number is real if it can be located on a number line.

Example 1

State whether each of the following is true:

a $3 \notin \mathbb{Q}$ **b** $\sqrt{9} \in \mathbb{Z}$ **c** $\pi \in \mathbb{Q}$
d $-6.9 \in \mathbb{Z}$ **e** $3.\overline{213} \in \mathbb{Q}$ **f** $\sqrt{-11} \in \mathbb{R}$

a False, as 3 can be written as $\frac{3}{1}$ and is therefore a rational number. We write $3 \in \mathbb{Q}$.
b True, as $\sqrt{9} = 3$, and $3 \in \mathbb{Z}$.
c False, as π is an irrational number. We write $\pi \notin \mathbb{Q}$.
d False, as $-6.9 = -6\frac{9}{10}$, which is not an integer. We write $-6.9 \notin \mathbb{Z}$.
e True, as $3.\overline{213} = 3\frac{213}{999} = 3\frac{71}{333} = \frac{1070}{333}$ which is rational. We write $3.\overline{213} \in \mathbb{Q}$.
f False, as $\sqrt{-11}$ is an imaginary number. $\sqrt{-11} \in \mathbb{C}$ but $\sqrt{-11} \notin \mathbb{R}$.

EQUALITY OF SETS

Two sets are **equal** if and only if they contain the same elements. The order in which elements in a set are listed is not important.

For example, for number sets, the set $\{a, b, c\}$ is the same set as $\{b, c, a\}$. The set $\{a, b, b, c\}$ is also equal to the previous two because repetitions of numerical elements are ignored.

Example 2

State whether the following pairs of sets are equal:

a $\{3, 5, 7\}$, $\{5, 7, 3\}$ **b** $\{2, 2, 3, 5\}$, $\{2, 3, 5\}$
c {vowels in the English alphabet}, {a, e, i, o, u}
d {prime numbers between 24 and 28 inclusive}, {prime numbers between 32 and 36 inclusive}
e {integers between -3 and 7 inclusive}, {natural numbers between -3 and 7 inclusive}

a The order of the elements in a set does not matter, so the sets are equal.
b Repetitions can be ignored, so the sets are equal.
c Both sets describe the same letters, so they are equal.
d Both sets are empty, so they are equal.
e The first set is $\{-3, -2, -1, 0, 1, 2, 3, 4, 5, 6, 7\}$ while the second is $\{0, 1, 2, 3, 4, 5, 6, 7\}$.
We see, for example, that -1 belongs to the first set but not the second set
$\therefore$ the sets are not equal.

When it is not feasible to list the elements of each set, we need an alternative method for determining whether or not two sets are equal. This will be discussed later.

THE EMPTY SET AND THE UNIVERSAL SET

There is exactly one set with no elements. It is called the **empty set** or **null set**, and is denoted $\varnothing$.

In any particular situation, the set containing all elements under consideration is called the **universal set**, $\mathbb{U}$. In statistics this is the population, and in probability it is the sample space.

EXERCISE A.1

1 List the elements of each set and state the number of elements in the set:

- **a** $\{a, b, c\}$
- **b** $\{x \mid x \text{ is a prime number less than ten}\}$
- **c** $\{x \mid x \in \mathbb{Z},\ x \in [3, 8[\}$
- **d** $\{x \mid x \in \mathbb{R},\ x^2 = -9\}$
- **e** $\{3, 4, 3, 4\}$
- **f** $\{\varnothing\}$

2 State whether the following sets are finite or infinite. If the set is finite, write down its cardinality.

- **a** $\{x \mid x \in \mathbb{Z},\ 0 < x < 100\}$
- **b** $\{x \mid x \in \mathbb{Q},\ 0 < x < 100\}$

3 State whether each of the following is true:

- **a** $7 \in \mathbb{Z}$
- **b** $\sqrt{13} \in \mathbb{Q}$
- **c** $e \in \mathbb{R}$
- **d** $-3.5 \in \mathbb{Q}$
- **e** $4.\overline{1} \in \mathbb{Z}^+$
- **f** $\sqrt{-2} \in \mathbb{C}$
- **g** $(\sqrt{3})^2 \in \mathbb{Z}$
- **h** $\pi^2 \in \mathbb{R}$

4 Which of the following pairs of sets are equal?

- **a** $\{1, 2, 3, 3\}$ and $\{1, 2, 3\}$
- **b** $\{1, m, n\}$ and $\{m, 1, n\}$
- **c** $\{x \mid x \in \mathbb{Z},\ x^2 = 4\}$ and $\{x \mid x \in \mathbb{R},\ |x| = 2\}$
- **d** {prime numbers of the form $2n,\ n \in \mathbb{N},\ n > 1$} and {negative numbers > 3}
- **e** $\{x \mid x \in \mathbb{R},\ x \in]2, 5[\}$ and $\{x \mid x \in \mathbb{R},\ x \in [2, 5]\}$

SUBSETS

If every element of set B is also an element of set A, then B is a **subset** of A.

In this case, for all $x \in B,\ x \in A$, and we write $B \subseteq A$.

We note that for any set A, $A \subseteq \mathbb{U}$.

The empty set $\varnothing$ is a subset of every set, and every set is a subset of itself.

So, for any set A, $\varnothing \subseteq A$ and $A \subseteq A$.

The subsets $\varnothing$ and A of set A are called the **trivial subsets** of A.

If a subset B of A is such that $B \neq A$ and $B \neq \varnothing$, then B is called a **proper subset** of A.

We write $B \subset A$.

The subsets of the set $\{a, b\}$ are $\varnothing$, $\{a\}$, $\{b\}$, and $\{a, b\}$.

Venn diagrams can be used to illustrate sets. The interior of a rectangle indicates the universal set $\mathbb{U}$, and the interiors of circles are used for other sets. In illustrations of large numbers of sets, other closed figures may be used.

The Venn diagram alongside illustrates $B \subseteq A$.

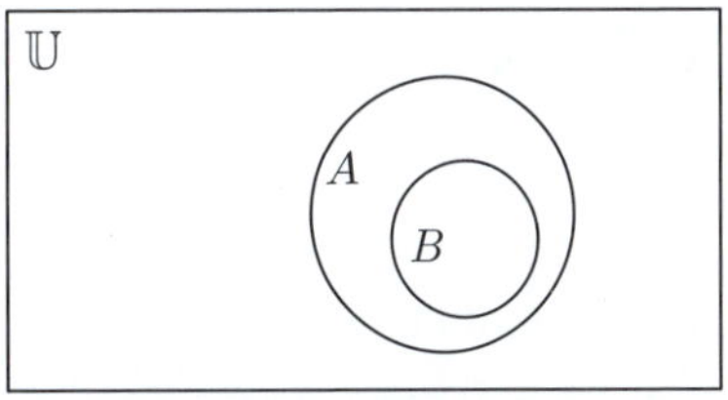

The set of subsets of a set A is called the **power set**, $P(A)$, of A.

The number of subsets of a set with m elements is 2^m. So, if $n(A) = m$, then $n(P(A)) = 2^m$.

Proof:

For every subset of A, there are two possibilities for each element $x \in A$: either x will be in the subset, or it will not.

$\therefore$ for all m elements there will be 2^m different choices in making a subset of A

$\therefore$ the number of subsets of A is 2^m.

Example 3

Find $P(A)$ if $A = \{p, q, r\}$.

Since $n(A) = 3$, there will be $2^3 = 8$ elements in $P(A)$.

$P(A) = \{\varnothing, \{p\}, \{q\}, \{r\}, \{p, q\}, \{p, r\}, \{q, r\}, \{p, q, r\}\}$

Two sets A and B are **equal** if $A \subseteq B$ and $B \subseteq A$.

This result provides us with the method to prove two sets A and B are equal, particularly when A and B are abstract sets for which it is not feasible to list their elements.

To prove two sets A and B are equal, we need to show that $x \in A \Leftrightarrow x \in B$.

For proofs involving an equivalence statement "if and only if" or iff or $\Leftrightarrow$, we need to perform the proof both ways.

So, if we are to prove that statement S_1 is true if and only if statement S_2 is true, then we have to do this both ways:

($\Rightarrow$) start by assuming statement S_1 and prove that statement S_2 is true, **and**

($\Leftarrow$) assume statement S_2 and prove that statement S_1 is true.

To show two sets A and B are equal, we need to show **both**:

($\Rightarrow$) For all $x \in A$, if $x \in A$ then $x \in B$. This establishes that $A \subseteq B$.

($\Leftarrow$) For all $y \in B$, if $y \in B$ then $y \in A$. This establishes that $B \subseteq A$.

Then, since $A \subseteq B$ and $B \subseteq A$, it follows that $A = B$.

EXERCISE A.2

1 Find the power set $P(A)$ for each of the following sets:

a $\{p, q\}$ **b** $\{1, 2, 3\}$ **c** $\{0\}$

2 For each of the following sets, state whether $A \subseteq B$ is true or false:

a $A = \{\text{vowels in the English alphabet}\}$, $B = \{\text{letters in the word 'sequoia'}\}$

b $A = \{0\}$, $B = \varnothing$

c $A = \{3, 5, 9\}$, $B = \{\text{prime numbers}\}$

d $A = \{x \mid x = a + b\sqrt{2},\ a, b \in \mathbb{Z}\}$, $B = \{\text{irrational numbers}\}$

3 Prove that $A = B$:

a $A = \{x \mid x = a + \frac{1}{2},\ a \in \mathbb{Z}\}$, $B = \{x \mid x = b - \frac{1}{2},\ b \in \mathbb{Z}\}$

b $A = \{x \mid x = \sqrt{y},\ y \in \mathbb{R}^+\}$, $B = \mathbb{R}^+$

4 Prove that $A \subseteq B$ but $A \neq B$:

a $A = \{y \mid y = (x+1)^2,\ x \in \mathbb{Z},\ x \text{ is even}\}$, $B = \{\text{odd integers}\}$

b $A = \{x \mid x = yy^*,\ y \in \mathbb{C}\}$, $B = \mathbb{R}$

5 Prove using mathematical induction that $n(P(A)) = 2^{n(A)}$.

ALGEBRA OF SETS

INTERSECTION

The set consisting of the elements common to both set A and set B is called the **intersection** of A and B, written $A \cap B$.

$$A \cap B = \{x \mid x \in A \text{ and } x \in B\}$$

In the Venn diagram, the shaded region is $A \cap B$.

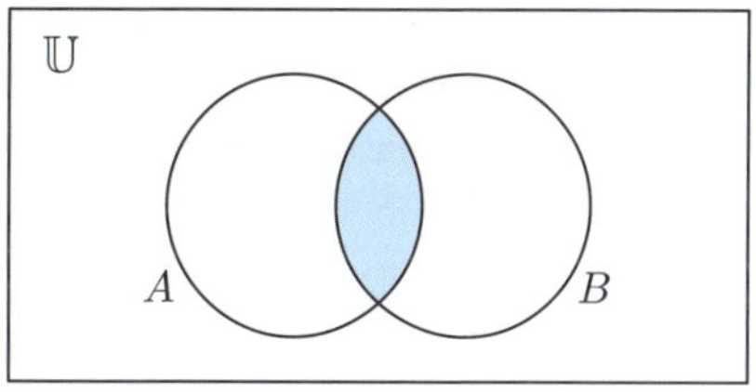

Example 4

Find $A \cap B$ if:

a $A = \{1, 2, 3, 4, 5, 6\}$ and $B = \{3, 5, 7, 9\}$

b $A = \{1, 2, 3, 4, 5, 6\}$ and $B = \{0, 7, 9\}$

a $A \cap B = \{3, 5\}$ **b** $A \cap B = \varnothing$

UNION

The set consisting of all the elements that are found in A *or* B (or both) is called the **union** of A and B, written $A \cup B$.

$$A \cup B = \{x \mid x \in A \text{ or } x \in B\}$$

In the Venn diagram, the shaded region is $A \cup B$.

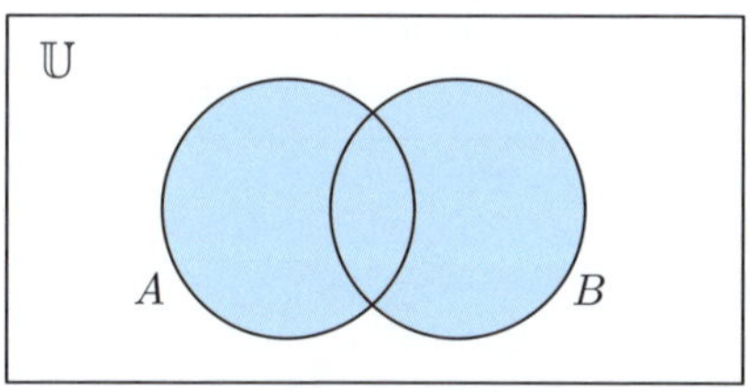

Example 5

Find $A \cup B$ if:

a $A = \{a, b, c, d, e\}$, $B = \{a, e, i, o, u\}$

b $A = \varnothing$, $B = \{1, 2, 3\}$

c $A = \{\text{even integers}\}$, $B = \{\text{odd integers}\}$

d $A = \{\text{prime integers}\}$, $B = \mathbb{N}$

a $A \cup B = \{a, b, c, d, e, i, o, u\}$

b $A \cup B = \{1, 2, 3\} = B$

c $A \cup B = \mathbb{Z}$

d $A \cup B = \mathbb{N}$

LAWS OF INTERSECTION AND UNION

- $A \cap B \subseteq A \cup B$
- If $A \cup B = A \cap B$, then $A = B$
- $A \cup B = A$ if and only if $B \subseteq A$
- $A \cap B = A$ if and only if $A \subseteq B$
- $A \cap A = A$ (*Idempotent Law*)
- $A \cup A = A$ (*Idempotent Law*)
- $A \cap \varnothing = \varnothing$ (*Identity Law*)
- $A \cup \varnothing = A$ (*Identity Law*)
- $A \cup \mathbb{U} = \mathbb{U}$ (*Identity Law*)
- $A \cap \mathbb{U} = A$ (*Identity Law*)

Example 6

Prove that $A \cup B = A$ if and only if $B \subseteq A$.

($\Rightarrow$) Suppose $A \cup B = A$.

If $B = \varnothing$ then we know $B \subseteq A$.

If $B \neq \varnothing$, then let $x \in B$

$\therefore\ x \in A \cup B$

$\therefore\ x \in A$ since $A \cup B = A$.

So, if $x \in B$ then $x \in A$. $\therefore$ $B \subseteq A$.

($\Leftarrow$) Now suppose $B \subseteq A$.

If $B = \varnothing$ then $B \cup A = \varnothing \cup A = A$.

If $B \neq \varnothing$ then if $x \in B$, $x \in A$. $\therefore$ $\{x\} \cup A = A$.

But this is true for all $x \in B$, so $B \cup A = A \cup B = A$.

Therefore $A \cup B = A$ if and only if $B \subseteq A$.

DISJOINT SETS

If $A \cap B = \varnothing$, we say that A and B are **disjoint**. A and B contain no common elements.

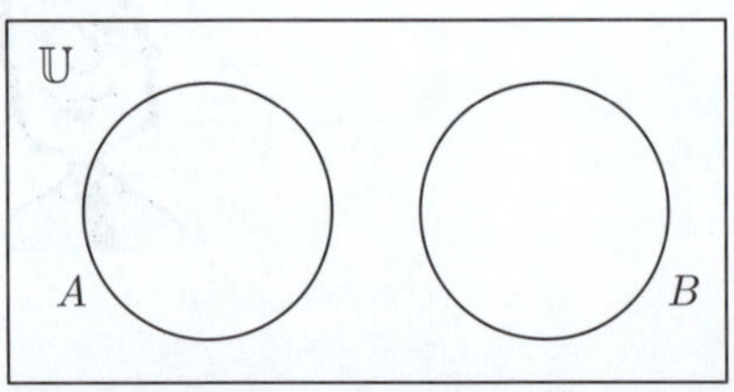

If $A \cap B = \varnothing$ and $A \cup B = \mathbb{U}$ we say that A and B **partition** $\mathbb{U}$.

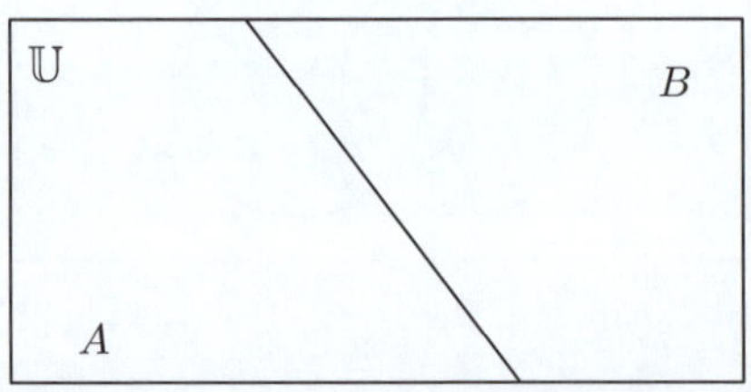

COMPLEMENT

The **complement** of A, written A', contains all elements of $\mathbb{U}$ which are not in A.

A' is sometimes called the **absolute complement**.

The shaded region in the diagram represents A'.

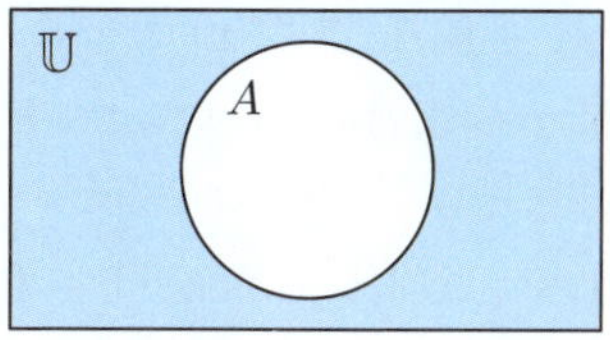

Notice that $A \cap A' = \varnothing$ and $A \cup A' = \mathbb{U}$, so A and A' partition $\mathbb{U}$.

EXERCISE A.3

1 $A = \{1, 3, 5, 7\}$, $B = \{0, 1, 2, 3, 4\}$, $C = \{6, 7, 8\}$, $\mathbb{U} = \{n \mid n \leqslant 9, n \in \mathbb{N}\}$
Find each of the following:

a $A \cup B$ **b** $A \cap C$ **c** $B \cap C$ **d** $A \cap (B \cup C)$
e $(A \cap B) \cup (A \cap C)$ **f** B' **g** $(A \cup B)'$ **h** $A' \cap B'$

2 Assuming A and B are non-empty sets, draw separate Venn diagrams to illustrate the following cases:

a $A \cap B = \varnothing$ **b** $A \cup B = A$ **c** $A \cap B' = A$ **d** $A \cup B = \varnothing$
e $A \cap B' = \varnothing$ **f** $A \cup B = A \cap B$ **g** $A \cup B = A \cap B'$

3 Consider the Venn diagram shown.
Show that if A and B are non-empty sets, then $A \cup B \neq A' \cap B$.

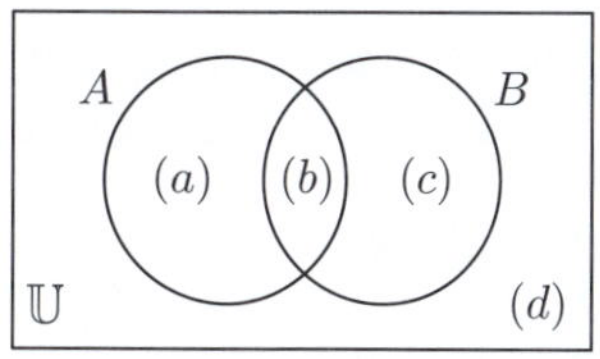

4 Prove that $A \cap B = A$ if and only if $A \subseteq B$.

5 Prove that if A and B are disjoint and $A \cup B = \mathbb{U}$, then $B = A'$.

6 **a** Prove that $n(A \cup B) = n(A) + n(B) - n(A \cap B)$

b In a class of 30 students, 16 play tennis and 15 play basketball. 6 students play neither of these games. How many students play both tennis and basketball?

7 Prove the **transitive property** of set inclusion: If $A \subseteq B$ and $B \subseteq C$, then $A \subseteq C$.

ASSOCIATIVE AND DISTRIBUTIVE PROPERTIES

Both union of sets and intersection of sets are **associative** operations:

$$(A \cup B) \cup C = A \cup (B \cup C) \quad \text{and} \quad (A \cap B) \cap C = A \cap (B \cap C)$$

The union of sets is also **distributive** over intersection, and intersection is **distributive** over union:

$$A \cup (B \cap C) = (A \cup B) \cap (A \cup C) \quad \text{and} \quad A \cap (B \cup C) = (A \cap B) \cup (A \cap C)$$

These laws can be easily illustrated using Venn diagrams. A formal proof for the first of the distributive laws is given in the following **Example**.

Example 7

For all sets A and B, prove that $A \cup (B \cap C) = (A \cup B) \cap (A \cup C)$.

$(\Rightarrow)$ Let $x \in A \cup (B \cap C)$.

$\therefore\ x \in A$ or $x \in B \cap C$

If $x \in A$, then $x \in A \cup B$ and $x \in A \cup C$

$\therefore\ x \in (A \cup B) \cap (A \cup C)$

If $x \in B \cap C$, then $x \in B$ and $x \in C$

$\therefore\ x \in A \cup B$ and $x \in A \cup C$

$\therefore\ x \in (A \cup B) \cap (A \cup C)$

This establishes that $A \cup (B \cap C) \subseteq (A \cup B) \cap (A \cup C)$ .... (1)

$(\Leftarrow)$ Now let $x \in (A \cup B) \cap (A \cup C)$.

$\therefore\ x \in A \cup B$ and $x \in A \cup C$

If $x \in A$, then $x \in A \cup (B \cap C)$

If $x \notin A$, then $x \in B$ and $x \in C$

$\therefore\ x \in B \cap C$

$\therefore\ x \in A \cup (B \cap C)$

This establishes that $(A \cup B) \cap (A \cup C) \subseteq A \cup (B \cap C)$ .... (2)

Together, (1) and (2) give: $A \cup (B \cap C) = (A \cup B) \cap (A \cup C)$

DE MORGAN'S LAWS

Two important laws in set algebra are known as **De Morgan's Laws**. These are:

$$(A \cup B)' = A' \cap B' \quad \text{and} \quad (A \cap B)' = A' \cup B'$$

Example 8

Prove that $(A \cup B)' = A' \cap B'$

$(\Rightarrow)$ If $x \in (A \cup B)'$, then $x \notin (A \cup B)$

$\therefore\ x \notin A$ and $x \notin B$

$\therefore\ x \in A'$ and $x \in B'$

$\therefore\ x \in A' \cap B'$

This establishes that $(A \cup B)' \subseteq A' \cap B'$ .... (1)

$(\Leftarrow)$ If $x \in A' \cap B'$, then $x \in A'$ and $x \in B'$

$\therefore\ x \notin A$ and $x \notin B$

$\therefore\ x \notin (A \cup B)$

$\therefore\ x \in (A \cup B)'$

This establishes that $A' \cap B' \subseteq (A \cup B)'$ .... (2)

Together, (1) and (2) give: $(A \cup B)' = A' \cap B'$

A summary of the laws of the algebra of sets is given below:

Idempotent Laws	$A \cup A = A$	$A \cap A = A$
Associative Laws	$(A \cup B) \cup C = A \cup (B \cup C)$	$(A \cap B) \cap C = A \cap (B \cap C)$
Commutative Laws	$A \cup B = B \cup A$	$A \cap B = B \cap A$
Distributive Laws	$A \cup (B \cap C) = (A \cup B) \cap (A \cup C)$	$A \cap (B \cup C) = (A \cap B) \cup (A \cap C)$
Identity Laws	$A \cup \varnothing = A, \quad A \cup \mathbb{U} = \mathbb{U}$	$A \cap \mathbb{U} = A, \quad A \cap \varnothing = \varnothing$
Complement Laws	$A \cup A' = \mathbb{U}, \quad (A')' = A$	$A \cap A' = \varnothing, \quad \mathbb{U}' = \varnothing, \quad \varnothing' = \mathbb{U}$
De Morgan's Laws	$(A \cup B)' = A' \cap B'$	$(A \cap B)' = A' \cup B'$

EXERCISE A.4

1 Suppose $P = \{o, n, u, a\}$, $M = \{c, n, a, e\}$, and the universal set is $\mathbb{U} = \{$letters in the word "conjugate"$\}$. Find:

- **a** $P \cup M$
- **b** $P \cap M$
- **c** P'
- **d** $P' \cup M'$
- **e** $(P \cap M)'$
- **f** $P \cap (M \cup P)$

2 For the Venn diagram shown, shade the region corresponding to:

- **a** $A \cup B'$
- **b** $A' \cap B$
- **c** $(A \cup B)'$
- **d** $(A \cap B)'$
- **e** $(A' \cap B')'$

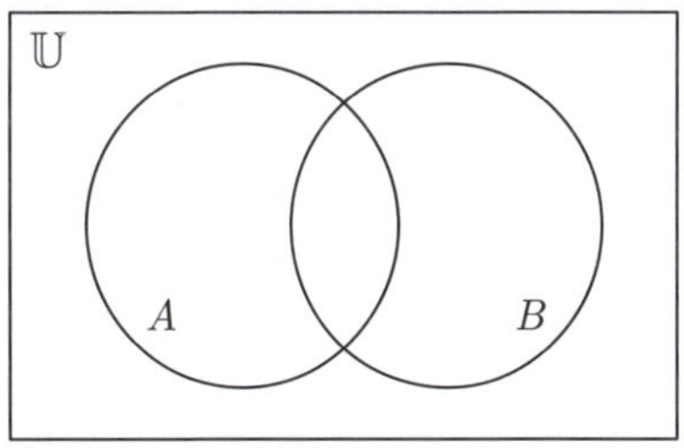

3 For all sets A and B, prove that:

- **a** $(A \cup B) \cup C = A \cup (B \cup C)$
- **b** $A \cap (B \cup C) = (A \cap B) \cup (A \cap C)$
- **c** $(A \cup B) \cap (A' \cup B) = B$
- **d** $(A \cap B)' = A' \cup B'$

DIFFERENCE

The **difference** between two sets A and B, sometimes called the **relative complement**, is defined to be

$$A \backslash B = \{x \mid x \in A \text{ and } x \notin B\}.$$

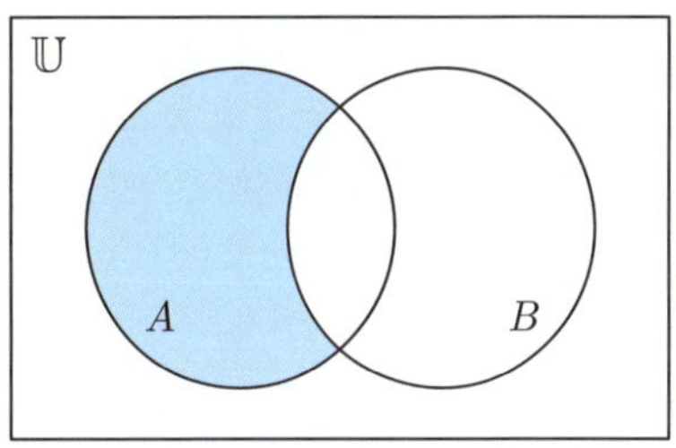

$A \backslash B$ consists of all those elements which are found in A but not in B, so

$$A \backslash B = A \cap B'$$

Set difference is *not* a commutative operation, which means that in general, $A \backslash B \neq B \backslash A$.

Example 9

For each pair of sets A and B, find: **i** $A\backslash B$ **ii** $B\backslash A$

a $A = \{1, 2, 3\}$, $B = \{4, 5\}$
b $A = \{a, b, c, d\}$, $B = \{b, d, e, f\}$
c $A = \{1, 2, 3, 4, 5\}$, $B = \{2, 4\}$

a **i** $A\backslash B = \{1, 2, 3\} = A$ **ii** $B\backslash A = \{4, 5\} = B$
b **i** $A\backslash B = \{a, c\}$ **ii** $B\backslash A = \{e, f\}$
c **i** $A\backslash B = \{1, 3, 5\}$ **ii** $B\backslash A = \varnothing$

SYMMETRIC DIFFERENCE

The **symmetric difference** between two sets A and B is defined by

$$A\Delta B = (A\backslash B) \cup (B\backslash A).$$

The symmetric difference of sets A and B is the set made up of all the elements which are in A or B but not both.

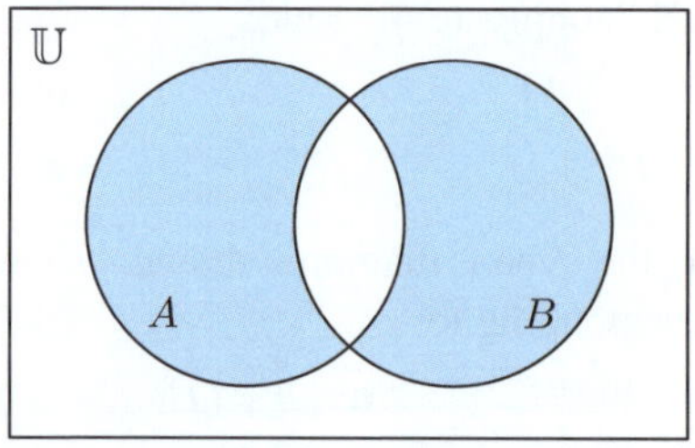

The symmetric difference has the following properties:

- $A\Delta B = B\Delta A$ **Commutative property**
- $A\Delta(B\Delta C) = (A\Delta B)\Delta C$ **Associative property**
- $A\Delta\varnothing = A$
- $A\Delta A = \varnothing$
- $A\Delta A' = \mathbb{U}$

Example 10

Find $A\Delta B$ for:

a $A = \{1, 2, 3\}$, $B = \{4, 5\}$
b $A = \{a, b, c, d\}$, $B = \{b, d, e, f\}$
c $A = \{1, 2, 3, 4, 5\}$, $B = \{2, 4\}$

a $A\Delta B = \{1, 2, 3, 4, 5\}$ **b** $A\Delta B = \{a, c, e, f\}$
c $A\Delta B = \{1, 3, 5\}$

EXERCISE A.5

1 For each pair of sets S and T, find: **i** $S\backslash T$ **ii** $T\backslash S$

a $S = \{1, 2, 3, 4\}$, $T = \{1, 3\}$ **b** $S = \mathbb{R}$, $T = \mathbb{Q}$
c $S = \{0, 1, 2, 3\}$, $T = \{2, 3, 4, 5\}$ **d** $S = \{2, 3, 4\}$, $T = \{0, 1, 5\}$

2 Find $A \Delta B$ for:

a $A = \{a, b, c, d, e\}$, $B = \{a, e\}$

b $A = \{1, 2, 3, 4\}$, $B = \{3, 4, 5\}$

c $A = \{2, 4, 6\}$, $B = \{1, 3, 5\}$

d $A = \{9, 11, 13\}$, $B = \varnothing$

3 For each of the Venn diagrams below, shade the region corresponding to:

i $A \cup B$ **ii** $A \cap B$ **iii** $A \backslash B$ **iv** $A \Delta B$

a

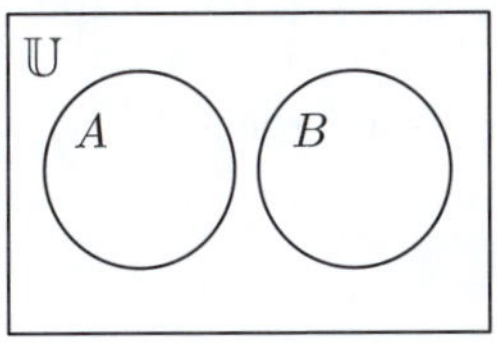

b

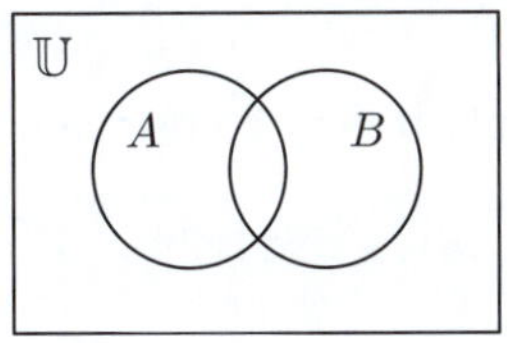

c

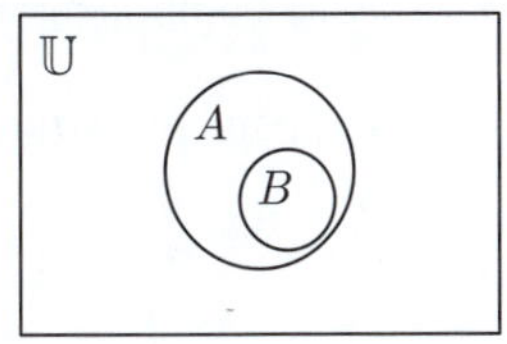

4 For the Venn diagram shown, shade the region corresponding to:

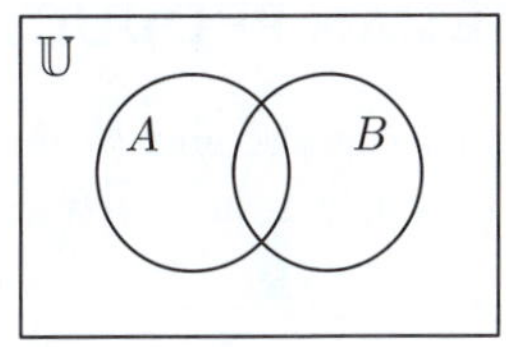

a $B \cup A'$

b $(A \cup B) \setminus (A \cap B)$

c $A \cap (B \cup A')$

d $A' \Delta B'$

5 Prove that $A \Delta B = A \cup B$ if and only if $A \cap B = \varnothing$.

6 For all sets A, B, and C, prove that $A \cap (B \backslash C) = (A \cap B) \setminus (A \cap C)$.

7 For all sets A and B, prove that $A \Delta B = A' \Delta B'$.

B ORDERED PAIRS

We are familiar with the concept of an ordered pair from locating points in the Cartesian plane. However, an ordered pair need not have numbers as elements.

An **ordered pair** (a, b) is defined to contain two components or coordinates: a first component a, and a second component b.

Two ordered pairs are **equal** if and only if their corresponding components are equal.

$$(a, b) = (c, d) \text{ if and only if } a = c \text{ and } b = d.$$

Thus $(a, b) = (b, a)$ if and only if $a = b$.

CARTESIAN PRODUCT

Given two sets A and B, the set which contains all the ordered pairs (a, b) such that $a \in A$ and $b \in B$ is called the **Cartesian product** of A and B, written $A \times B$.

$$A \times B = \{(a, b) \mid a \in A, \ b \in B\}$$

For example, $\{1, 2, 3\} \times \{5, 6\} = \{(1, 5), (1, 6), (2, 5), (2, 6), (3, 5), (3, 6)\}$.

The **Cartesian plane** is $\mathbb{R} \times \mathbb{R}$, sometimes written $\mathbb{R}^2$.

In general, $A \times B \neq B \times A$. The exceptions are when $A = B$, or when either A or B is the empty set, in which case $A \times B$ and $B \times A$ both equal the empty set.

The number of elements in $A \times B$ is found by multiplying the number of elements in each of A and B:

$$n(A \times B) = n(A) \times n(B)$$

Example 11

Prove that the Cartesian product is distributive over set intersection:

$$A \times (B \cap C) = (A \times B) \cap (A \times C)$$

($\Rightarrow$) Let $(x, y) \in A \times (B \cap C)$
$\therefore$ $x \in A$ and $y \in B \cap C$
$\therefore$ $x \in A$, $y \in B$, and $y \in C$
$\therefore$ $(x, y) \in A \times B$ and $(x, y) \in A \times C$
$\therefore$ $(x, y) \in (A \times B) \cap (A \times C)$
$\therefore$ $A \times (B \cap C) \subseteq (A \times B) \cap (A \times C)$ .... (1)

($\Leftarrow$) Let $(x, y) \in (A \times B) \cap (A \times C)$
$\therefore$ $(x, y) \in A \times B$ and $(x, y) \in A \times C$
$\therefore$ $x \in A$, $y \in B$, and $y \in C$
$\therefore$ $x \in A$ and $y \in B \cap C$
$\therefore$ $(x, y) \in A \times (B \cap C)$
$\therefore$ $(A \times B) \cap (A \times C) \subseteq A \times (B \cap C)$ .... (2)

From (1) and (2), $A \times (B \cap C) = (A \times B) \cap (A \times C)$

EXERCISE B.1

1 For each pair of sets A and B, find: **i** $A \times B$ **ii** $B \times A$

a $A = \{1, 2\}$ and $B = \{3, 4, 5\}$ **b** $A = \{a\}$ and $B = \{a, b\}$

c $A = \{1, 2, 3\}$ and $B = \varnothing$

2 For each pair of sets A and B, graph $A \times B$ on the Cartesian plane.

a $A = \{-2, 0, 2\}$, $B = \{-1, 0, 1\}$

b $A = \{x \mid 2 \leqslant x < 5,\ x \in \mathbb{R}\}$, $B = \{x \mid -1 \leqslant x < 4,\ x \in \mathbb{R}\}$

3 Prove that the Cartesian product is distributive over set union: $A \times (B \cup C) = (A \times B) \cup (A \times C)$.

RELATIONS

A **relation** is any set of ordered pairs.

Any subset of the Cartesian product of two sets A and B is a relation.

If R is a relation and $(x, y) \in R$, then we sometimes write xRy.

If $R \subseteq A \times B$, then R is said to be "a relation from A to B".

If $R = X \times Y$, then X is called the **domain** of R and Y is called the **range** of R.

The **domain** consists of all possible first components of the ordered pairs of the relation.

The **range** contains all possible second components of the ordered pairs of the relation.

If R is a relation from A to B then the domain of R is a subset of A, and the range of R is a subset of B.

If $R \subseteq A \times A$, we say that R is "a relation in A".

The following are examples of relations:

$R = \{(1, 3), (2, 4), (3, 1), (3, 4)\}$ is a relation in $\mathbb{N}$ or $\mathbb{Z}^+$

$R = \{(1, 2.5), (2, 3.7), (4, 2), (3, 7.3)\}$ is a relation from $\mathbb{N}$ to $\mathbb{Q}$

$R = \{(x, y) \mid x^2 + y^2 = 9,\ x, y \in \mathbb{R}\}$ is a relation in $\mathbb{R}$

$R = \{(x, (y, z)) \mid y^2 + z^2 = x^2,\ x, y, z \in \mathbb{R}\}$ is a relation from $\mathbb{R}$ to $\mathbb{R}^2$.

REFLEXIVE RELATIONS

A relation R in a set S is said to be **reflexive** if, for all $a \in S$, aRa.

R is a reflexive relation on the set $\{1, 2, 3, 4\}$ if and only if $\{(1, 1), (2, 2), (3, 3), (4, 4)\} \subseteq R$

Example 12

Which of the following relations are reflexive?

a The relation R in a set of school students, where xRy if and only if x and y attend the same school.

b The relation in children in a family, "is the brother of".

c The relation R in $\mathbb{Z}$, where xRy if and only if $x \leqslant y$.

d The relation R in $\{1, 2, 3\}$, where $R = \{(1, 1), (1, 2), (3, 2), (3, 3)\}$.

e The relation R in $\mathbb{R}$, where xRy if and only if $x = y$.

a The relation is reflexive since a student always goes to the same school as him or herself.

b The relation is not reflexive since you are not your own brother, especially if you are a girl.

c The relation is reflexive since $x \leqslant x$ for all $x \in \mathbb{Z}$.

d The relation is not reflexive since $(2, 2) \notin R$.

e The relation is reflexive by definition.

SYMMETRIC RELATIONS

A relation R in a set S is said to be **symmetric** if, for all $a, b \in S$, aRb implies bRa.

So, a relation R is symmetric if, whenever $(a, b) \in R$, then also $(b, a) \in R$.

Example 13

Which of the following are symmetric relations?

a The relation R in $\{1, 2, 3, 4\}$, where $R = \{(1, 2), (2, 1), (3, 3), (4, 2), (2, 4)\}$

b The relation in a set of people, "is the sibling of".

c The relation in a set of people, "is the brother of".

d The relation in $\mathbb{Z}$, where xRy if and only if $x \leqslant y$.

e The relation in $\mathbb{R}$, where xRy if and only if $x = y$.

a The relation is symmetric.

b The relation is symmetric. In a set of people, not every person will have a sibling. All that is required here is that if a is the brother or sister of b, then b will be the brother or sister of a.

c The relation is not symmetric. For example, Paul may be the brother of Anne, but Anne is not the brother of Paul.

d The relation is not symmetric. For example, $3 \leqslant 7$ and so $(3, 7) \in R$, but $7 \nleqslant 3$ and so $(7, 3) \notin R$.

e The relation is symmetric.

Note that when a relation is not symmetric, we describe it as *non-symmetric* or just *not symmetric.* The term *anti-symmetric* is reserved for a special case of non-symmetric relations.

A relation R is *anti-symmetric* if, for all xRy, it is never true that yRx unless $x = y$.

For example: $\{(1, 2), (2, 1), (3, 2), (2, 3)\}$ is symmetric.
$\{(1, 2), (2, 1), (3, 2)\}$ is non-symmetric but not anti-symmetric.
$\{(1, 2), (2, 3), (3, 3)\}$ is non-symmetric and anti-symmetric.

TRANSITIVE RELATIONS

A relation R in a set S is said to be **transitive** if, for all $a, b, c \in S$, aRc whenever aRb and bRc.

In a transitive relation R, if (a, b) and (b, c) are both elements of R, then so must (a, c). Establishing this can often be a time consuming process. It may be useful to make a list of all possibilities and check each one.

Example 14

Which of the following relations are transitive?

a The relation R in $\{1, 2, 3, 4\}$, where $R = \{(1, 1), (1, 2), (2, 3), (1, 3)\}$.
b The relation in a set of buildings, "is older than".
c The relation in a set of people, "is the father of".
d The relation R in $\mathbb{Z}$, where xRy if and only if $x \leqslant y$.
e The relation in $\mathbb{R}$, where xRy if and only if $x = y$.

a The relation is transitive. For example, since $(1, 2) \in R$ and $(2, 3) \in R$, $(1, 3)$ must be in R, which is true.
b The relation is transitive. If building a is older than building b, and building b is older than building c, then building a is older than building c.
c The relation is not transitive. If a fathers b and b fathers c, then a is the grandfather of c, not the father.
d The relation is transitive. If $a \leqslant b$ and $b \leqslant c$, then $a \leqslant c$.
e The relation is transitive. If $a = b$ and $b = c$, then $a = c$.

EXERCISE B.2

1 State the domain and range of each of the following relations:

a $\{(0, 5), (1, 3), (2, 2)\}$
b $\{(x, y) \mid x^2 + y^2 = 9,\ x \in \mathbb{Z}\}$
c $\{(x, y) \mid y = \sin x,\ x \in \mathbb{R}\}$
d $\{(x, (y, z)) \mid y^2 + z^2 = x^2,\ x, y, z \in \mathbb{Z}^+,\ x \leqslant 10\}$

2 $A = \{2, 3, 4, 5\}$ and $B = \{5, 6, 7, 8\}$. Suppose R is a relation from A to B. Write R as a set of ordered pairs if:

a $xRy \Leftrightarrow x$ is a factor of y
b $xRy \Leftrightarrow y = x + 3$
c $xRy \Leftrightarrow y = 2x$
d $xRy \Leftrightarrow y > 2x$

3 Determine whether each of the following relations is:

i reflexive **ii** symmetric **iii** transitive.

a xRy if y is the brother of x
b xRy if y is older than x
c xRy if x and y live in the same country
d xRy if x and y have the same mother.

4 Let R be a relation in $\mathbb{N}$ defined by xRy if x and y are co-prime. Determine whether R is:

a reflexive **b** symmetric **c** transitive.

x and y are **co-prime** if they share no common factors except 1.

5 Let R be a relation in a family of sets. In each of the following cases, determine whether R is:

i reflexive **ii** symmetric **iii** transitive.

a $ARB \Leftrightarrow A$ and B are disjoint **b** $ARB \Leftrightarrow A \subseteq B$

c $ARB \Leftrightarrow n(A) = n(B)$

EQUIVALENCE RELATIONS

A relation in a set S which is reflexive, symmetric, *and* transitive is said to be an **equivalence relation** in S.

We saw in the previous section that the relation of equality is reflexive, symmetric, and transitive. It is therefore an equivalence relation.

If we graph a relation on the Cartesian plane, then the following apply:

- If R is reflexive, all possible points on the line $y = x$ are included. For example, if $S = \{-2, -1, 0, 1\}$ then $(-2, -2), (-1, -1), (0, 0), (1, 1) \in R$ and therefore these points all appear on the graph.
- If R is symmetric then the graph is symmetric about the line $y = x$.

Note that since an equivalence relation in S is reflexive, $(a, a) \in S$ for all $a \in S$, and so the domain and range are both S.

THE EMPTY RELATION

A relation R in a set is a set of ordered pairs, so any subset of a set of ordered pairs will be a relation. This includes the empty set, which is referred to as the **empty relation**.

For example, if $A = \{1, 2, 3\}$, examples of relations in A are: $R_1 = \{(1, 3), (2, 1), (1, 1)\}$

$R_2 = \{(1, 2)\}$

$R_3 = \varnothing$ which is the empty relation.

For the empty relation $R = \varnothing$ in a non-empty set S, the following are both vacuously true statements:

- for all $a, b \in S$, if aRb then bRa
- for all $a, b, c \in S$, if aRb and bRc then aRc.

In mathematics, a **vacuous truth** is a statement asserting something about all members of an empty class, in this case the empty relation R.

Since there are no $a, b \in S$ such that aRb, the empty relation is symmetric and transitive by default.

Also, since $R = \varnothing$, $(a, a) \notin R$ for all $a \in S$. Hence $R = \varnothing$ is not reflexive unless $S = \varnothing$.

We conclude that:

- The empty relation in a non-empty set is symmetric and transitive but is not reflexive. It is hence not an equivalence relation.
- The empty relation in an empty set is reflexive, symmetric, and transitive, and is therefore an equivalence relation.

Note that the empty relation in a non-empty set S is not the only instance of a relation which is symmetric and transitive but not reflexive.

For example, consider the relation R in $A = \{a, b, c, d\}$, where

$$R = \{(a, a), (a, b), (b, a), (b, b), (a, c), (c, a), (c, c), (c, b), (b, c)\}.$$

Since $(d, d) \notin R$, R is not reflexive. However, we can see that R is both symmetric and transitive.

EQUIVALENCE CLASSES

If a set S is separated into subsets which are disjoint and such that their union is S, then we say S has been **partitioned**. An equivalence relation in S partitions S into subsets which are called **equivalence classes**. This result is proven with **Theorem 1** below.

Examples:

1 Define the relation R in $\mathbb{Z}$ by

$aRb \Leftrightarrow a$ and b have the same remainder on division by 2, where $a, b \in \mathbb{Z}$.

This relation partitions $\mathbb{Z}$ into two equivalence classes: the set of odd integers and the set of even integers.

2 Let P be the set of polygons.

Define the relation R in P by

$aRb \Leftrightarrow a$ and b have the same number of sides, where $a, b \in P$.

R partitions P into an infinite number of equivalence classes: the set of triangles, the set of quadrilaterals, the set of pentagons, and so on.

Theorem 1

If R is an equivalence relation in a set S, then the equivalent classes defined by R are disjoint subsets of S which partition S.

Proof:

Let R be an equivalence relation in a non-empty set S.

For $a \in S$, $S_a = \{b \in S \mid aRb\}$ is the **equivalence class defined by a**. Since R is reflexive, aRa and therefore $a \in S_a$.

Each element of S therefore lies in an equivalence class, and therefore S is the union of all equivalence classes defined by R.

It remains to prove that distinct equivalence classes are disjoint.

Suppose for $a, b \in S$, that S_a and S_b are two equivalence classes which are not disjoint. We prove that $S_a = S_b$.

Let $c \in S_a \cap S_b$, since $S_a \cap S_b \neq \varnothing$.

(continued next page)

($\Rightarrow$) If $x \in S_a$, then aRx and so xRa {symmetric property of an equivalence relation}.

Since $c \in S_a$, aRc.

Now xRa and aRc implies xRc {transitive property of an equivalence relation}.

Since $c \in S_b$, bRc and $\therefore$ cRb {symmetric property}.

We also have xRc and cRb, so xRb {transitive property}
$\therefore$ bRx {symmetric property}

Since bRx, $x \in S_b$.

x was any element of S_a, so $S_a \subseteq S_b$.

($\Leftarrow$) Similarly, we can show $S_b \subseteq S_a$.

Together, $S_a \subseteq S_b$ and $S_b \subseteq S_a$ give $S_a = S_b$.

Thus, if two equivalence classes are not disjoint, then they are identical. Hence distinct equivalence classes are disjoint.

Since S is the union of all equivalence classes defined by R, the set of equivalence classes in S are disjoint subsets of S which partition S.

The number of distinct equivalence classes may range from one in the case $R = S \times S$, to $n(S)$ in the case where each equivalence class contains only one element (for example for an equivalence relation R defined by $aRb \Leftrightarrow a = b$ in a finite set S).

Example 15

Let $A = \{1, 2, 3, 4\}$ and define a relation R by: $xRy \Leftrightarrow x + y$ is even.

a Show that R is an equivalence relation. **b** Find the equivalence classes.

a *Reflexive*: $x + x = 2x$
Now $2x$ is even for all $x \in A$, so xRx for all $x \in A$.

Symmetric: If xRy then $x + y$ is even.
Now $x + y = y + x$ for all $x, y \in A$
$\therefore$ $y + x$ is also even, which means yRx.
So, if xRy, then yRx.

Transitive: Suppose xRy and yRz
$\therefore$ $x + y$ is even and $y + z$ is even.
$\therefore$ $x + y = 2m$ and $y + z = 2n$ where $m, n \in \mathbb{Z}$
$\therefore$ $x + y + y + z = 2m + 2n$
$\therefore$ $x + z = 2m + 2n - 2y$
$\therefore$ $x + z = 2(m + n - y)$
Since $m, n, y \in \mathbb{Z}$, $m + n - y \in \mathbb{Z}$ also
$\therefore$ $x + z$ is even, which means xRz.

So, if xRy and yRz then xRz.
Since R is reflexive, symmetric, and transitive, R is an equivalence relation in A.

b Now $R = \{(1, 1), (1, 3), (3, 3), (3, 1), (2, 2), (2, 4), (4, 4), (4, 2)\}$
The first four ordered pairs contain only the elements 1 and 3 from A, and the remaining four ordered pairs contain only the elements 2 and 4.

So, there are two equivalence classes: $\{1, 3\}$ and $\{2, 4\}$.
R can be graphed on the Cartesian plane as shown.

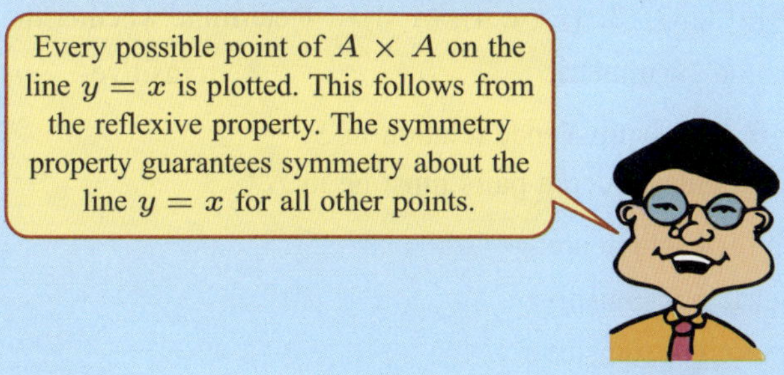

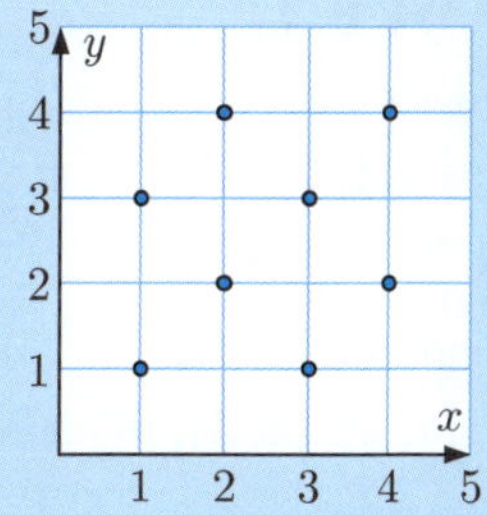

Example 16

Let S be the set of all triangles. Define the relation R such that if $x, y \in S$, then $xRy \Leftrightarrow x$ is similar to y.

Show that R is an equivalence relation, and describe the equivalence classes.

Reflexive: A triangle is similar to itself since, for any triangle ABC, $\frac{\text{AB}}{\text{AB}} = \frac{\text{BC}}{\text{BC}} = \frac{\text{AC}}{\text{AC}}$.
Therefore xRx for all $x \in S$.

Symmetric: If x is similar to y, then its corresponding angles are equal.
$\therefore$ y is also similar to x.
Hence for all $x, y \in S$, if xRy then yRx.

Transitive: Consider triangles $x, y, z \in S$.
If x is similar to y, then the corresponding angles of x and y are equal.
Also, if y is similar to z, the corresponding angles of y and z are equal.
Therefore, the corresponding angles of x and z must also be equal, and so x is similar to z.
$\therefore$ for all $x, y, z \in S$, if xRy and yRz then xRz.

Since R is reflexive, symmetric, and transitive, R is an equivalence relation in S.

The equivalence classes are sets of triangles, each set containing all triangles which are similar to each other.

In this instance there are infinitely many equivalence classes, each with an infinite number of members.

Example 17

Consider the relation R in $\mathbb{R}$, where for all $x, y \in \mathbb{R}$, xRy if $x > y$.

Show that R is not an equivalence relation.

The relation is not reflexive since, for example, 5 is not greater than itself. This is sufficient to establish that R is not an equivalence relation.

However, we note that the relation is also not symmetric since, for example, $7 > 2$, but $2 \not> 7$.

The relation *is* transitive since, if $x > y$ and $y > z$, then $x > z$.

EXERCISE B.3

1 Consider the relation R in $\{1, 2, 3, 4\}$ where $R = \{(1, 1), (1, 2), (2, 2), (2, 3), (3, 3), (3, 4), (4, 4), (4, 3)\}$. Determine whether the relation is:

a reflexive **b** symmetric **c** transitive.

2 Suppose $S = \{1, 2, 3, 4\}$ and R is an equivalence relation in S.
If $(1, 2), (2, 3), (4, 4) \in R$, what other ordered pairs must be in R?

3 If $A = \{a, b, c\}$, find relations in A which are:

- **a** reflexive but neither symmetric nor transitive
- **b** symmetric but neither reflexive nor transitive
- **c** transitive but neither reflexive nor symmetric
- **d** reflexive and symmetric but not transitive
- **e** reflexive and transitive but not symmetric
- **f** symmetric and transitive but not reflexive.

Example 18

R is a relation in $\mathbb{Z} \times \mathbb{Z}$ such that for $(a, b), (x, y) \in \mathbb{Z} \times \mathbb{Z}$, $(a, b)R(x, y)$ if and only if $x + 5y = a + 5b$.

a Show that R is an equivalence relation.

b Describe how R partitions $\mathbb{Z} \times \mathbb{Z}$ and state the equivalence classes.

a *Reflexive*: Letting $a = x$ and $b = y$,
$x + 5y = x + 5y$ which is true for all $(x, y) \in \mathbb{Z} \times \mathbb{Z}$
Hence $(x, y)R(x, y)$ for all $(x, y) \in \mathbb{Z} \times \mathbb{Z}$.

Symmetric: If $(a, b)R(x, y)$ then $x + 5y = a + 5b$
$\therefore\ a + 5b = x + 5y$
$\therefore\ (x, y)R(a, b)$ for all $(a, b), (x, y) \in \mathbb{Z} \times \mathbb{Z}$
Hence if $(a, b)R(x, y)$ then $(x, y)R(a, b)$ for all $(a, b), (x, y) \in \mathbb{Z} \times \mathbb{Z}$.

Transitive: Suppose $(a, b)R(x, y)$ and $(x, y)R(c, d)$
$\therefore\ x + 5y = a + 5b$ and $c + 5d = x + 5y$
$\therefore\ c + 5d = a + 5b$
$\therefore\ (a, b)R(c, d)$ for all $(a, b), (c, d) \in \mathbb{Z} \times \mathbb{Z}$
Hence if $(a, b)R(x, y)$ and $(x, y)R(c, d)$ then $(a, b)R(c, d)$ for all $(a, b), (c, d), (x, y) \in \mathbb{Z} \times \mathbb{Z}$.

Since R is reflexive, symmetric, and transitive, it is an equivalence relation.

b For any $(x, y) \in \mathbb{Z} \times \mathbb{Z}$, we know that $x + 5y$ is some integer $c \in \mathbb{Z}$.
$\therefore$ the relation R partitions $\mathbb{Z} \times \mathbb{Z}$ into an infinite number of equivalence classes, each equivalence class containing the different points (x, y) that result in $x + 5y$ being a particular value. Each class is therefore the set of points with integer coordinates that lie on the line $x + 5y = c$, $c \in \mathbb{Z}$.

For example:
$\{...., (0, 0), (5, -1), (10, -2), ...\}$ is the equivalence class corresponding to $x + 5y = 0$.
$\{...., (1, 0), (6, -1), (11, -2), ...\}$ is the equivalence class corresponding to $x + 5y = 1$.

4 R is a relation in the family of lines in the Euclidean plane such that $xRy \Leftrightarrow x$ and y have the same gradient.

a Show that R is an equivalence relation. b Describe the equivalence classes.

5 Show that R is an equivalence relation in $\mathbb{N}$ if $xRy \Leftrightarrow x - y$ is divisible by 7.

6 Let S be the set of regular polygons. Define the relation R such that if $x, y \in S$, then $xRy \Leftrightarrow x$ is similar to y.

a Show that R is an equivalence relation. b Describe the equivalence classes.

c Explain how the equivalence classes partition S.

7 Consider the relation R in $\mathbb{Z}$, where for all $x, y \in \mathbb{Z}$, xRy if $x \leqslant y$. Show that R is not an equivalence relation.

8 R is a relation in $\mathbb{Z} \times \mathbb{Z}$ such that for $(a, b), (x, y) \in \mathbb{Z} \times \mathbb{Z}$, $(a, b)R(x, y) \Leftrightarrow x = a$.

a Show that R is an equivalence relation.

b Describe how R partitions $\mathbb{Z} \times \mathbb{Z}$, and state the equivalence classes.

9 R is a relation in $\mathbb{R} \times \mathbb{R} \setminus \{(0, 0)\}$ such that for $(a, b), (x, y) \in \mathbb{R} \times \mathbb{R} \setminus \{(0, 0)\}$, $(a, b)R(x, y)$ if and only if $ay = bx$.

a Show that R is an equivalence relation.

b Describe how R partitions $\mathbb{R} \times \mathbb{R} \setminus \{(0, 0)\}$, and state the equivalence classes.

10 R is a relation in $\mathbb{R} \times \mathbb{R}$ such that for $(a, b), (x, y) \in \mathbb{R} \times \mathbb{R}$, $(a, b)R(x, y)$ if and only if $y - b = 3x - 3a$.

a Show that R is an equivalence relation.

b Describe how R partitions $\mathbb{R} \times \mathbb{R}$, and state the equivalence classes.

11 Let R be the relation on $\mathbb{Z}$ defined by $aRb \Leftrightarrow 3ab \geqslant 0$ for $a, b \in \mathbb{Z}$.

a Determine whether R is:

i reflexive ii symmetric iii transitive.

b Is R an equivalence relation? Explain your answer.

12 The relation R on $\mathbb{Z} \times \mathbb{Z}$ is defined by

$$(a, b)R(c, d) \Leftrightarrow a - c \text{ is a multiple of 2 and } b - d \text{ is a multiple of 3}$$

for $(a, b), (c, d) \in \mathbb{Z} \times \mathbb{Z}$.

a Prove that R is an equivalence relation.

b Find explicitly the equivalence class containing: i $(0, 0)$ ii $(1, 3)$

c List the remaining (distinct) equivalence classes.

RESIDUE CLASSES

The multiples of 3 such as 3, 6, and 9, each give remainder 0 on division by 3.

The integers 1, 4, and 7 each give remainder 1 on division by 3.

The integers 2, 5, and 8 each give remainder 2 on division by 3.

In fact, by the **Division Algorithm**, for any integer a, there exist a unique pair of integers q and r such that $a = 3q + r$ and $r \in \{0, 1, 2\}$.

The value r is the **remainder on division of a by 3**.

For example, for negative integers:

$$-6 = 3 \times -2 + 0, \quad \text{so } -6 \text{ has remainder 0 on division by 3}$$
$$-5 = 3 \times -2 + 1, \quad \text{so } -5 \text{ has remainder 1 on division by 3}$$
$$-13 = 3 \times -5 + 2, \quad \text{so } -13 \text{ has remainder 2 on division by 3.}$$

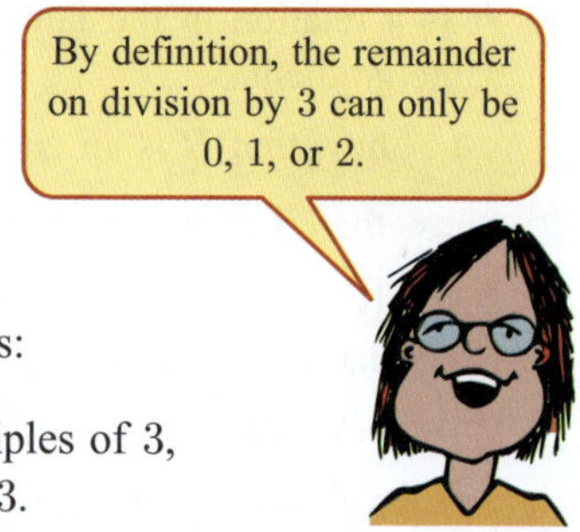

In this way, the set $\mathbb{Z}$ of all integers is partitioned into three disjoint sets:

$[0] = \{...., -12, -9, -6, -3, 0, 3, 6, 9, 12, ...\}$ is the set of multiples of 3, which are the integers which have remainder 0 on division by 3.

$[1] = \{...., -8, -5, -2, 1, 4, 7, 10, ...\}$ is the set of integers which have remainder 1 on division by 3.

$[2] = \{...., -7, -4, -1, 2, 5, 8, 11, ...\}$ is the set of integers which have remainder 2 on division by 3.

The subsets $[0]$, $[1]$, $[2]$ of $\mathbb{Z}$ are the **residue classes modulo 3**.

Notice that any two integers from the same residue class differ by a multiple of 3.

This property generalises as follows:

For any fixed integer $n \in \mathbb{Z}^+$ and any integer $a \in \mathbb{Z}$:

1. The **remainder on division of a by n** will be one of $0, 1, 2, ..., n-1$.
2. The integers $\mathbb{Z}$ will be partitioned by the n disjoint **residue classes modulo n** which are denoted by $[0], [1], [2], ..., [n-1]$.
   - $[0]$ is the set of multiples of n,
   - $[1]$ is the set of integers which have remainder 1 on division by n,
   - ⋮
   - $[n-1]$ is the set of integers which have remainder $n-1$ on division by n.
3. Any two integers in the same residue class modulo n will differ by a multiple of n.

CONGRUENCE

Let $n \in \mathbb{Z}^+$ and let $a, b \in \mathbb{Z}$. We say **a is congruent to b modulo n**, written $a \equiv b \pmod{n}$, if and only if $a - b$ is a multiple of n.

If a is not congruent to b modulo n, we write $a \not\equiv b \pmod{n}$.

For example: Consider the integers 10 and 4.

- Since $10 - 4 = 6$, which is a multiple of 3, $10 \equiv 4 \pmod{3}$.
  Also, since $4 - 10 = -6$, which is a multiple of 3, $4 \equiv 10 \pmod{3}$.
  Hence congruence modulo 3 is symmetric.
- Both 10 and 4 have the same remainder 1 on division by 3, so 10 and 4 both belong to the same residue class modulo 3.

These results can be generalised in the following theorem.

Theorem 2

Suppose $n \in \mathbb{Z}^+$ and $a, b \in \mathbb{Z}$. The following are equivalent statements:

1 $a \equiv b \pmod{n}$

2 $b \equiv a \pmod{n}$

3 $a - b \equiv 0 \pmod{n}$

4 a and b have the same remainder on division by n

5 a and b belong to the same residue class modulo n

6 a and b differ by a multiple of n, which means $a = b + sn$ for some $s \in \mathbb{Z}$.

Proof:

We need to prove that one statement is true if and only if each other statement is true.

1 ⇔ 2

$a \equiv b \pmod{n}$

$\Leftrightarrow$ $a - b = sn$ for some $s \in \mathbb{Z}$ {by definition of congruence modulo n}

$\Leftrightarrow$ $b - a = -sn$, where $-s \in \mathbb{Z}$

$\Leftrightarrow$ $b \equiv a \pmod{n}$

1 ⇔ 3

$a \equiv b \pmod{n}$

$\Leftrightarrow$ $a - b = sn$ for some $s \in \mathbb{Z}$ {by definition of congruence modulo n}

$\Leftrightarrow$ $(a - b) - 0 = -sn$ for some $s \in \mathbb{Z}$

$\Leftrightarrow$ $a - b \equiv 0 \pmod{n}$ {by definition of congruence modulo n}

1 ⇔ 4 Firstly, $a \equiv b \pmod{n}$

$\Leftrightarrow$ $a - b = sn$ for some $s \in \mathbb{Z}$ .... ($*$)

($\Rightarrow$) Suppose a and b have remainders r_1 and r_2 respectively, on division by n.

$\therefore$ $a = nq_1 + r_1$ and $b = nq_2 + r_2$ for some $q_1, q_2 \in \mathbb{Z}$ and such that $r_1, r_2 \in \{0, 1, 2, ..., n-1\}$

$$\therefore \ a - b = (nq_1 + r_1) - (nq_2 + r_2)$$
$$= n(q_1 - q_2) + (r_1 - r_2)$$

From ($*$), $a - b = sn$

$$\therefore \ sn = n(q_1 - q_2) + (r_1 - r_2)$$
$$\therefore \ (r_1 - r_2) = n(s - q_1 + q_2),$$ which is a multiple of n.

But since $r_1, r_2 \in \{0, 1, 2, ..., n-1\}$ the only solution is $r_1 - r_2 = 0$.

Therefore $r_1 = r_2$, and so a and b have the same remainder on division by n.

($\Leftarrow$) Suppose a and b have the same remainder r on division by n.

$\therefore$ $a = nq_1 + r$ and $b = nq_2 + r$ for some $q_1, q_2 \in \mathbb{Z}$ and $r \in \{0, 1, ..., n-1\}$.

$$\therefore \ a - b = n(q_1 - q_2) + r - r$$
$$= n(q_1 - q_2)$$

$\therefore$ $a - b$ is a multiple of n, and hence by the definition of congruence modulo n, $a \equiv b \pmod{n}$.

(continued next page)

4 $\Leftrightarrow$ **5** by definition of residue class modulo n.

5 $\Leftrightarrow$ **6** $a = q_1 n + r$ and $b = q_2 n + r$ for some $q_1, q_2 \in \mathbb{Z}$ and $r \in \{0, 1, ..., n-1\}$

$$\begin{aligned} \Leftrightarrow\ a - b &= (q_1 n + r) - (q_2 n + r) \\ &= n(q_1 - q_2) + r - r \\ &= n(q_1 - q_2) \quad \text{where } (q_1 - q_2) \in \mathbb{Z} \end{aligned}$$

$\Leftrightarrow$ a and b differ by a multiple of n.

For $n \in \mathbb{Z}^+$, it can be proved that congruence modulo n is an equivalence relation in $\mathbb{Z}$ with equivalence classes $[0], [1], ..., [n-1]$, the n residue classes modulo n. We prove the case $n = 5$ in the following example.

Example 19

Suppose R is a relation in $\mathbb{Z}$ such that xRy if and only if $x \equiv y \pmod 5$.
Show that R is an equivalence relation and describe the equivalence classes.

Reflexive: $x - x = 0$ which is a multiple of 5
$\therefore$ by definition, $x \equiv x \pmod 5$ for all $x \in \mathbb{Z}$.

Hence xRx for all $x \in \mathbb{Z}$.

Symmetric: For $x, y \in \mathbb{Z}$, if xRy then $x \equiv y \pmod 5$
$\therefore$ $x - y = 5s$ for some $s \in \mathbb{Z}$
$\therefore$ $y - x = 5(-s)$ where $-s \in \mathbb{Z}$
$\therefore$ $y - x$ is a multiple of 5
$\therefore$ $y \equiv x \pmod 5$ {by definition of congruence modulo 5}
$\therefore$ yRx.

Hence if xRy then yRx for all $x, y \in \mathbb{R}$.

Transitive: For $x, y, z \in \mathbb{Z}$, suppose xRy and yRz.

$\therefore$ $x \equiv y \pmod 5$ and $y \equiv z \pmod 5$
$\therefore$ $x - y = 5s$ and $y - z = 5t$ for some $s, t \in \mathbb{Z}$.

$$\begin{aligned} \therefore\ x - z &= x - y + y - z \\ &= 5s + 5t \\ &= 5(s + t) \quad \text{where } (s + t) \in \mathbb{Z} \end{aligned}$$

$\therefore$ $x - z$ is a multiple of 5
$\therefore$ $x \equiv z \pmod 5$ {by definition of congruence modulo 5}
$\therefore$ xRz.

Hence, if xRy and yRz then xRz for all $x, y, z \in \mathbb{Z}$.

Since R is reflexive, symmetric, and transitive, it is an equivalence relation.

Equivalence classes: If $a \in \mathbb{Z}$, then the other elements of the equivalence class to which a belongs will be $a \pm 5$, $a \pm 10$, $a \pm 15$, ...

There will be 5 such classes corresponding to the 5 possible remainders on division by 5: 0, 1, 2, 3, 4.

The 5 residue classes modulo 5 are:

$[0] = \{...., -10, -5, 0, 5, 10, ...\}$ is the set of multiples of 5.

$[1] = \{...., -9, -4, 1, 6, 11, ...\}$ is the set of integers which leave remainder 1 on division by 5.

$[2] = \{...., -8, -3, 2, 7, 12, ...\}$ is the set of integers which leave remainder 2 on division by 5.

$[3] = \{...., -7, -2, 3, 8, 13, ...\}$ is the set of integers which leave remainder 3 on division by 5.

$[4] = \{...., -6, -1, 4, 9, 14, ...\}$ is the set of integers which leave remainder 4 on division by 5.

In the above example, it can be seen that each integer belongs to one and only one residue class. These sets are therefore pair-wise disjoint and their union is $\mathbb{Z}$.

The set of residue classes modulo 5 is called $\mathbb{Z}_5$ and is written $\{[0], [1], [2], [3], [4]\}$ or just $\{0, 1, 2, 3, 4\}$.

In general,

$\mathbb{Z}_n = \{0, 1, 2, ..., n-2, n-1\}$ is the set of residues modulo n, $n \in \mathbb{Z}^+$.

EXERCISE B.4

1 If $a \equiv b \pmod{n}$ and $c \equiv d \pmod{n}$, prove that:

a $a + c \equiv b + d \pmod{n}$ **b** $ac \equiv bd \pmod{n}$

2 Find the smallest positive integer x that is a solution of the congruence $ax \equiv 1 \pmod{11}$ for each of the values $a = 1, 2, 3, 4, 5, 6, 7, 8, 9, 10$.

3 Suppose R is a relation in $\mathbb{Z}$ such that xRy if and only if $x \equiv y \pmod{n}$, $n \in \mathbb{Z}^+$. Show that R is an equivalence relation and describe the equivalence classes.

4 Determine whether the relation R in $\mathbb{N}$ is an equivalence relation if $xRy \Leftrightarrow x^2 \equiv y^2 \pmod{3}$.

5 **a** If $a \equiv b \pmod{n}$, show that $a^2 \equiv b^2 \pmod{n}$.

b If $a^2 \equiv b^2 \pmod{n}$, show that it is not necessarily true that $a \equiv b \pmod{n}$.

GRAPHICS CALCULATOR INSTRUCTIONS FOR CONGRUENCE

C FUNCTIONS

The work in this section follows on from **Chapter 2** of the Core HL text.

A relation f from set A to set B, is said to be a **function** from A to B if, for each $x \in A$, there is at most one element $y \in B$ such that $(x, y) \in f$.

Functions are sometimes referred to as **mappings**.
The **domain** of the function is A or a subset of A.
B is called the **codomain**. The **range** of f will be a subset of B.

Rather than write $(x, y) \in f$ or xfy, the standard notation used is $y = f(x)$ or $f : x \mapsto y$.

Example 20

The diagrams below map relations from $A = \{1, 2, 3, 4\}$ to $B = \{1, 2, 3, 4\}$.
Determine whether each relation is a function. If it is, state the domain, codomain, and range.

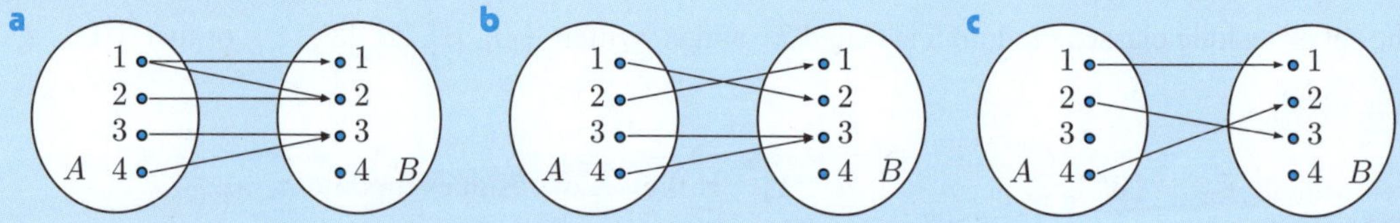

a The element 1 in A is mapped to two elements, 1 and 2, in B.
$\therefore$ the relation is not a function.

b Each element in A is mapped to exactly one element in B, so the relation is a function.
The domain of the function is $\{1, 2, 3, 4\}$, the codomain is also $\{1, 2, 3, 4\}$, and the range is $\{1, 2, 3\}$.

c Each element in A is mapped to at most one element in B, so the relation is a function.
The domain is $\{1, 2, 4\}$, the codomain is $\{1, 2, 3, 4\}$, and the range is $\{1, 2, 3\}$.

Example 21

Determine whether each of the following relations is a function:

a the relation in $\mathbb{N}$, $\{(1, 3), (2, 5), (2, 3), (3, 7)\}$

b the relation in $\mathbb{R}$ defined by $\{(x, y) \mid y > x\}$

c the relation R from $A = \{1, 2, 3, 4\}$ to $B = \{1, 2, 3, 4\}$ where $R = \{(1, 4), (2, 4), (3, 4), (4, 1)\}$

d the relation $f : \mathbb{R} \to \mathbb{R}$ where $f(x) = 2x^2 - 3$.

a The relation is not a function, as 2 is mapped to two different elements, 5 and 3.

b The relation is not a function, as each element x is mapped to an infinite number of elements y.

c The relation is a function since, for each different first component of the ordered pairs, there is only one possible second component.

d The relation is a function since, for each value of x, there is only one value of $2x^2 - 3$.

A test for functions which can be graphed in the Cartesian plane is the **vertical line test**.

If we draw all possible vertical lines on the graph of a relation, the relation:

- is a function if each line cuts the graph no more than once
- is not a function if at least one line cuts the graph more than once.

Example 22

For each of the following, graph the relation and use the vertical line test to determine (if possible) whether or not the relation is a function.

a $R = \{(0, 0), (1, -1), (1, 1), (4, -2), (4, 2)\} \subseteq \mathbb{Z} \times \mathbb{Z}$

b $R = \left\{(x, y) \mid y = \dfrac{x^3 - 2x^2 - x + 2}{x}\right\} \subseteq \mathbb{R} \times \mathbb{R}$

a

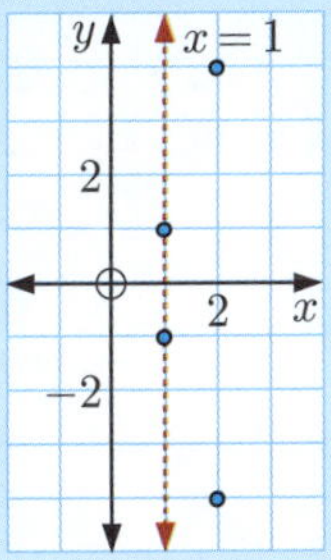

The vertical line $x = 1$ intersects the graph of R in two points $(1, -1)$ and $(1, 1)$.

$\therefore$ by the vertical line test, R is not a function.

b

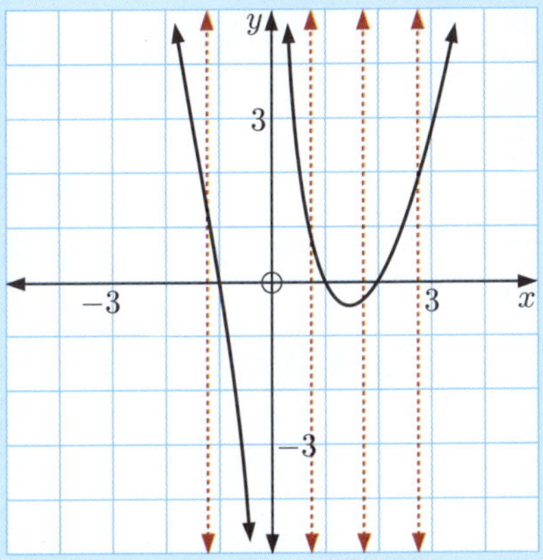

It *appears* that every vertical line meets the graph at most once. Note that the vertical asymptote $x = 0$ does not meet the graph at all. However, only a portion of the graph can be shown, as the domain of the relation is $\mathbb{R}$.

To *prove* this relation is a function we would need to show algebraically *for all* $x, y_1, y_2 \in \mathbb{R}$, that if $(x, y_1), (x, y_2) \in R$, then $y_1 = y_2$.

The above example shows that the vertical line test can only be used to prove that a relation is *not* a function.

INJECTIONS

If a function f is such that each element in the range corresponds to only one element in the domain, then f is said to be **one-to-one** or an **injection**.

To show that a function is an injection, it is sufficient to prove that $f(x_1) = f(x_2)$ implies $x_1 = x_2$.

Alternatively, if f is differentiable then showing that either $f'(x) > 0$ for all x in the domain, or $f'(x) < 0$ for all x in the domain, will prove that f is an injection.

Example 23

Is the illustrated function from $A = \{1, 2, 3\}$ to $B = \{1, 2, 3, 4\}$ an injection?

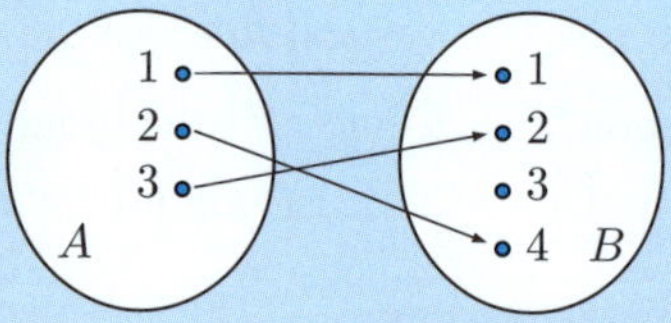

The function is an injection since each element in the range corresponds to only one element in the domain. In other words, no two elements in the domain are mapped to the same element in the range.

Example 24

Prove that the function $f : \mathbb{Z}^+ \to \mathbb{Z}^+$ where $f(x) = x^2$ is an injection.

Suppose there is an element in the range which corresponds to two distinct elements x_1 and x_2 in the domain, where $x_1 \neq x_2$.

$$\begin{aligned} \therefore\ f(x_1) &= f(x_2) \\ \therefore\ x_1^2 &= x_2^2 \\ \therefore\ x_1^2 - x_2^2 &= 0 \\ \therefore\ (x_1 - x_2)(x_1 + x_2) &= 0 \\ \therefore\ x_1 &= \pm x_2 \\ \therefore\ x_1 &= x_2 \quad \{\text{since } x_1, x_2 \in \mathbb{Z}^+\} \end{aligned}$$

This is a contradiction, so f is an injection.

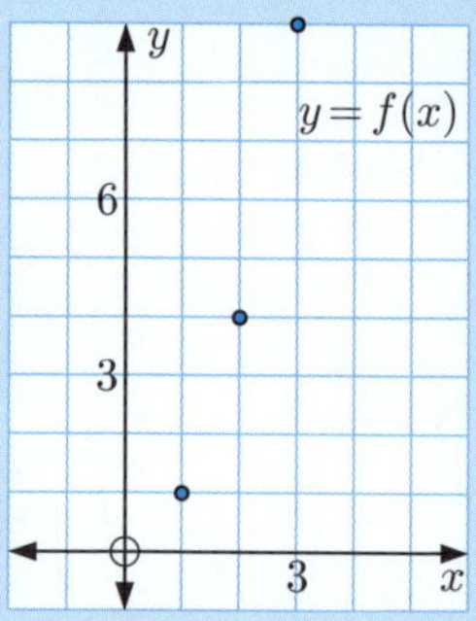

If a function can be graphed in the Cartesian plane, then the **horizontal line test** may be used to show a function is **not** an injection.

If any horizontal line intersects the graph of a function more than once, then the function is not an injection.

Example 25

Use the horizontal line test to show the function $f : \mathbb{R} \to \mathbb{R}$ such that $f(x) = x^2$, is not an injection.

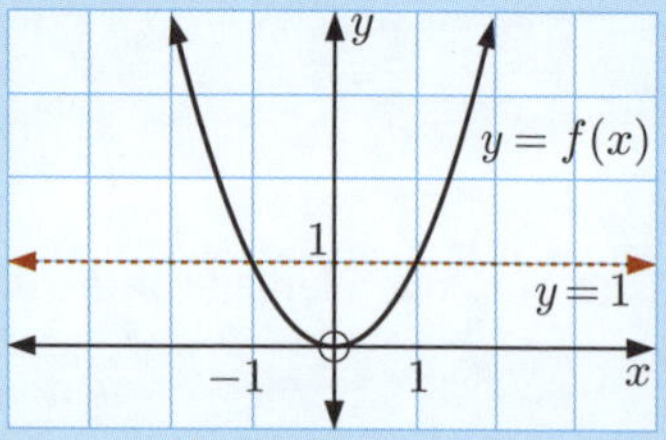

The horizontal line $y = 1$ meets the graph in the two distinct points $(-1, 1)$ and $(1, 1)$. Hence $f(-1) = f(1) = 1$, and so f is not an injection.

SURJECTIONS

For a function f from A to B, f is said to be **onto** or a **surjection** if the range of f is B.

Every element in B will be the image of an element in A, so the codomain is the same as the range.

Example 26

Is the illustrated function from $A = \{1, 2, 3, 4\}$ to $B = \{1, 2, 3\}$ a surjection?

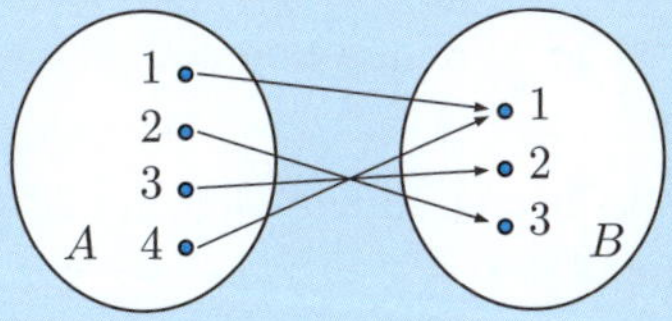

The function is a surjection, as every element of B is the image of some element of A.

Example 27

Determine whether each of the following functions is a surjection:

a $f : \mathbb{R} \to \mathbb{R}^+ \cup \{0\}$ where $f(x) = x^2$ **b** $f : \mathbb{Z}^+ \to \mathbb{Z}^+$ where $f(x) = 2x$.

a f is a surjection because every non-negative real number is the square of a real number.

b If we take any positive integer and double it, we get an even positive integer.
$\Rightarrow$ no elements of $\mathbb{Z}^+$ will map to an odd positive integer.
$\Rightarrow$ not all elements in the codomain correspond to elements in the domain.
$\Rightarrow$ f is not a surjection.

BIJECTIONS

A function which is both an injection *and* a surjection is called a **bijection**.

Example 28

Is the illustrated function from $A = \{1, 2, 3, 4\}$ to $B = \{1, 2, 3, 4\}$ a bijection?

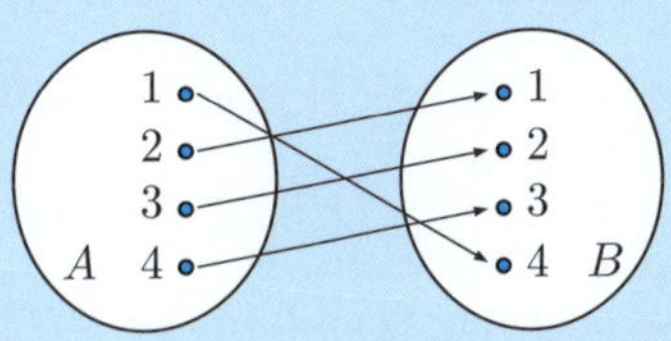

Each element in the domain maps to at most one element in the range, so the mapping is indeed a function.

Each element of the codomain is the image of at most one element from the domain, so the function is an injection (one-to-one).

Each element in the codomain is in the range, so the function is a surjection (onto).
$\therefore$ since the function is both an injection and a surjection, it is a bijection.

Example 29

Determine whether each of the following functions is a bijection:

a $f: \mathbb{R} \to \mathbb{R}$ where $f(x) = x^3$

b $f: \mathbb{R} \to \mathbb{R}$ where $f(x) = x^2$

c $f: \mathbb{R}^+ \to \mathbb{R}^+$ where $f(x) = x^2$.

a Each real number has a unique real cube root, so f is an injection.
Each real number is the cube of a unique real number, so f is a surjection.
$\therefore$ f is a bijection.

b f is not an injection since up to two elements of the domain can map to the same element of the range. For example, $f(-2) = f(2) = 4$.
Also, no negative real number is the square of a real number, so the range is not the whole of the codomain. The function is therefore also not a surjection.
$\therefore$ f is not a bijection.

c f is an injection since each element of the range is the square of only one element in the domain.
f is also a surjection since each real positive number is the square of a real positive number.
$\therefore$ f is a bijection.

EXERCISE C.1

1 The diagrams below map relations from $A = \{1, 2, 3\}$ to $B = \{1, 2, 3\}$.
Determine whether each relation is a function. If it is, state the domain, codomain, and range.

a

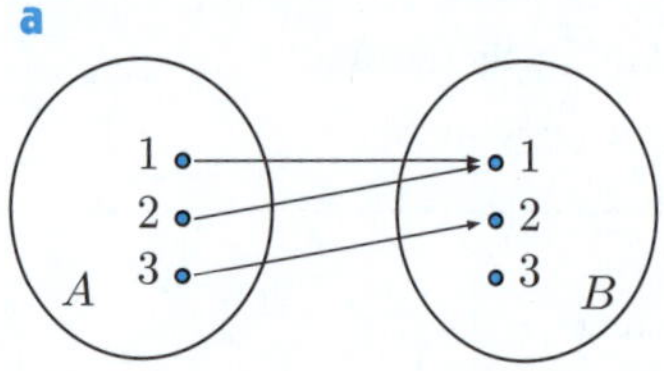

b

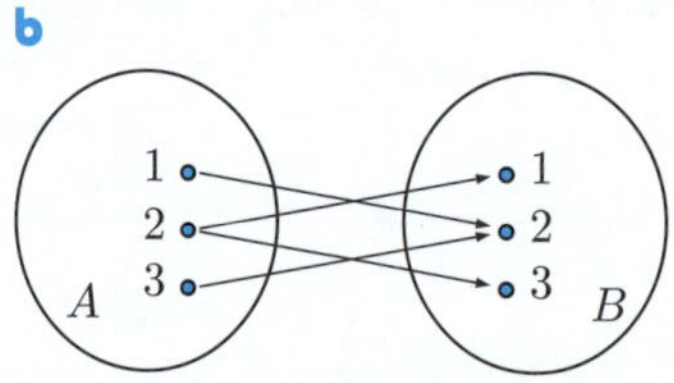

c

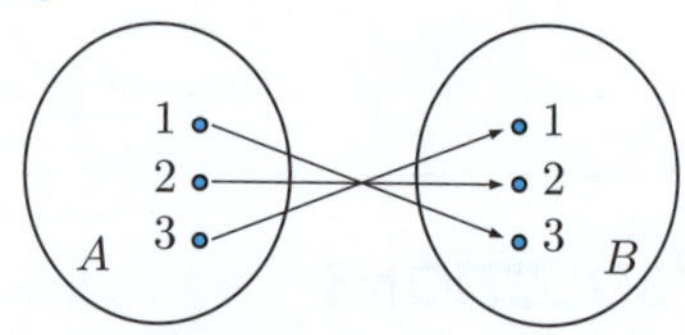

2 Determine whether each of the following relations is a function:

a the relation in $\mathbb{Z}$, $\{(0, -2), (1, 0), (2, 2), (3, 4)\}$

b the relation in $\mathbb{R}$ defined by $\{(x, y) \mid x > y\}$

c the relation in $\mathbb{R}$ defined by $\{(x, y) \mid y = \frac{2}{x^2},\ x \neq 0\}$

d the relation in $\mathbb{Z}$ defined by $\{(3, 2), (2, 2), (1, 2), (2, -1), (1, -1), (0, -1)\}$.

3 The diagrams below map functions from $A = \{1, 2, 3, 4\}$ to $B = \{1, 2, 3, 4\}$. In each case, determine whether the function is:

i an injection **ii** a surjection **iii** a bijection.

a

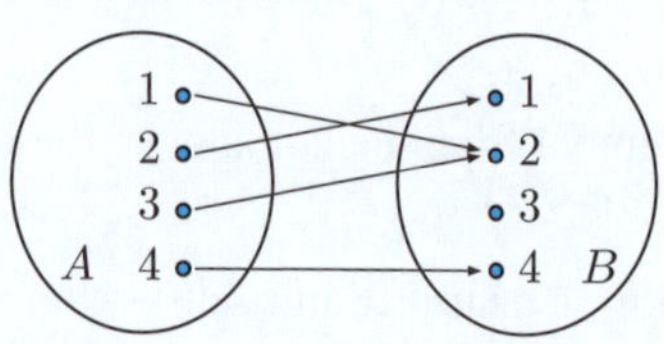

b

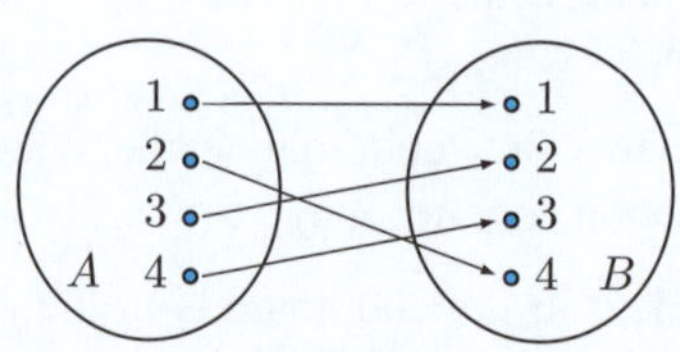

4 For each of the following, graph the relation and use the vertical line test to determine (if possible) whether or not the relation is a function.

a $R = \{(1, 2), (1, 3), (2, 4), (3, 5)\} \subseteq \mathbb{Z}^+ \times \mathbb{Z}^+$

b $R = \{(x, y) \mid y = \sin x\} \subseteq \mathbb{R} \times \mathbb{R}$

c $R = \{(x, y) \mid x^2 + 2y^2 = 1\} \subseteq \mathbb{R} \times \mathbb{R}$.

5 Use true and false to complete the following table for the given relations from $\{1, 2, 3, 4, 5\}$ to $\{1, 2, 3, 4, 5\}$.

	Relation	*Function*	*Injection*	*Bijection*
a	$\{(1, 2), (2, 4), (3, 5), (1, 3), (4, 1), (5, 2)\}$			
b	$\{(1, 5), (2, 4), (3, 5), (4, 5), (5, 3)\}$			
c	$\{(1, 3), (2, 4), (3, 5), (4, 2), (5, 1)\}$			

6 If possible, use the horizontal line test to determine whether each function is not an injection.

a $f : \mathbb{R} \to \mathbb{R}$ where $f(x) = x^3$

b $f : \mathbb{R} \to \mathbb{R}$ where $f(x) = x^3 - x$

c $f : \mathbb{R}^+ \to \mathbb{R}$ where $f(x) = x^2 + 2x - 24$

7 State whether each of the following relations is a function, and if so, determine whether it is:

i an injection

ii a surjection

iii a bijection.

a The relation R from $\{0, 1, 2\}$ to $\{1, 2\}$ where $R = \{(0, 1), (1, 2), (2, 2)\}$

b The relation R from $\{0, 1, 2\}$ to $\{1, 2\}$ where $R = \{(0, 1), (1, 1), (2, 1)\}$

c The relation R from $\{0, 1, 2\}$ to $\{1, 2\}$ where $R = \{(0, 1), (1, 1), (1, 2), (2, 2)\}$

d The relation from $\mathbb{Z}$ to $\mathbb{Z}^+$ defined by $\{(x, y) \mid y = x^2 + 1\}$

e The relation from $\mathbb{R}^2$ to $\mathbb{R}$ defined by $(x, y)Rz$ if and only if $z = x^2 + y^2$.

f The relation from $\mathbb{Z} \times \mathbb{Z}$ to $\mathbb{Z} \times \mathbb{Z}$ where $(a, b)R(x, y)$ if and only if $y = a$ and $x = b$.

8 Determine whether each of the following functions is a bijection. Give reasons for your answers.

a $f : \mathbb{R} \to \mathbb{R}, \quad f(x) = 2x - 1$

b $f : \mathbb{R} \to \mathbb{Z}, \quad f(x) = \lfloor x \rfloor$, where $\lfloor x \rfloor$ means "the greatest integer less than or equal to x"

c $f : \mathbb{Z} \to \mathbb{Z}^+ \cup \{0\}, \quad f(x) = |x|$

d $f : \mathbb{Q}^+ \to \mathbb{Q}^+, \quad f(x) = x^2$

e $f : \left[0, \frac{\pi}{2}\right] \to [0, 1], \quad f(x) = \sin x$

f $f : \mathbb{Z}^+ \to \mathbb{Z}^+, \quad f(x) = 2x$

9 Consider a function $f : S \to S$. Let $f(A) = \{f(x) \mid x \in A\}$ and $f(B) = \{f(x) \mid x \in B\}$. Prove that if $A \subseteq B \subseteq S$ then $f(A) \subseteq f(B)$.

10 Find an example of a function $f : \mathbb{R} \to \mathbb{R}$ which is:

a one-to-one but not onto

b onto but not one-to-one

c one-to-one and onto

d neither one-to-one nor onto.

11 **a** Let $f : S \to S$ be a function with domain $S = \{1, 2, 3\}$.
Find, if possible, an example of a function f which is:
  i onto but not one-to-one **ii** one-to-one but not onto.

b Complete the following:
Let $f : S \to S$ be a function with domain S, where S is any *finite* set.
f is one-to-one $\Leftrightarrow$ f is ....

12 Let $P = \{p(x) \mid p(x)$ is a polynomial of degree n, $n \in \mathbb{Z}^+ \cup \{0\}$ with coefficients in $\mathbb{R}\}$.
Define $f : P \to P$ where $f(p(x)) = p'(x)$, the derivative of $p(x)$. Determine whether f is:
a a function **b** an injection **c** a surjection **d** a bijection.

COMPOSITION OF FUNCTIONS

Suppose f is a function from A to B and g is a function from B to C.

We can define the **composite function** $g(f(x))$ or $g \circ f$ from a subset of A to C provided the domain of g contains the range of f.

In general, the domain of $g \circ f$ is $\{x \mid x \in \text{domain of } f \textit{ and } f(x) \in \text{domain of } g\}$.

Example 30

Let f map $\{1, 2, 3, 4\}$ to $\{5, 6, 7\}$ where $f = \{(1, 6), (2, 6), (3, 5), (4, 7)\}$.

Let g map $\{5, 6, 7\}$ to $\{8, 9\}$ where $g = \{(5, 8), (6, 9), (7, 8)\}$.

Find, if possible: **a** $g \circ f$ **b** $f \circ g$

a $(g \circ f)(1) = g(f(1)) = g(6) = 9$
$(g \circ f)(2) = g(f(2)) = g(6) = 9$
$(g \circ f)(3) = g(f(3)) = g(5) = 8$
$(g \circ f)(4) = g(f(4)) = g(7) = 8$
$\therefore\ g \circ f = \{(1, 9), (2, 9), (3, 8), (4, 8)\}$

b $f \circ g$ is not defined because the domain of f does not contain the range of g.

Example 31

Let $f : \mathbb{R} \to \mathbb{R}$ and $g : \mathbb{R} \to \mathbb{R}$ where $f(x) = x + 2$ and $g(x) = x^3$.
Find: **a** $(g \circ f)(x)$ **b** $(f \circ g)(x)$

a $(g \circ f)(x) = g(f(x)) = g(x + 2) = (x + 2)^3$

b $(f \circ g)(x) = f(g(x)) = f(x^3) = x^3 + 2$

INVERSE FUNCTIONS

If f is a bijection from A to B such that $f : x \mapsto y$, then the function f is said to be **invertible**, and it is possible to define a new function such that y is mapped to x. This function is called the **inverse** of f, denoted f^{-1}, and is a function from B to A.

If f is written as a set of ordered pairs, then the inverse function f^{-1} is obtained by reversing the order of the components in each pair. Note that since f is a bijection, f^{-1} is also a bijection.

It follows that:

For f a bijection $f : A \to B$, the **inverse of *f*** is the function $f^{-1} : B \to A$ which satisfies:

- $(f^{-1} \circ f)(x) = x$ for all x in the domain of f
- $(f \circ f^{-1})(x) = x$ for all x in the domain of f^{-1}.

The domain of f equals the range of f^{-1}.
The range of f equals the domain of f^{-1}.

Example 32

Find the inverse of the bijection from $A = \{1, 2, 3, 4\}$ to $B = \{1, 2, 3, 4\}$ where $f = \{(1, 3), (2, 2), (3, 4), (4, 1)\}$

Since f is a bijection, f^{-1} exists.

Swapping the order of coordinates in each pair we find $f^{-1} = \{(3, 1), (2, 2), (4, 3), (1, 4)\}$.

Example 33

Find the inverse of $f : \mathbb{R} \to \mathbb{R}$ where $f(x) = 2x^3 + 1$.

First, we note that f is both an injection and a surjection, so f is a bijection and therefore has an inverse. Next, we let $y = 2x^3 + 1$.

We then interchange x and y, which has the effect of reversing the order of the components of each ordered pair of the function.

$\therefore$ for the inverse function, $x = 2y^3 + 1$

$$\therefore\ 2y^3 = x - 1$$

$$\therefore\ y^3 = \frac{x-1}{2}$$

$$\therefore\ y = \sqrt[3]{\frac{x-1}{2}}$$

So, $f^{-1}(x) = \sqrt[3]{\dfrac{x-1}{2}}$.

Example 34

Let $f : [5, \infty[\to \mathbb{R}$ be defined by $f(x) = \sqrt{x-5}$.

a Find the inverse f^{-1} of f.

b State the domain and range of f.

c State the domain and range of f^{-1}.

d Graph both functions on the same set of axes.

a f is both an injection and a surjection, so f is a bijection and therefore has an inverse.

Let $y = \sqrt{x-5}$.

Interchanging x and y, $x = \sqrt{y-5}$

$$\therefore\ x^2 = y - 5$$
$$\therefore\ y = x^2 + 5$$
$$\therefore\ f^{-1}(x) = x^2 + 5.$$

b The domain of f is $[5, \infty[$ and the range of f is $[0, \infty[$.

This can be seen from the graph of $y = f(x)$:

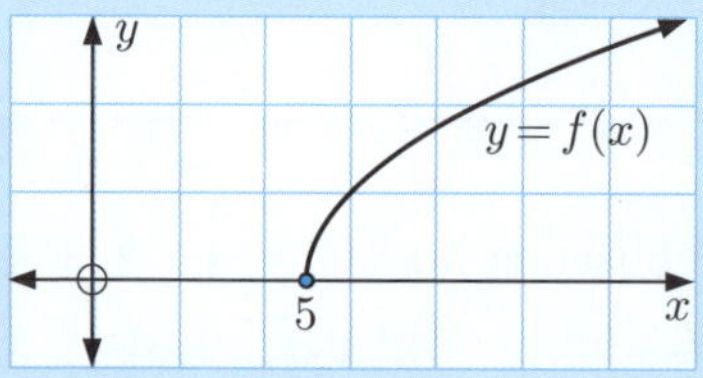

c The domain of f^{-1} = the range of f = $[0, \infty[$.

The range of f^{-1} = the domain of f = $[5, \infty[$.

d

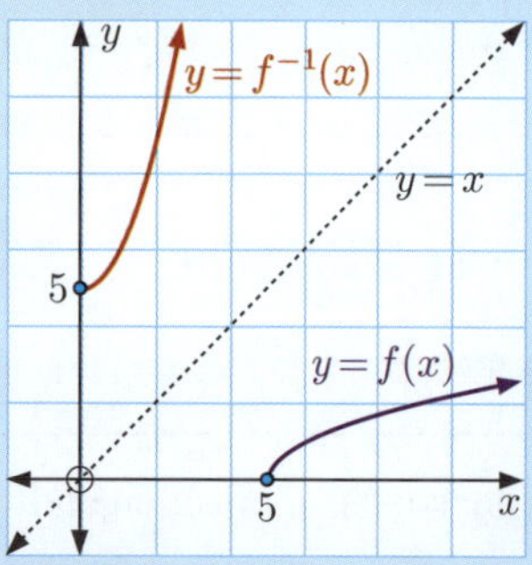

EXERCISE C.2

1 Let f map $\{1, 2, 3\}$ to $\{4, 5, 6\}$ where $f = \{(1, 5), (2, 6), (3, 4)\}$.
Let g map $\{4, 5, 6\}$ to $\{0, 1\}$ where $g = \{(4, 1), (5, 0), (6, 1)\}$.
Find, if possible: **a** $g \circ f$ **b** $f \circ g$

2 Suppose $A = \{0, 1, 2, 3\}$. Let f and g be functions mapping A to A, where
$f = \{(0, 1), (1, 2), (2, 0), (3, 3)\}$ and $g = \{(0, 2), (1, 3), (2, 0), (3, 1)\}$.

a Find:

i $(f \circ g)(1)$ **ii** $(g \circ f)(1)$ **iii** $(f \circ g)(3)$ **iv** $(g \circ f)(3)$

b Find:

i f^{-1} **ii** g^{-1} **iii** $(g \circ f)^{-1}$ **iv** $(f^{-1} \circ g^{-1})$

3 Find the inverse of:

a $f : \mathbb{Z} \to \mathbb{Z}$ where $f(x) = 1 - x$

b $f : \mathbb{R} \to \mathbb{R}$ where $f(x) = x^3 - 2$.

4 Let f be defined by $f(x) = \sqrt{1 - x}$.

a State the domain and range of f.

b Graph $y = f(x)$.

c Find the inverse f^{-1} of f.

d State the domain and range of f^{-1}.

e Graph $y = f^{-1}$ on the same set of axes used in **b**.

5 f and g are functions with domain $\mathbb{R}^+$ such that: $f(x) = \ln(x + 1)$ and $g(x) = x^2$.
Find each of the following:

a $(g \circ f)(x)$

b $(f \circ g)(x)$

c $f^{-1}(x)$

d $(g \circ f)^{-1}(x)$

e $(f^{-1} \circ g^{-1})(x)$

6 Let $f : \mathbb{R} \to \mathbb{R}$ and $g : \mathbb{R} \to \mathbb{R}$ be invertible functions. Prove that $g \circ f$ is an invertible function and that $(g \circ f)^{-1} = f^{-1} \circ g^{-1}$.

7 For each of the following real-valued functions, determine whether f is invertible, and if it is, find f^{-1}.

a $f(x) = e^x + 3e^{-x}$

b $f(x) = e^x - 3e^{-x}$

8 Let $f : \mathbb{R}^+ \times \mathbb{R}^+ \to \mathbb{R}^+ \times \mathbb{R}^+$ be defined by $f(x, y) = \left(\frac{y}{x}, xy\right)$.

a Prove that f is a bijection.

b Find a formula for $f^{-1} : \mathbb{R}^+ \times \mathbb{R}^+ \to \mathbb{R}^+ \times \mathbb{R}^+$.

D BINARY OPERATIONS

Given a non-empty set S, a **binary operation** on S is a rule for combining any two elements $a, b \in S$ to give a unique result c, where c is not necessarily an element of S.

Many binary operations are familiar from operations on number. Addition, subtraction, multiplication, and division are all examples of binary operations.

For example, given the set of real numbers $\mathbb{R}$, the binary operation of addition with 3 and 5 gives 8, and we write $3+5=8$.

Less familiar binary operations between two elements in a set are often defined using a symbol such as $*$.

Example 35

Let a binary operation $*$ on $\mathbb{Z}$ be defined by $a*b=a+2b-3$

Find:

a $3*5$ **b** $3*0$ **c** $0*3$ **d** $-5*0$

a $3*5=3+2\times 5-3$
$=10$

b $3*0=3+2\times 0-3$
$=0$

c $0*3=0+2\times 3-3$
$=3$

d $-5*0=-5+2\times 0-3$
$=-8$

CLOSURE

A set S is said to be **closed under the binary operation** $*$, or the binary operation $*$ is said to be **closed on S**, if $a*b\in S$ for all $a, b \in S$.

A closed binary operation on a set S is a function with domain $S\times S$ and codomain S.

For example:

- Given the set of integers $\mathbb{Z}$, the binary operation of addition is closed. The sum of any two integers is an integer.
- Given the set of natural numbers $\mathbb{N}$, the binary operation of subtraction is *not* closed. For example, $5-7=-2$, and this result does not belong to $\mathbb{N}$. By contrast, the set of integers $\mathbb{Z}$ is closed under subtraction because the result of subtracting any integer from another integer is always an integer.

Note that some definitions of a binary operation include closure as a property. The definition used here does not, and so closure must not be assumed.

Example 36

Determine whether the following binary operations are closed on $\mathbb{Z}$:

a $a * b = \dfrac{a+b}{a^2}$ **b** $a * b = 2^{a+b}$ **c** $a * b = a + b - 3ab$

a Consider $a = 2$ and $b = 3$.
$2 * 3 = \dfrac{2+3}{4} = \frac{5}{4} \notin \mathbb{Z}$
$\therefore$ the binary operation is not closed on $\mathbb{Z}$.

b Consider $a = -2$ and $b = 0$.
$-2 * 0 = 2^{-2+0} = \frac{1}{4} \notin \mathbb{Z}$
$\therefore$ the binary operation is not closed on $\mathbb{Z}$.

c Since a and b are in $\mathbb{Z}$, their sum $a + b$ and product ab are also in $\mathbb{Z}$.
$\therefore$ $a + b - 3ab$ is also in $\mathbb{Z}$
$\therefore$ $a * b \in \mathbb{Z}$
$\therefore$ the binary operation is closed on $\mathbb{Z}$.

EXERCISE D.1

1 Define two binary operations in $\mathbb{Q}$ by $a * b = a - b + 1$ and $a \diamondsuit b = ab - a$.

a Find:

i $3 * 4$ **ii** $4 * 3$ **iii** $(-2) \diamondsuit 3$ **iv** $6 \diamondsuit 0$
v $0 \diamondsuit 7$ **vi** $4 * ((-5) \diamondsuit 2)$ **vii** $(4 * (-5)) \diamondsuit 2$

b Solve for x:

i $4 * x = 7$ **ii** $x \diamondsuit 3 = -2$

2 Copy and complete the table for closure using true (T) and false (F):

	$+$	$-$	$\times$	$\div$
$\mathbb{Z}^+$				
$\mathbb{Z}$				
$\mathbb{Q}^+$				
$\mathbb{Q}$				
$\mathbb{R}$				

3 Determine whether each of the following sets is closed under multiplication:

a $\{a + bi \mid a, b \in \mathbb{Q},\ b \neq 0\}$ **b** $\{a + bi \mid a, b \in \mathbb{Q},\ a \neq 0\}$

c $\{a + bi \mid a, b \in \mathbb{Q},\ a$ and b not both equal to zero$\}$

4 State whether each of the following sets is closed under the given operation:

a The set of even positive integers $\{2, 4, 6, ...\}$ under addition

b The set of even positive integers $\{2, 4, 6, ...\}$ under multiplication

c The set of odd positive integers $\{1, 3, 5, ...\}$ under addition

d The set of odd positive integers $\{1, 3, 5, ...\}$ under multiplication

5 Determine whether the following binary operations are closed on: **i** $\mathbb{Z}$ **ii** $\mathbb{Q}$

a $a * b = a^2 - b$ **b** $a * b = \dfrac{a+b}{a}$ **c** $a * b = \sqrt{a^2b^2}$ **d** $a * b = \sqrt{|ab|}$

ASSOCIATIVE LAW

Consider the following examples of repeated use of the binary operation multiplication on $\mathbb{Z}$:

$$\begin{aligned} &3 \times (2 \times 5) \\ &= 3 \times 10 \\ &= 30 \end{aligned} \qquad\qquad \begin{aligned} &(3 \times 2) \times 5 \\ &= 6 \times 5 \\ &= 30 \end{aligned}$$

We observe that the order of grouping the terms makes no difference. This is true for multiplication of all real numbers, so we say that multiplication in $\mathbb{R}$ is **associative**.

A binary operation $*$ on a set S is said to be **associative** if
$a * (b * c) = (a * b) * c$ for all $a, b, c \in S$.

For example:

- Addition in $\mathbb{R}$ is associative.
- $8 - (3 - 5) \neq (8 - 3) - 5$ so subtraction in $\mathbb{R}$ is not associative.
- $12 \div (6 \div 2) \neq (12 \div 6) \div 2$ so division in $\mathbb{R}$ is not associative.

If a binary operation is associative on a set, then the associativity will also hold in any subset of the set.

By contrast, note that a binary operation is not necessarily closed on a subset of a set. Thus care must be taken, as not all properties of an operation on a set are transferable to a subset.

Example 37

Determine whether the following binary operations on $\mathbb{R}$ defined below are associative.

a $a * b = 2a + 3b$ **b** $a * b = a + b + ab$

a $(a * b) * c = (2a + 3b) * c$
$= 2(2a + 3b) + 3c$
$= 4a + 6b + 3c$

$a * (b * c) = a * (2b + 3c)$
$= 2a + 3(2b + 3c)$
$= 2a + 6b + 9c$
$\neq (a * b) * c$ in general.

In fact, noting that $(1 * 0) * 2 = 10$
whereas $1 * (0 * 2) = 20$
is enough to prove $*$ is not associative.

b $(a * b) * c = (a + b + ab) * c$
$= (a + b + ab) + c + (a + b + ab)c$
$= a + b + ab + c + ac + bc + abc$

$a * (b * c) = a * (b + c + bc)$
$= a + (b + c + bc) + a(b + c + bc)$
$= a + b + c + bc + ab + ac + abc$
$= (a * b) * c$

$\therefore$ since $a * (b * c) = (a * b) * c$ for all $a, b, c \in \mathbb{R}$, the operation $*$ is associative.

As shown in the above example, to prove a law holds, the full general proof is required. To prove a law does *not* hold, it is sufficient to give one example (called a **counter-example**) for which the law fails. For more information, see the **Appendix** on **Methods of Proof**.

NOTATION CONVENTIONS

Although multiplication and addition of real numbers are binary operations, since they are associative we can write statements such as $3+6+17$ or $2\times5\times7$ without any need for grouping the terms into pairs.

This is true in general for associative operations. So, if $*$ is associative then there is no ambiguity in writing $a*b*c$ rather than $(a*b)*c$ or $a*(b*c)$.

We will also follow the convention of writing $\underbrace{a*a*a*....*a}_{n \text{ times}}$ as a^n, so care must be taken here to not assume that this operation is simply the multiplication of real numbers.

The familiar index laws apply for associative operations.

For example:

- $$a^m * a^n = \underbrace{\underbrace{a*a*a*....*a}_{m \text{ times}} * \underbrace{a*a*a*....*a}_{n \text{ times}}}_{m+n \text{ times}} = a^{m+n}$$
- $$\begin{aligned}(a^m)^n &= \underbrace{a^m * a^m * .... * a^m}_{n \text{ times}} \\ &= \underbrace{(\underbrace{a*a*....*a}_{m \text{ times}}) * (\underbrace{a*a*....*a}_{m \text{ times}}) * .... * (\underbrace{a*a*....*a}_{m \text{ times}})}_{n \text{ times}} \\ &= \underbrace{a*a*....*a}_{mn \text{ times}} \qquad \{\text{brackets can be removed since } * \text{ is associative}\} \\ &= a^{mn}\end{aligned}$$

If $*$ is associative, then $(a^m)^n = a^{mn}$ and $a^m * a^n = a^{m+n}$.

COMMUTATIVE LAW

A binary operation $*$ on a set S is said to be **commutative** if $a*b=b*a$ for all $a, b \in S$.

Multiplication and addition in $\mathbb{R}$ are examples of commutative operations. Subtraction in $\mathbb{R}$ is not commutative.

If $*$ is both associative and commutative then $(a*b)^n = a^n * b^n$.

Example 38

If $*$ is both associative and commutative on a set S, show that $(a*b)^2 = a^2*b^2$.

$$\begin{aligned}(a*b)^2 &= (a*b)*(a*b) \\ &= a*(b*a)*b \qquad \{\text{Associative Law}\} \\ &= a*(a*b)*b \qquad \{\text{Commutative Law}\} \\ &= (a*a)*(b*b) \qquad \{\text{Associative Law}\} \\ &= a^2*b^2\end{aligned}$$

Example 39

Determine whether the following operations on $\mathbb{R}$ are commutative:

a $a * b = 2a + b$ **b** $a * b = 3^{a+b}$

a $3 * 2 = 2 \times 3 + 2 = 8$

$2 * 3 = 2 \times 2 + 3 = 7 \neq 3 * 2$

$\therefore$ the operation $*$ is not commutative.

b $b * a = 3^{b+a}$

$= 3^{a+b}$ {addition on $\mathbb{R}$ is a commutative operation}

$= a * b$

$\therefore$ the operation $*$ is commutative.

DISTRIBUTIVE LAW

Given two binary operations $*$ and $\circ$ on a set S, $*$ is said to be **distributive over** $\circ$ if $a * (b \circ c) = (a * b) \circ (a * c)$ for all $a, b, c \in S$.

For example:

- In $\mathbb{R}$, multiplication is distributive over addition, since $a(b + c) = ab + ac$ for all $a, b, c \in \mathbb{R}$.
- In $\mathbb{R}$, addition is not distributive over multiplication, since in general $a + (bc) \neq (a + b)(a + c)$. For example, $1 + (2 \times 3) = 7$ whereas $(1 + 2) \times (1 + 3) = 12$.

Example 40

$*$ and $\circ$ are binary operations on $\mathbb{R}$ defined by $a * b = a + 2b$ and $a \circ b = 2ab$.

a Is $*$ distributive over $\circ$? **b** Is $\circ$ distributive over $*$?

a $a * (b \circ c) = a * (2bc) = a + 4bc$

and $(a * b) \circ (a * c) = (a + 2b) \circ (a + 2c) = 2(a + 2b)(a + 2c) = 2a^2 + 4ac + 4ab + 8bc \neq a * (b \circ c)$ in general.

For example, $0 * (1 \circ 2) = 8$ and $(0 * 1) \circ (0 * 2) = 16$

$\therefore$ $0 * (1 \circ 2) \neq (0 * 1) \circ (0 * 2)$

Therefore $*$ is not distributive over $\circ$.

b $a \circ (b * c) = a \circ (b + 2c) = 2a(b + 2c) = 2ab + 4ac$

and $(a \circ b) * (a \circ c) = (2ab) * (2ac) = 2ab + 4ac = a \circ (b * c)$

Therefore $\circ$ is distributive over $*$.

EXERCISE D.2

1 Determine whether the following binary operations on $\mathbb{R}$ are:

i commutative **ii** associative

a $a * b = a + 2b$ **b** $a * b = a^2 + b^2$ **c** $a * b = ab - a - b$ **d** $a * b = \dfrac{1}{a+b}$

2 Prove by mathematical induction:
If $*$ is both associative and commutative on a set S, then $(a * b)^n = a^n * b^n$ for all $n \in \mathbb{Z}^+$.

3 Suppose $*$ and $\diamondsuit$ are binary operations on $\mathbb{R}$ defined by $a * b = a - b$ and $a \diamondsuit b = ab$.
Determine whether:

a $*$ is distributive over $\diamondsuit$ **b** $\diamondsuit$ is distributive over $*$.

4 Suppose $*$ and $\diamondsuit$ are binary operations on $\mathbb{R}^+$ defined by $a * b = 3a + b$ and $a \diamondsuit b = 2ab$.
Determine whether:

a $*$ is distributive over $\diamondsuit$ **b** $\diamondsuit$ is distributive over $*$.

5 Using suitable Venn diagrams, determine if each statement is true or false.

a Set difference is not associative. **b** Symmetric difference of sets is associative.

IDENTITY

Consider a binary operation $*$ on a set S. If there exists an element $e \in S$ such that $e * x = x * e = x$ for all $x \in S$, then e is said to be an **identity** element for $*$ on S.

If an identity e for operation $*$ exists in S, we define using index notation $x^0 = e$ for all $x \in S$.

For addition in $\mathbb{R}$, the identity element is the number 0.

For multiplication in $\mathbb{R}$, the identity element is 1.

Subtraction in $\mathbb{R}$ does not have an identity element because, although $x - 0 = x$ for all $x \in \mathbb{R}$, $0 - x \neq x$ for all $x \in \mathbb{R} \setminus \{0\}$. For example, $0 - 1 \neq 1$.

There is no identity for division in $\mathbb{R}$.

For a commutative binary operation $*$ on S, to establish that there is an identity element $e \in S$ it is sufficient to check that just one of $e * x = x$ or $x * e = x$ is true for all $x \in S$.

Theorem 3

If a binary operation on a set has an identity element, then it is unique.

Proof:

Suppose a binary operation $*$ on a set S has more than one identity element.

Let e and f be two such identity elements, where $e \neq f$.

$\Rightarrow$ for all $x \in S$, $e * x = x * e = x$ .... (1) and $f * x = x * f = x$ .... (2).

Since $f \in S$, we can replace x by f in (1), so $e * f = f * e = f$.

Similarly, since $e \in S$, we can replace x by e in (2), so $f * e = e * f = e$.

$\therefore$ $e = f$, which contradicts the original assumption.

$\therefore$ if it exists, the identity element is unique.

Example 41

For each of the following operations, determine whether an identity element exists in $\mathbb{R}$:

a $a * b = 3ab$ **b** $a * b = 3a + b$

a Suppose e is an identity element for the binary operation $*$ on $\mathbb{R}$.

$$\therefore\ a * e = a$$
$$\therefore\ 3ae = a$$
$$\therefore\ 3ae - a = 0$$
$$\therefore\ a(3e - 1) = 0$$

$\Rightarrow$ $a * e = a$ is satisfied by $e = \frac{1}{3}$ for all $a \in \mathbb{R}$.

We must now *either* show that $*$ is commutative *or* that $e * a = a$ for all $a \in \mathbb{R}$ and $e = \frac{1}{3}$.

Here we do the latter: $e * a = \frac{1}{3} * a = 3(\frac{1}{3})a = a$

$\therefore$ an identity element e exists, and $e = \frac{1}{3}$.

b Suppose e is an identity element for the binary operation $*$ on $\mathbb{R}$.

$$\therefore\ a * e = a$$
$$\therefore\ 3a + e = a$$
$$\therefore\ e = -2a, \quad \text{but this is not unique for all } a \in \mathbb{R}.$$

$\therefore$ an identity element does not exist for $*$.

INVERSE

Consider a binary operation $*$ on a set S with an identity element $e \in S$. For $a \in S$, an **inverse element** $a^{-1} \in S$ exists for a if and only if $a^{-1} * a = a * a^{-1} = e$.

For addition in $\mathbb{R}$, each element $a \in \mathbb{R}$ has inverse $-a$, since $a + (-a) = (-a) + a = 0$, where 0 is the additive identity.

No inverse exists for addition on $\mathbb{Z}^+$ for any element in $\mathbb{Z}^+$.

For multiplication in $\mathbb{R}$, each element $a \in \mathbb{R}\backslash\{0\}$ has inverse $\frac{1}{a}$, since $a \times \frac{1}{a} = \frac{1}{a} \times a = 1$, the multiplicative identity.

0 is the unique element in $\mathbb{R}$ which does not have a multiplicative inverse, since there is no element $x \in \mathbb{R}$ for which $0 \times x = x \times 0 = 1$.

Theorem 4

Let $*$ be an *associative* binary operation on a set S with identity element e.

If an element $a \in S$ has an inverse, then it is unique.

Proof:

Suppose an element $a \in S$ has more than one inverse.

Let two of these inverses be x and y, where $x \neq y$.

$\Rightarrow \quad x * a = a * x = e \quad .... (1) \quad$ and $\quad y * a = a * y = e \quad .... (2)$

$$\begin{aligned} \text{Using (1),} \quad (x * a) * y &= e * y = y \\ \therefore \quad x * (a * y) &= y \quad \{\text{Associative Law}\} \\ \therefore \quad x * e &= y \quad \{\text{from (2)}\} \\ \therefore \quad x &= y \end{aligned}$$

This contradicts the original assumption, so the inverse element (if it exists) must be unique.

The contrapositive of this theorem can be useful:

If the inverse of an element is not unique then associativity does not hold.

However, uniqueness of an inverse does not ensure that associativity holds.

Example 42

Consider the binary operation $*$ on $\mathbb{R}$ defined by $a * b = 3ab$.
Determine the values of $a \in \mathbb{R}$ for which $*$ has an inverse, and the value of the inverse in each case.

From **Example 41**, $*$ has identity element $\frac{1}{3}$.

If an element a has inverse b, then

$$\begin{aligned} b * a &= a * b = \tfrac{1}{3} \\ \therefore \quad 3ba &= 3ab = \tfrac{1}{3} \\ \therefore \quad b &= \frac{1}{9a} \quad \text{provided } a \neq 0 \end{aligned}$$

$\therefore$ each element $a \in \mathbb{R} \setminus \{0\}$ has inverse $\dfrac{1}{9a}$, and element 0 has no inverse.

EXERCISE D.3

1 Where one exists, state the identity element for each of the following:

- **a** $\mathbb{R}$ under addition
- **b** $\mathbb{Z}$ under multiplication
- **c** $\mathbb{R}$ under $*$ where $a * b = a$
- **d** $\mathbb{R}$ under $*$ where $a * b = 5ab$
- **e** $\mathbb{R}$ under $*$ where $a * b = 2a + ab + 2b$
- **f** $\mathbb{R}$ under division.

2 For each of the following, state the identity, if it exists. If an identity exists, determine whether each element in the set has an inverse. Whenever it can be found, state the inverse.

- **a** $\mathbb{Q}$ under addition
- **b** $\mathbb{Q}$ under multiplication
- **c** $\mathbb{Z}^+$ under multiplication
- **d** $\mathbb{R}$ under $*$ where $a * b = 2ab$

3 Let $S = \{2, 4, 6, 8\}$ under the binary operation $\times_{10}$, which is multiplication modulo 10, or $a \times_{10} b = (a \times b) (\text{mod } 10)$. For example, $4 \times_{10} 8 = 4 \times 8 (\text{mod } 10) = 32 (\text{mod } 10) = 2$.

- **a** Show that S is closed under $\times_{10}$.
- **b** Find, with justification, the identity element.

Example 43

a Explain why the set operations *union* and *intersection* are binary operations.

b For union of sets:

i is there an identity element

ii does each set have an inverse?

c For intersection of sets:

i is there an identity element

ii does each set have an inverse?

---

a For sets A and B, set union $A \cup B$ and set intersection $A \cap B$ are both binary operations as they are operations on two sets and they have unique results.

b **i** Consider a set E such that $A \cup E = E \cup A = A$ for all sets A.
$\therefore$ $E \subseteq A$ for all sets A.
$E = \varnothing$, the empty set, is the unique set with this property.
$\therefore$ for the union of sets, the identity element is the empty set $\varnothing$.

ii For set union, a set A has inverse B if and only if $A \cup B = B \cup A = \varnothing$, the identity for set union.
But $A \cup B = \varnothing \;\Leftrightarrow\; A = B = \varnothing$.
$\therefore$ no non-empty set has an inverse under the union of sets. However, $\varnothing$ is its own inverse.

c **i** Consider a set E such that $A \cap E = E \cap A = A$ for all sets A.
$\therefore$ $A \subseteq E$ for all sets A.
$\therefore$ $E = \mathbb{U}$, since the universal set $\mathbb{U}$ is the only set with this property.
$\therefore$ for the intersection of sets, the identity element is the universal set $\mathbb{U}$.

ii For set intersection, a set A has inverse B if and only if $A \cap B = B \cap A = \mathbb{U}$, the identity for set intersection.
But $A \cap B = \mathbb{U}$ only when $A = B = \mathbb{U}$.
$\therefore$ no set other than $\mathbb{U}$ has an inverse under set intersection, and the set $\mathbb{U}$ is its own inverse.

---

4 **a** Let $\diamondsuit$ be a binary operation in $\mathbb{Q} \setminus \{1\}$ such that $a \diamondsuit b = a - ab + b$.

i Show that $\mathbb{Q} \setminus \{1\}$ is closed under $\diamondsuit$.

ii Prove that $\diamondsuit$ is associative in $\mathbb{Q} \setminus \{1\}$.

iii Find an identity element or show that one does not exist.

iv Does each element have an inverse?

b Are the results in **a** the same for $\diamondsuit$ in $\mathbb{R} \setminus \{1\}$?

5 A binary operation $*$ is defined on the set $\mathbb{R}^2$ by $(a, b) * (c, d) = (ac - bd, ad + bc)$.

a Is $*$ associative?

b Is $*$ commutative?

c Is there an identity element in $\mathbb{R}^2$? If so, state it.

d Does each element in $\mathbb{R}^2$ have an inverse?

e Whenever it exists, find the inverse of (a, b).

6 Let $P = \{p(x) \mid p(x)$ is a polynomial of degree $n \in \mathbb{Z}^+ \cup \{0\}$, with coefficients in $\mathbb{R}\}$.
Consider the set P under the operation addition. Determine, where possible:

a if P is closed under addition

b if the operation is commutative

c if the operation is associative

d the identity in P

e the inverse of each element in P.

CAYLEY TABLES

The possible results of a binary operation on a finite set can be set out in a **Cayley table**, named after **Arthur Cayley** (1821 - 1895).

For a binary operation $*$ on a finite set S, the Cayley table is a square array. Each element of S appears once to the left of a row, and once heading a column. The result $a * b$ is entered at the intersection of the row corresponding to a and the column corresponding to b.

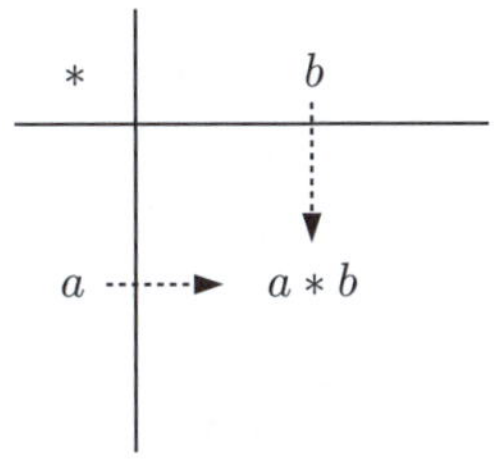

Example 44

Let a binary operation on $S = \{0, 1, 2, 3\}$ be defined by $a * b = a^2 + ab$.

a Construct the Cayley table for $*$.

b Is the operation closed on S?

c Is the operation commutative?

a The Cayley table is:

$*$	0	1	2	3
0	0	0	0	0
1	1	2	3	4
2	4	6	8	10
3	9	12	15	18

b From the table, $*$ is not closed on S.
For example, $3 * 2 = 15 \notin S$.

c The lack of symmetry about the leading diagonal indicates that $*$ is not commutative.
For example, $3 * 2 = 15$ whereas $2 * 3 = 10$.

When we study **groups**, we will see the significance of when a Cayley table is a **Latin square**.

A **Latin square** of order n, $n \in \mathbb{Z}^+$, is an $n \times n$ array using n distinct symbols, where each row and column contains each symbol exactly once.

For example:

$$\begin{matrix} 1 & 2 & 3 \\ 3 & 1 & 2 \\ 2 & 3 & 1 \end{matrix}$$

is a Latin square of order 3.

$$\begin{matrix} 1 & 2 & 3 \\ 3 & 2 & 1 \\ 2 & 1 & 2 \end{matrix}$$

is not a Latin square since, for example, row 3 contains the element 2 twice.

A Latin square of infinite order can be similarly defined.

EXERCISE D.4

1 Construct a Cayley table for multiplication modulo 5 on $\{1, 2, 3, 4\}$.

a Use the table to solve the following for x in modulo 5.

i $2x = 1$ **ii** $4x = 3$

iii $3x = 4$ **iv** $4x + 3 = 4$

b State the identity, if it exists.

c Verify that $3^{-1} = 2$.

d State the inverse of each other element.

2 Each of the following Cayley tables describes a different closed binary operation $*$ in $S = \{a, b, c\}$. For each operation:

i Find an identity element, if it exists.

ii Find an inverse for each element, if one exists.

iii State whether the operation is commutative.

iv State whether the operation is associative.

v State whether the Cayley table is a Latin square.

a

$*$	a	b	c
a	a	b	c
b	b	c	a
c	c	a	b

b

$*$	a	b	c
a	a	a	a
b	a	b	c
c	a	c	b

c

$*$	a	b	c
a	a	c	b
b	c	b	a
c	b	a	c

d

$*$	a	b	c
a	c	a	b
b	a	b	c
c	b	c	c

e

$*$	a	b	c
a	b	c	a
b	a	b	c
c	c	a	b

3 a Let $U_4 = \{1, i, -i, -1\} \subseteq \mathbb{C}$ and consider the operation of multiplication in U_4.

i Construct a Cayley table for the operation.

ii Is the operation: **A** commutative **B** associative?

iii State the identity, if it exists.

iv Find the inverse of each element, if possible.

b Now consider $U_n = \{1, \alpha, \alpha^2, ..., \alpha^{n-1}\}$, $n \in \mathbb{Z}^+$, where $\alpha = \text{cis}\left(\frac{2\pi}{n}\right)$.

i Describe the set U_n. **ii** Show that when $n = 4$, this is the set $\{1, i, -i, -1\}$.

E GROUPS

A set with one or more operations defined on it is called an **algebraic structure**.

Within the set of algebraic structures there is a hierarchy:

An algebraic structure with one closed binary operation defined is referred to as a **groupoid**.

If the associative law is obeyed, the groupoid qualifies as a **semigroup**.

A semigroup with an identity element is known as a **monoid**.

In some of these monoids, each element will have an inverse and this leads us to **groups**.

A non-empty set G on which a binary operation $*$ is defined is said to be a **group**, written $\{G, *\}$, if each of the following four axioms hold:

- G is **closed** under $*$.
  So, for all $a, b \in G$, $a * b \in G$.
- $*$ is **associative** on G.
  So, for all $a, b, c \in G$, $(a * b) * c = a * (b * c)$.
- $*$ has an **identity** element in G.
  So, there exists an element $e \in G$ such that $a * e = e * a = a$ for all $a \in G$.
- Each element of G has an **inverse** under $*$.
  So, for each $a \in G$, there exists an element $a^{-1} \in G$ such that $a^{-1} * a = a * a^{-1} = e$.

For a group $\{G, *\}$, if it is clear which operation is the relevant group operation, the group will sometimes be referred to simply as G.

A group is called **finite** if it has a finite number of elements, or **infinite** if it has an infinite number of elements.

By **Theorems 3** and **4**, the identity element in a group is **unique**, and each element in the group has a **unique** inverse. These results can be proved directly from the four **group axioms** which are in the definition of a group.

In general, we may assume the closure of the set of real numbers $\mathbb{R}$ and the set of integers $\mathbb{Z}$ under the operations $+$, $-$, and $\times$. $\mathbb{R}\setminus\{0\}$ is closed under $\div$.

For example:

- We have seen that $\mathbb{Z} \subseteq \mathbb{Q} \subseteq \mathbb{R} \subseteq \mathbb{C}$. Each of these sets, together with the operation of addition, forms a group with identity 0. So, $\{\mathbb{Z}, +\}$, $\{\mathbb{Q}, +\}$, $\{\mathbb{R}, +\}$, and $\{\mathbb{C}, +\}$ are all infinite groups. In each case, the inverse of an element x is $x^{-1} = -x$, the negative of x, since $x + (-x) = (-x) + x = 0$, the identity. These are all infinite groups.
- The integers modulo 3, $\mathbb{Z}_3 = \{0, 1, 2\}$ with addition modulo 3, form the finite group $\{\mathbb{Z}_3, +_3\}$ with three elements. The identity is 0, since $0 +_3 0 = 0$, $1 +_3 0 = 0 +_3 1 = 1$ and $2 +_3 0 = 0 +_3 2 = 2$.

  In this case, the inverses are:

  $1^{-1} = 2$, since $1 +_3 2 = 2 +_3 1 = 0$, the identity,
  $2^{-1} = 1$, since $2 +_3 1 = 1 +_3 2 = 0$, and
  $0^{-1} = 0$.
- It can be shown that for $\mathbb{Z}_n = \{0, 1, 2, ..., n-1\}$, $n \in \mathbb{Z}^+$, and $+_n$ addition modulo n, $\{\mathbb{Z}_n, +_n\}$ is a (finite) group.

- The set $\mathbb{R}$ is closed under multiplication $\times$, the operation $\times$ is associative, and an identity 1 exists in $\mathbb{R}$ for multiplication. However $\{\mathbb{R}, \times\}$ is not a group since the element $0 \in \mathbb{R}$ has no inverse under $\times$. It can be shown that $\{\mathbb{R} \setminus \{0\}, \times\}$ is a group.

Example 45

The given Cayley table is for the operation $*$ on the set $S = \{e, a, b, c, d, x\}$.

Show that:

a S is closed under $*$

b there is an identity element for $*$ in S

c each element of S has a unique inverse

d $*$ is not associative.

$*$	e	a	b	c	d	x
e	e	a	b	c	d	x
a	a	e	c	d	x	b
b	b	d	e	x	c	a
c	c	x	a	e	b	d
d	d	b	x	a	e	c
x	x	c	d	b	a	e

a For all $a, b \in S$, $a * b \in S$. $\therefore$ S is closed under $*$.

b For all $y \in S$, $e * y = y * e = y$. $\therefore$ since $e \in S$, the identity is e.

c For all $y \in S$, $y * y = e$. $\therefore$ each element has a unique inverse, which is itself.

d $a * (b * c) = a * x = b$

$(a * b) * c = c * c = e \neq a * (b * c)$

Thus $*$ is not an associative operation, and so S does *not* form a group under $*$.

Notice in this example that each element has a unique inverse. So, while associativity implies that each inverse is unique (see **Theorem 4**), the converse of this result is not true. When verifying that a set is a group, the property of associativity must therefore be checked.

However, if the Cayley table indicates the inverse is not unique, we can conclude that the operation is not associative.

CANCELLATION LAWS

The group axioms lead to the following cancellation laws. As commutativity is not a group axiom, it is necessary to consider both left and right cancellation laws.

Theorem 5

Given a group $\{G, *\}$, the following apply for all $a, b, c \in G$:

Left cancellation law If $a * b = a * c$ then $b = c$.

Right cancellation law If $b * a = c * a$ then $b = c$.

Proof of right cancellation law:

$b * a = c * a$

$\Rightarrow$ $(b * a) * a^{-1} = (c * a) * a^{-1}$ {multiply on the right by a^{-1}, where $a^{-1} \in G$ is the inverse of a, which exists since G is a group}

$\Rightarrow$ $b * (a * a^{-1}) = c * (a * a^{-1})$ {Associative Law}

$\Rightarrow$ $b * e = c * e$ {where $e \in G$ is the identity}

$\Rightarrow$ $b = c$

A similar proof establishes the left cancellation law.

CAYLEY TABLES FOR GROUPS

Cayley tables for groups have the property of being **Latin squares**, as described in the following theorem:

Theorem 6

> If $\{G, *\}$ is a group, then each element of G will appear exactly once in every row and exactly once in every column of its Cayley table.

Proof:

Let $a, p \in G$.

As $\{G, *\}$ is a group, $a^{-1} \in G$ where a^{-1} is the inverse of a

$\Rightarrow$ $a^{-1} * p \in G$ and $p * a^{-1} \in G$ for all a, p. {Closure}

$$\begin{aligned} \text{Now} \quad a * (a^{-1} * p) &= (a * a^{-1}) * p && \{\text{Associative}\} \\ &= e * p && \{e \text{ is the identity element}\} \\ &= p \end{aligned}$$

Therefore for any p and a it is always possible to find an element $x = a^{-1} * p$ of G such that $a * x = p$.

Hence p must be on the row corresponding to a. This means that every element must appear on every row.

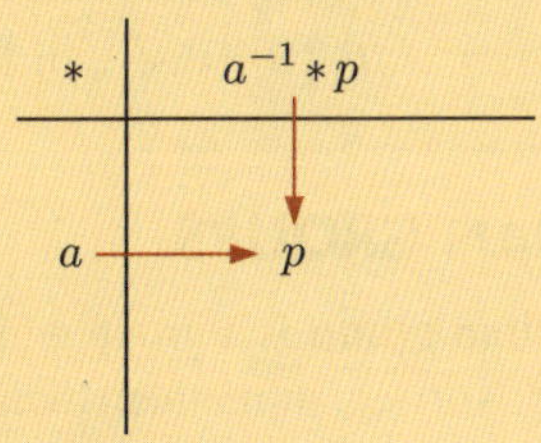

Similarly, we can show that an element $y = p * a^{-1}$ of G can be found such that $y * a = p$, so p will appear in every column.

Now we need to show that the elements appear *only* once in each row and column.

For finite groups, we could note that there are only n spaces to fill in each row and column. Since each element must appear at least once, it can appear *only* once.

However more generally, suppose that x_1 and x_2 are such that $a * x_1 = p$ and $a * x_2 = p$. Then $a * x_1 = a * x_2$, and so $x_1 = x_2$. {left cancellation law}

We can argue similarly for each column.

Hence p must appear exactly once in every row and column.

The converse of **Theorem 6** is not true. That is, a Latin square need not give rise to a group.

For example, Cayley table

$*$	a	b	c
a	a	b	c
b	c	a	b
c	b	c	a

is a Latin square, but $\{\{a, b, c\}, *\}$ is not a group,

since for example there is no identity element:

a is not an identity since $b * a = c \neq b$

b is not an identity since $b * a = c \neq a$

c is not an identity since $c * a = b \neq a$.

Example 46

a Show that $\mathbb{Z}_4 \setminus \{0\}$, that is $\{1, 2, 3\}$, does not form a group under $\times_4$, multiplication modulo 4.

b Prove that if n is composite, then $\mathbb{Z}_n \setminus \{0\}$ does not form a group under $\times_n$, multiplication modulo n.

a The Cayley table for $\mathbb{Z}_4 \setminus \{0\}$ under $\times_4$ is:

$\times_4$	1	2	3
1	1	2	3
2	2	0	2
3	3	2	1

$\mathbb{Z}_4 \setminus \{0\}$ is not closed under $\times_4$ as $2 \times_4 2 = 0$ and $0 \notin \mathbb{Z}_4 \setminus \{0\}$.

$\therefore$ $\mathbb{Z}_4 \setminus \{0\}$ does not form a group under $\times_4$.

b If n is composite then $n = pq$ for some $p, q \in \mathbb{Z}^+$ with $1 < p, q < n$

Thus $p, q \in \mathbb{Z}_n$ and $p \times q = n \equiv 0 \pmod{n}$

$\therefore$ $p \times_n q = 0$.

But $0 \notin \mathbb{Z}_n \setminus \{0\}$

$\therefore$ $\mathbb{Z}_n \setminus \{0\}$ is not closed under $\times_n$

$\therefore$ $\mathbb{Z}_n \setminus \{0\}$ does not form a group under $\times_n$.

ABELIAN GROUPS

An **Abelian group** is a group which has the commutativity property. The set of Abelian groups is named after the Norwegian mathematician **Niels Henrik Abel** (1802 - 1829).

A group $\{G, *\}$ is **Abelian** if $a * b = b * a$ for all $a, b \in G$.

If the Cayley table for a group G is symmetric about the main diagonal, then $a * b = b * a$ for all $a, b \in G$, and the group is Abelian.

Example 47

a Show that the set $\mathbb{Z}_5 \setminus \{0\}$, that is $\{1, 2, 3, 4\}$, forms a group under $\times_5$, multiplication modulo 5.

b Is this group Abelian?

a *Closure*: The Cayley table is:

$\times_5$	1	2	3	4
1	1	2	3	4
2	2	4	1	3
3	3	1	4	2
4	4	3	2	1

From the table $a \times_5 b \in \mathbb{Z}_5 \setminus \{0\}$ for all $a, b \in \mathbb{Z}_5 \setminus \{0\}$.

Associative: This follows from the associativity of multiplication of integers.

Identity: The element $1 \in \mathbb{Z}_5 \setminus \{0\}$ is such that $a \times_5 1 = 1 \times_5 a = a$ for all $a \in \mathbb{Z}_5 \setminus \{0\}$. Therefore 1 is the multiplicative identity element for $\mathbb{Z}_5 \setminus \{0\}$.

Inverse: $1 \times_5 1 = 1$ and $4 \times_5 4 = 1$, so each of 1 and 4 is its own inverse. $3 \times_5 2 = 2 \times_5 3 = 1$, so 2 and 3 are inverses of each other.

Thus for each element $a \in \mathbb{Z}_5 \setminus \{0\}$ there is an inverse $a^{-1} \in \mathbb{Z}_5 \setminus \{0\}$, for the operation $\times_5$ with identity 1.

Therefore $\{\mathbb{Z}_5 \setminus \{0\}, \times_5\}$ forms a group, since it satisfies all four group axioms.

b Since the multiplication of integers is commutative, it follows that $a \times_5 b = b \times_5 a$ for all $a, b \in \mathbb{Z}_5 \setminus \{0\}$. Therefore the group is Abelian.

Example 48

Show that the set of bijections on a set A under the operation *composition of functions* forms a group. Is the group Abelian?

Closure: If $f : A \mapsto A$ and $g : A \mapsto A$, then $g \circ f : A \mapsto A$.
The composition of two bijections is a bijection, so closure applies.

Associative: The composition of functions is associative, since

$$\begin{aligned}(h \circ g) \circ f &= (h \circ g)(f(x)) \\ &= h(g(f(x))) \\ &= h((g \circ f)(x)) \\ &= h \circ (g \circ f)\end{aligned}$$

$\therefore$ the composition of bijections is also associative.

Identity: The identity function $e : x \mapsto x$ for $x \in A$, is a bijection.
For all functions f, $e \circ f = f \circ e = f$
$\therefore$ there is an identity in the set of bijections under composition of functions.

Inverse: Every bijection f on A has an inverse f^{-1} such that $f \circ f^{-1} = f^{-1} \circ f = e$.

Therefore the set of bijections on a set A forms a group under the operation composition of functions.

In general, $f \circ g \neq g \circ f$, so the group is not Abelian. {see **Example 31**}

Example 49

Show that the set $\mathbb{R}$ with the binary operation $+$ is an Abelian group.

Closure: When two real numbers are added, the result is always a real number. Therefore $\mathbb{R}$ is closed under addition.

Associative: For all $a, b, c \in \mathbb{R}$, $a + (b + c) = (a + b) + c = a + b + c$.
Therefore $+$ is an associative operation on $\mathbb{R}$.

Identity: There exists an element $0 \in \mathbb{R}$ such that for all $a \in \mathbb{R}$, $a + 0 = 0 + a = a$.
Therefore there is an identity element in $\mathbb{R}$ for $+$.
0 is called the **additive identity** in $\mathbb{R}$.

Inverse: If $a \in \mathbb{R}$, then $-a \in \mathbb{R}$ and $a + (-a) = (-a) + a = 0$.
Therefore each element of $\mathbb{R}$ has an inverse in $\mathbb{R}$.

Therefore, $\{\mathbb{R}, +\}$ is a group, and is an example of an infinite group.

Since $a + b = b + a$ for all $a, b \in \mathbb{R}$, $\{\mathbb{R}, +\}$ is an Abelian group.

EXERCISE E.1

1 The given Cayley table is for the operation $*$ on the set $S = \{e, a, b, c, d\}$.

$*$	e	a	b	c	d
e	e	a	b	c	d
a	a	e	c	d	b
b	b	d	e	a	c
c	c	b	d	e	a
d	d	c	a	b	e

a Show that:

i S is closed under $*$

ii there is an identity element for $*$ in S

iii each element of S has a unique inverse.

b How many different checks must be made to show that $*$ is associative?

2 Prove the left cancellation law:

Consider a group $\{G, *\}$. For all $a, b, c \in G$, if $a * b = a * c$ then $b = c$.

3 Determine, giving reasons, whether each of the following is a group. If it is, state whether the group is Abelian.

a $\mathbb{Q}$ under multiplication.

b $\mathbb{Q} \setminus \{0\}$ under multiplication.

c The set of odd integers under multiplication.

d $\{3^n \mid n \in \mathbb{Z}\}$ under multiplication.

e $\left\{1, -\frac{1}{2} + i\frac{\sqrt{3}}{2}, -\frac{1}{2} - i\frac{\sqrt{3}}{2}\right\}$ under multiplication.

f $\{3n \mid n \in \mathbb{Z}\}$ under addition.

g $\{3n \mid n \in \mathbb{Z}\}$ under multiplication.

h $\mathbb{C}$ under addition.

i $\mathbb{C}$ under multiplication.

j $\{a + bi \mid a, b \in \mathbb{R}, \ |a + bi| = 1\}$ under multiplication.

k $\mathbb{C} \setminus \{0\}$ under multiplication.

4 Let $G = \mathbb{Q}^+$ with binary operation $*$ such that $a * b = \dfrac{ab}{4}$ for all $a, b \in \mathbb{Q}^+$.

Show that $\{G, *\}$ is a group.

5 Determine whether the following statement is true or false, giving reasons for your answer:

"If the Cayley table of a set G with closed binary operation $*$ is a Latin square, then $\{G, *\}$ is a group."

6 **a** Verify that $\{\mathbb{Z}_7 \setminus \{0\}, \times_7\}$ is a group, where $\times_7$ is multiplication modulo 7.

b Show that $\{\mathbb{Z}_9 \setminus \{0\}, \times_9\}$ is not a group.

7 Let F be the set of real-valued functions on $\mathbb{R}$, so $F = \{f \mid f : \mathbb{R} \to \mathbb{R}$ is a function$\}$.

Define addition in F by $(f + g)(x) = f(x) + g(x)$, $x \in \mathbb{R}$.

Verify that $\{F, +\}$ is a group.

8 Use the axioms of a group to prove each of the following:

a A group $\{G, *\}$ has a *unique* identity element.

b In a group $\{G, *\}$, each element has a *unique* inverse.

9 Prove: If $\{G, *\}$ is a group, then for any fixed elements $a, b \in G$, the equations $a * x = b$ and $y * a = b$ each have a unique solution in G.

10 Let $S = \mathbb{R} \setminus \{-2\}$ and define the binary operation $*$ on S by $a * b = a + b + \frac{ab}{2}$.

a Verify that $\{S, *\}$ is an Abelian group.

b Find the solution(s) in S to the equation:

i $2 * x * 5 = 11$ **ii** $x * 3 * 8 = 12$.

11 Suppose G is a set with a binary operation $*$ such that:

1 $*$ is closed and associative in G.

2 G has a **left identity** element e, so for each $a \in G$, $e * a = a$.

3 Each element in G has a **left inverse**, so for each $a \in G$ there exists $a_L^{-1} \in G$ such that $a_L^{-1} * a = e$.

Prove that $\{G, *\}$ is a group.

12 Suppose G is a set with a closed, associative binary operation $*$, and an identity element e. Prove that if the Cayley table for G contains e once in every row and once in every column, then $\{G, *\}$ is a group.

13 Let S be the set of all subsets of a universal set $\mathbb{U}$. Show that $\{S, \Delta\}$ is a group. You may assume that the symmetric difference Δ of sets is associative from **Exercise D.2** question **5**.

ORDER

In a group $\{G, *\}$ where $a \in G$, we define:

$$a^m = \underbrace{a * a * .... * a}_{m \text{ times}}$$

a^m is unique since $*$ is closed and associative in G.

We define $a^0 = e$, where e is the identity in G.

The **order of a group** $\{G, *\}$ is the number of elements in G, which is $n(G)$ or $|G|$.

The **order of an element** a of a group $\{G, *\}$ is the smallest *positive* integer m for which $a^m = e$, where e is the identity in G. We write $|a| = m$ to denote element a has finite order m. If no such $m \in \mathbb{Z}^+$ exists, then element a has **infinite order** in $\{G, *\}$.

In any group, the **order of the identity element** is 1.

For example:

- In the group $\{\mathbb{Z}_3, +_3\}$, the identity is 0 and the element 2 has order 3 since:

  $2^1 = 2 \neq 0$

  $2^2 = 2 +_3 2 = 1 \neq 0$

  $2^3 = 2 +_3 2 +_3 2 = 0$, the identity

  $\therefore$ 3 is the smallest positive integer such that $2^3 = e = 0$.

$a^m = \underbrace{a * a * .... * a}_{m \text{ times}}$ where $*$ is $+_3$ in this example.

- In the group $\{\mathbb{R}, +\}$, the identity is 0 and the element 2 has infinite order since:

$$2^1 = 2 \neq 0$$
$$2^2 = 2 + 2 = 4 \neq 0$$
$$\vdots$$
$$2^m = \underbrace{2 + 2 + ... + 2}_{m \text{ times}} = 2m \neq 0 \quad \text{for all } m \in \mathbb{Z}^+$$

$\therefore$ there exists no $m \in \mathbb{Z}^+$ such that $2^m = 2m = 0$

$\therefore$ 2 has infinite order in $\{\mathbb{R}, +\}$.

A **finite group** is a group with finite order, which is a group containing a finite number of elements.

An **infinite group** is a group with infinite order, which is a group containing an infinite number of elements.

Theorem 7

In a finite group $\{G, *\}$, each element has finite order.

Proof:

If $a \in G$, then $a^2, a^3, a^4, ..., a^m, ... \in G$ for all $m \in \mathbb{Z}^+$, since G is closed under $*$.

Since G is a finite group, this list of elements cannot be infinite. There must exist $m_1, m_2 \in \mathbb{Z}^+$ such that $a^{m_1} = a^{m_2}$.

Suppose $m_1 < m_2$ (without loss of generality). The inverse of a, element a^{-1}, exists since G is a group. Multiplying $a^{m_1} = a^{m_2}$ on the left by a^{-1} m_1 times,

$$\underbrace{a^{-1} * a^{-1} * .... * a^{-1}}_{m_1 \text{ times}} * a^{m_1} = \underbrace{a^{-1} * .... * a^{-1}}_{m_1 \text{ times}} * a^{m_2}$$

{We omit brackets since in a group $*$ is associative}

$$\therefore \underbrace{a^{-1} * a^{-1} * .... * a^{-1}}_{m_1 - 1 \text{ times}} * \underbrace{a^{-1} * a}_{e} * \underbrace{a * .... * a}_{m_1 - 1 \text{ times}} = \underbrace{a^{-1} * .... * a^{-1}}_{m_1 \text{ times}} * \underbrace{a * .... * a}_{m_2 \text{ times}}$$

$$\therefore \underbrace{a^{-1} * a^{-1} * .... * a^{-1}}_{m_1 - 1 \text{ times}} * \underbrace{a * .... * a}_{m_1 - 1 \text{ times}} = \underbrace{a^{-1} * .... * a^{-1}}_{m_1 - 1 \text{ times}} * (a^{-1} * a) * \underbrace{a * .... * a}_{m_2 - 1 \text{ times}}$$

{since $a^{-1} * a = e$}

$$\vdots$$

$$\therefore e = \underbrace{a * .... * a}_{m_2 - m_1 \text{ times}} \qquad \{\text{since } m_2 > m_1\}$$

$$\therefore e = a^{m_2 - m_1} \quad \text{where } m_2 - m_1 \in \mathbb{Z}^+.$$

Hence the element a has finite order.

Theorem 8

Consider a group $\{G, *\}$ with identity e. For any $a \in G$:

1 $(a^{-1})^{-1} = a$

2 $(a^n)^{-1} = (a^{-1})^n$

Proof:

1 By definition of the inverse of a, $a * a^{-1} = a^{-1} * a = e$

$\therefore \ (a^{-1})^{-1} = a.$

2 For any $a \in G$, $a^{-1} \in G$ since G is a group.

Also a^n, $(a^{-1})^n \in G$ since $*$ is closed in G.

Consider
$$\begin{aligned}(a^{-1})^n * a^n &= \underbrace{a^{-1} * a^{-1} * \dots * a^{-1}}_{n \text{ times}} * \underbrace{a * \dots * a}_{n \text{ times}} \\ &= \underbrace{a^{-1} * a^{-1} * \dots * a^{-1}}_{n-1 \text{ times}} * e * \underbrace{a * \dots * a}_{n-1 \text{ times}} \quad \{\text{associativity and } a^{-1} * a = e\} \\ &\ \ \vdots \\ &= e\end{aligned}$$

Similarly $a^n * (a^{-1})^n = e$

$\therefore \ (a^n)^{-1} = (a^{-1})^n$

Example 50

a Show that $\mathbb{Z}_4 = \{0, 1, 2, 3\}$ under the operation of $+_4$, addition modulo 4, is a group.

b Is the group Abelian?

c State the order of each element of the group.

---

a The Cayley table is:

$+_4$	0	1	2	3
0	0	1	2	3
1	1	2	3	0
2	2	3	0	1
3	3	0	1	2

Closure: For all $a, b \in \mathbb{Z}_4$, $a +_4 b \in \mathbb{Z}_4$.
$\therefore$ $\mathbb{Z}_4$ is closed under addition modulo 4.

Associative: Associativity follows from the associative property of addition in $\mathbb{Z}$.

Identity: For all $a \in \mathbb{Z}_4$, $0 +_4 a = a +_4 0 = a$.
$\therefore$ since $0 \in \mathbb{Z}_4$, it is the identity element in $\mathbb{Z}_4$ for $+_4$.

Inverse: From the table, $0 +_4 0 = 0$, $2 +_4 2 = 0$, and $1 +_4 3 = 3 +_4 1 = 0$.
$\therefore$ each of 0 and 2 is its own inverse, while 1 and 3 are inverses of each other.

Therefore $\{\mathbb{Z}_4, +_4\}$ is a group, since the four group axioms hold.

b From the symmetry of the Cayley table about the main diagonal, $a + b = b + a$ for all $a, b \in \mathbb{Z}_4$.

Therefore, $\{\mathbb{Z}_4, +_4\}$ is an Abelian group.

c 0 is the identity which has order 1.

1 has order 4, since $1^1, 1^2, 1^3 \not\equiv 0 \pmod 4$
whereas $1^4 = 1 +_4 1 +_4 1 +_4 1 = 0$.

2 has order 2, since $2^1 \not\equiv 0 \pmod 4$
whereas $2^2 = 2 +_4 2 = 0$.

3 has order 4, since $3^1, 3^2, 3^3 \not\equiv 0 \pmod 4$
whereas $3^4 = 3 +_4 3 +_4 3 +_4 3 = 0$.

Example 51

Let $\alpha = \frac{1}{2} + i\frac{\sqrt{3}}{2}$ and let $G = \{\alpha^n \mid n \in \mathbb{Z}^+\}$.

a Show that G is an Abelian group under multiplication in $\mathbb{C}$ and construct the Cayley table.

b What is the order of G?

c Find the order of:

i α **ii** α^2 **iii** α^3 **iv** α^4.

Notice that $\alpha = \frac{1}{2} + i\frac{\sqrt{3}}{2} = \text{cis}\left(\frac{\pi}{3}\right)$

By De Moivre's theorem:

$\alpha^2 = \text{cis}\left(\frac{2\pi}{3}\right)$

$\alpha^3 = \text{cis}\left(\frac{3\pi}{3}\right) = \text{cis}(\pi) = -1$

$\alpha^4 = \text{cis}\left(\frac{4\pi}{3}\right)$

$\alpha^5 = \text{cis}\left(\frac{5\pi}{3}\right)$

$\alpha^6 = \text{cis}(2\pi) = 1$

$\alpha^7 = \text{cis}\left(\frac{7\pi}{3}\right) = \text{cis}\left(\frac{\pi}{3}\right) = \alpha$

$\alpha^8 = \alpha^2$, and so on.

$\therefore\ G = \{\alpha, \alpha^2, \alpha^3 = -1, \alpha^4, \alpha^5, \alpha^6 = 1\}$ is the set of six sixth roots of unity in $\mathbb{C}$.

a

$\times$	1	α	α^2	α^3	α^4	α^5
1	1	α	α^2	α^3	α^4	α^5
α	α	α^2	α^3	α^4	α^5	1
α^2	α^2	α^3	α^4	α^5	1	α
α^3	α^3	α^4	α^5	1	α	α^2
α^4	α^4	α^5	1	α	α^2	α^3
α^5	α^5	1	α	α^2	α^3	α^4

Closure: Each element in the table is in G $\therefore\ a \times b \in G$ for all $a, b \in G$.

Associative: Multiplication is associative in $\mathbb{C}$ $\therefore$ it is associative in G.

Identity: The element $1 \in G$ is the identity, since $1 \times \alpha^i = \alpha^i \times 1 = \alpha^i$ for all $i = 1, ..., 6$.

Inverse: Since $\alpha^6 = 1$ from the table, we obtain:

$(\alpha)^{-1} = \alpha^5$, $(\alpha^2)^{-1} = \alpha^4$, $(\alpha^3)^{-1} = \alpha^3$, $(\alpha^4)^{-1} = \alpha^2$,

$(\alpha^5)^{-1} = \alpha$, $1^{-1} = 1$

Hence $\{G, \times\}$ is a group.

Since $\times$ is commutative in $\mathbb{C}$, $\{G, \times\}$ is commutative in G, and G is therefore an Abelian group.

b G has 6 elements, so G has order 6.

c **i** From above, $m = 6$ is the smallest positive integer such that $\alpha^m = 1$

$\therefore\ \alpha$ has order 6

ii $\alpha^2 \neq 1$, $(\alpha^2)^2 = \alpha^4 \neq 1$, and $(\alpha^2)^3 = \alpha^6 = 1$

$\therefore\ \alpha^2$ has order 3

iii $\alpha^3 \neq 1$ and $(\alpha^3)^2 = \alpha^6 = 1$

$\therefore\ \alpha^3$ has order 2

iv $\alpha^4 \neq 1$, $(\alpha^4)^2 = \alpha^8 = \alpha^6\alpha^2 = 1 \times \alpha^2 = \alpha^2 \neq 1$, and $(\alpha^4)^3 = \alpha^{12} = (\alpha^6)^2 = 1^2 = 1$

$\therefore\ \alpha^4$ has order 3

EXERCISE E.2

1 Let $G = \{2, 4, 6, 8\}$ under the binary operation $*$, where $*$ denotes multiplication modulo 10.

- **a** Write down the Cayley table for $\{G, *\}$.
- **b** Does $\{G, *\}$ have an identity element?
- **c** Show that $\{G, *\}$ is a group.
- **d** Find the order of each element in G.
- **e** Find all solution pairs (x, y) where $x, y \in G$, to the equation $y = 2 * x * 4$.

2 Let $A = \{1, 3, 5, 7\}$ under the binary operation $*$, where $*$ denotes multiplication modulo 8.

- **a** Verify that $\{A, *\}$ is a group.
- **b** Is the group Abelian?
- **c** Find the order of each element in $\{A, *\}$.
- **d** Let $H = \{1, 3\}$. Verify that $\{H, *\}$ is a group.

3
- **a** For $n \in \mathbb{Z}^+$ any fixed positive integer, let $U_n = \{\alpha \mid \alpha^n = 1, \ \alpha \in \mathbb{C}\}$ be the nth roots of unity in $\mathbb{C}$. Verify that $\{U_n, \times\}$ is a group under multiplication in $\mathbb{C}$.
- **b** Write down the elements of U_4.
- **c** State the order of each element in U_4.

4 Let $\{G, *\}$ be a group with identity e, and suppose $x \in G$ has finite order m.
Prove that $x^n = e \Leftrightarrow n$ is a multiple of m.

5 Let $g, f \in G$ be elements of the group $\{G, *\}$. Show that:

- **a** g has finite order $\Leftrightarrow g^{-1}$ has finite order
- **b** $|g| = |g^{-1}|$, which means g and g^{-1} have the same order
- **c** if $|fg| = m \in \mathbb{Z}^+$ then $|gf| = m$.

INVESTIGATION — GROUPS OF ORDER $n = 1, 2, 3$, OR 4

We can use the definition of a group and our established properties of groups to construct groups of order 1, 2, 3, and 4. By doing this we investigate for each order how many such structures are possible.

Case $n = 1$

Consider a group $\{G, *\}$ for which $|G| = 1$ and the identity element is e. Since e is the only element of G, $G = \{e\}$.

1. Construct a Cayley table for the binary operation $*$.
2. Show that $*$ is associative and that e is its own unique inverse.
3. Explain why each of the following is a group of order 1:
   - **a** $\{\{1\}, \times\}$
   - **b** $\{\{0\}, +\}$
   - **c** $\{\{6\}, \times_{10}\}$

Case $n = 2$

Consider a group $\{G, *\}$ for which $|G| = 2$ and the identity element is e. Since $|G| = 2$, there exists an element $a \in G$ such that $a \neq e$.

1. Show that the Cayley table for any such group G must be

$*$	e	a
e	e	a
a	a	e

2. Show that $*$ is associative.
3. Show that each element in G has a unique inverse.
4. Explain why each of the following is a group of order 2:
   - **a** $\{\mathbb{Z}_2, +_2\} = \{\{0, 1\}, +_2\}$
   - **b** $\{\mathbb{Z}_3 \backslash \{0\}, \times_3\} = \{\{1, 2\}, \times_3\}$

These groups are different but they have the same **structure** as determined by the Cayley table. We say the groups are **isomorphic**.

Case $n = 3$

Consider a group $\{G, *\}$ for which $|G| = 3$ and the identity element is e. Suppose the other distinct elements are a and b, so $G = \{e, a, b\}$.

1. Since $*$ is closed in a group, the Cayley table for this group is a Latin square with elements e, a, and b. Thus each row and column must contain e, a, b each exactly once. Remembering that e is the identity element, show that there is only one possible Cayley table.
2. Verify that:
   - **a** $a^2 = a * a = b$
   - **b** $b^2 = b * b = a$
   - **c** $a * b = a^3 = e$
   - **d** $b * a = b^3 = e$
3. Find the inverse of each element in G.
4. Explain why $\{\mathbb{Z}_3, +_3\} = \{\{0, 1, 2\}, +_3\}$ is a group of order 3.

All groups of order 3 are isomorphic.

Case $n = 4$

Consider a group $\{G, *\}$ for which $|G| = 4$, the identity element is e, and $G = \{e, a, b, c\}$.

1 Remembering that each row and column must contain e, a, b, and c each exactly once, show that there are *two* possible structures for the Cayley table:

$*$	e	a	b	c
e	e	a	b	c
a	a	b	c	e
b	b	c	e	a
c	c	e	a	b

  which corresponds to the **cyclic group** of order 4 generated by a, $\{G, *\} = \{\{e, a, a^2, a^3\}, *\}$.

$*$	e	a	b	c
e	e	a	b	c
a	a	e	c	b
b	b	c	e	a
c	c	b	a	e

  which corresponds to the **Klein 4-group** $\{V_4, *\} = \{\{e, a, b, c \mid a^2 = b^2 = c^2 = e\}, *\}$.

2 Show that the Cayley table

$*$	e	a	b	c
e	e	a	b	c
a	a	c	e	b
b	b	e	c	a
c	c	b	a	e

can be rewritten as the cyclic group of order 4 by relabelling the elements.

3 State the order of each element in:

a the cyclic group of order 4

b the Klein 4-group.

4 Show that:

a $\{\{1, 2, 3, 4\}, +_4\}$ is a cyclic group of order 4

b $\{\{1, 3, 5, 7\}, \times_8\}$ is a Klein 4-group

c $\{\{2, 4, 6, 8\}, \times_{10}\}$ is a cyclic group of order 4.

For finite groups of order $n \geqslant 5$, there are often many essentially different possible structures and it becomes unwieldy to consider the different possible Cayley tables. We will instead investigate the different possible group structures using **isomorphisms** later in the course.

F PERMUTATION GROUPS

A **permutation** is a bijection from a non-empty set to itself.

For example, consider the illustrated mapping of $f : S \to S$ where $S = \{1, 2, 3, 4\}$.

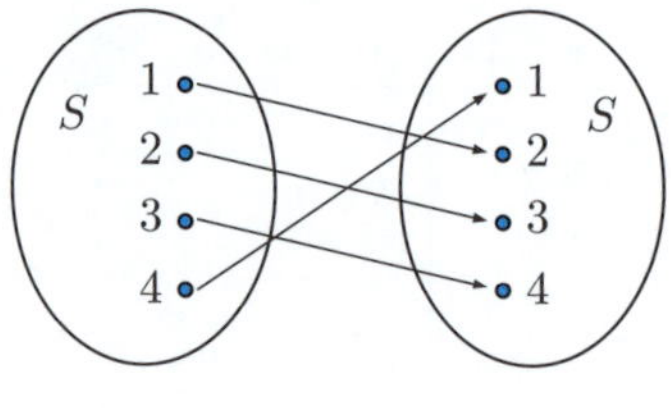

We observe that: $f : 1 \mapsto f(1) = 2$
$f : 2 \mapsto f(2) = 3$
$f : 3 \mapsto f(3) = 4$
$f : 4 \mapsto f(4) = 1$

The ordered pairs of the bijection are $\{(1, 2), (2, 3), (3, 4), (4, 1)\}$, and the permutation is also commonly written as:

$$p_a = \begin{pmatrix} 1 & 2 & 3 & 4 \\ f(1) & f(2) & f(3) & f(4) \end{pmatrix} = \begin{pmatrix} 1 & 2 & 3 & 4 \\ 2 & 3 & 4 & 1 \end{pmatrix}$$

More generally, a permutation p of the elements of $S = \{1, 2, 3, ..., n\}$, $n \in \mathbb{Z}^+$, is a bijection $f : S \to S$ which we write as $p = \begin{pmatrix} 1 & 2 & 3 & .... & n \\ f(1) & f(2) & f(3) & .... & f(n) \end{pmatrix}$.

A permutation can be viewed as a particular arrangement of the elements in a set.

Let S_n be the set of all permutations of the n distinct elements $1, 2, 3, ..., n$, where $n \in \mathbb{Z}^+$.

Since there are $n!$ possible arrangements of n distinct elements, there are $n!$ distinct permutations in S_n. We write $|S_n| = n!$.

COMPOSITION OF PERMUTATIONS

The **composition of two permutations** is also called **combining**, **multiplying**, or **finding the product**.

Let two permutations on $S = \{1, 2, 3, 4\}$ be $p_a = \begin{pmatrix} 1 & 2 & 3 & 4 \\ 2 & 3 & 4 & 1 \end{pmatrix}$ and $p_b = \begin{pmatrix} 1 & 2 & 3 & 4 \\ 3 & 1 & 2 & 4 \end{pmatrix}$.

Consider the composition of functions where p_a is followed by p_b, as shown in the diagram:

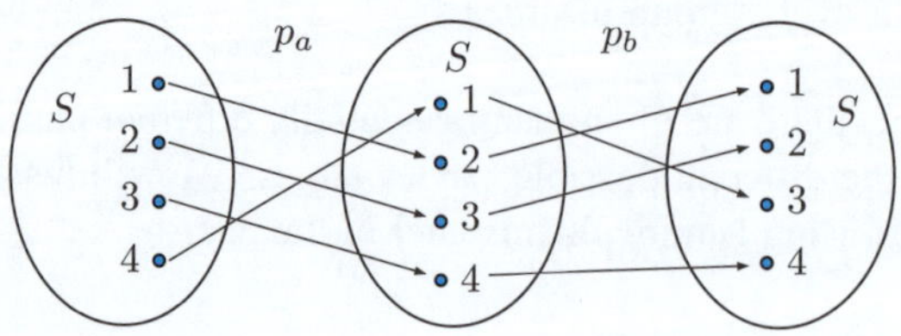

By following the arrows through, we find the resulting permutation

$$p_bp_a = \begin{pmatrix} 1 & 2 & 3 & 4 \\ 1 & 2 & 4 & 3 \end{pmatrix}$$

p_bp_a can also be found by writing the combined permutation in the order

$$\begin{pmatrix} 1 & 2 & 3 & 4 \\ 3 & 1 & 2 & 4 \end{pmatrix}\begin{pmatrix} 1 & 2 & 3 & 4 \\ 2 & 3 & 4 & 1 \end{pmatrix} = \begin{pmatrix} 1 & 2 & 3 & 4 \\ 1 & 2 & 4 & 3 \end{pmatrix} = p_bp_a$$

Note that we work from right to left when combining permutations. This is consistent with compositions of functions:

$$(p_b p_a)(x) = p_b(p_a(x)) = p_b \circ p_a(x)$$

However, not all texts follow this convention.

Composition of functions is, in general, not commutative, and this is therefore also the case for composition of permutations.

For example:
$$\begin{aligned} p_a p_b &= \begin{pmatrix} 1 & 2 & 3 & 4 \\ 2 & 3 & 4 & 1 \end{pmatrix}\begin{pmatrix} 1 & 2 & 3 & 4 \\ 3 & 1 & 2 & 4 \end{pmatrix} \\ &= \begin{pmatrix} 1 & 2 & 3 & 4 \\ 4 & 2 & 3 & 1 \end{pmatrix} \\ &\neq p_b p_a \end{aligned}$$

However, the composition of bijective functions of the form $f : S \to S$ is associative (see **Example 48**), so the process of composition of permutations can be used for more than two permutations.

For example:

$$p_4 p_3 p_2 p_1 = \begin{pmatrix} \cdot & \boxed{2} & \cdot & \cdot \\ \cdot & \boxed{3} & \cdot & \cdot \end{pmatrix}\begin{pmatrix} \cdot & \cdot & \cdot & \boxed{4} \\ \cdot & \cdot & \cdot & \boxed{2} \end{pmatrix}\begin{pmatrix} \cdot & \boxed{2} & \cdot & \cdot \\ \cdot & \boxed{4} & \cdot & \cdot \end{pmatrix}\begin{pmatrix} \boxed{1} & \cdot & \cdot & \cdot \\ \boxed{2} & \cdot & \cdot & \cdot \end{pmatrix} = \begin{pmatrix} 1 & \cdot & \cdot & \cdot \\ 3 & \cdot & \cdot & \cdot \end{pmatrix}$$

This process generalises for permutations in S_n. If we rearrange n distinct elements and then rearrange them again, the result is another arrangement of the same original n elements. It follows that S_n is **closed** under the binary and **associative** operation of **composition of permutations**.

IDENTITY

The **identity permutation** in S_n is $e = \begin{pmatrix} 1 & 2 & 3 & .... & n \\ 1 & 2 & 3 & .... & n \end{pmatrix}$, and $ep = pe = p$ for any permutation $p \in S_n$.

INVERSES

Intuitively, if n distinct objects can be rearranged by a permutation, then they can be put back in their original order by a corresponding inverse permutation.

Formally, since each permutation p is a bijection $f : S \to S$ on $S = \{1, 2, ..., n\}$, $n \in \mathbb{Z}^+$, there exists an inverse bijection $f^{-1} : S \to S$ such that $f \circ f^{-1} = f^{-1} \circ f = e$ in the group of bijections on S under composition of functions (see **Example 48**).

The permutation $p = \begin{pmatrix} 1 & 2 & 3 & .... & n \\ f(1) & f(2) & f(3) & .... & f(n) \end{pmatrix}$ has **inverse permutation**

$$p^{-1} = \begin{pmatrix} 1 & 2 & 3 & .... & n \\ f^{-1}(1) & f^{-1}(2) & f^{-1}(3) & .... & f^{-1}(n) \end{pmatrix}$$

such that $pp^{-1} = p^{-1}p = e$, the identity permutation in S_n.

For a given permutation $p = \begin{pmatrix} 1 & 2 & 3 & .... & n \\ f(1) & f(2) & f(3) & .... & f(n) \end{pmatrix}$,

we find p^{-1} by writing $p^{-1} = \begin{pmatrix} f(1) & f(2) & .... & f(n) \\ 1 & 2 & .... & n \end{pmatrix}$

and then rearranging the columns in p^{-1} so that the top row is in ascending order.

For example, in S_4 let $p = \begin{pmatrix} 1 & 2 & 3 & 4 \\ 3 & 1 & 4 & 2 \end{pmatrix}$.

$$\therefore \ p^{-1} = \begin{pmatrix} 3 & 1 & 4 & 2 \\ 1 & 2 & 3 & 4 \end{pmatrix} \qquad \{\text{swap rows in } p\}$$

$$= \begin{pmatrix} 1 & 2 & 3 & 4 \\ 2 & 4 & 1 & 3 \end{pmatrix} \qquad \{\text{rearrange the columns to obtain } 1\ 2\ 3\ 4 \text{ in the top row}\}$$

$$\text{On checking, } pp^{-1} = \begin{pmatrix} 1 & 2 & 3 & 4 \\ 3 & 1 & 4 & 2 \end{pmatrix}\begin{pmatrix} 1 & 2 & 3 & 4 \\ 2 & 4 & 1 & 3 \end{pmatrix} = e$$

$$\text{and } p^{-1}p = \begin{pmatrix} 1 & 2 & 3 & 4 \\ 2 & 4 & 1 & 3 \end{pmatrix}\begin{pmatrix} 1 & 2 & 3 & 4 \\ 3 & 1 & 4 & 2 \end{pmatrix} = e$$

We have verified the axioms of a group, so we are now able to state:

Theorem 9

The set S_n of all permutations of the n distinct elements $\{1, 2, ..., n\}$, $n \in \mathbb{Z}^+$, is a group of order $n!$ under the operation of composition of permutations.

We can also define:

For $n \in \mathbb{Z}^+$, S_n is called the **symmetric group of degree n**.

Note that S_4 is not an Abelian group, since we showed above $p_a p_b \neq p_b p_a$ for the given permutations $p_a, p_b \in S_4$.

Example 52

Consider the symmetric group of degree 3, which is the set S_3 of all possible permutations on $S = \{1, 2, 3\}$.

a Find all elements of S_3.

b Show, by direct verification of the group axioms, that S_3 is a group under composition of permutations.

c Is S_3 Abelian?

---

a We know that there are $3! = 6$ different permutations in S_3.

The identity $e = \begin{pmatrix} 1 & 2 & 3 \\ 1 & 2 & 3 \end{pmatrix}$.

Letting $\alpha = \begin{pmatrix} 1 & 2 & 3 \\ 2 & 3 & 1 \end{pmatrix}$, we find $\alpha^2 = \begin{pmatrix} 1 & 2 & 3 \\ 2 & 3 & 1 \end{pmatrix}\begin{pmatrix} 1 & 2 & 3 \\ 2 & 3 & 1 \end{pmatrix} = \begin{pmatrix} 1 & 2 & 3 \\ 3 & 1 & 2 \end{pmatrix}$

which is another permutation. We notice that $\alpha\alpha^2 = e$ and $\alpha^2\alpha = e$.

Letting $\beta = \begin{pmatrix} 1 & 2 & 3 \\ 1 & 3 & 2 \end{pmatrix}$, we find $\beta^2 = e$.

Letting $\gamma = \begin{pmatrix} 1 & 2 & 3 \\ 3 & 2 & 1 \end{pmatrix}$, we find $\gamma^2 = e$.

Letting $\delta = \begin{pmatrix} 1 & 2 & 3 \\ 2 & 1 & 3 \end{pmatrix}$, we find $\delta^2 = e$.

So, the six permutations on S are e, α, α^2, β, γ, and δ.

b $\alpha\beta = \begin{pmatrix} 1 & 2 & 3 \\ 2 & 3 & 1 \end{pmatrix}\begin{pmatrix} 1 & 2 & 3 \\ 1 & 3 & 2 \end{pmatrix} = \begin{pmatrix} 1 & 2 & 3 \\ 2 & 1 & 3 \end{pmatrix} = \delta$

$\alpha\gamma = \begin{pmatrix} 1 & 2 & 3 \\ 2 & 3 & 1 \end{pmatrix}\begin{pmatrix} 1 & 2 & 3 \\ 3 & 2 & 1 \end{pmatrix} = \begin{pmatrix} 1 & 2 & 3 \\ 1 & 3 & 2 \end{pmatrix} = \beta$

$\alpha\delta = \begin{pmatrix} 1 & 2 & 3 \\ 2 & 3 & 1 \end{pmatrix}\begin{pmatrix} 1 & 2 & 3 \\ 2 & 1 & 3 \end{pmatrix} = \begin{pmatrix} 1 & 2 & 3 \\ 3 & 2 & 1 \end{pmatrix} = \gamma$

Continuing in this way enables us to construct the Cayley table for the composition of permutations in S_3:

$*$	e	α	α^2	β	γ	δ
e	e	α	α^2	β	γ	δ
α	α	α^2	e	δ	β	γ
α^2	α^2	e	α	γ	δ	β
β	β	γ	δ	e	α	α^2
γ	γ	δ	β	α^2	e	α
δ	δ	β	γ	α	α^2	e

Closure: For all $a, b \in S_3$, $a * b \in S_3$. Therefore S_3 is closed under the operation.

Associative: Composition of functions is an associative operation, so the composition of permutations on S is also associative.

Identity: From the table, $a * e = e * a = a$ for all $a \in S_3$.
$\therefore$ e is indeed the identity element in S_3 for $*$.

Inverse: e is its own inverse.

$\alpha\alpha^2 = \alpha^2\alpha = e$ so $\alpha^{-1} = \alpha^2$ and $(\alpha^2)^{-1} = \alpha$

$\beta^2 = e$ so $\beta^{-1} = \beta$

$\gamma^2 = e$ so $\gamma^{-1} = \gamma$

$\delta^2 = e$ so $\delta^{-1} = \delta$

Hence each element in S_3 has an inverse in S_3.

We have directly verified the group axioms
$\therefore$ $\{S_3, *\}$ is a group.

c The Cayley table is not symmetric about the main diagonal.
For example, $\beta\alpha^2 = \delta$ but $\alpha^2\beta = \gamma \neq \delta$.
$\therefore$ S_3 is not an Abelian group.

Consistent with our previous definition of order of an element, we define:

The **order m of a permutation** $p \in S_n$ is the least positive integer m such that $p^m = \underbrace{ppp....p}_{m \text{ times}} = e$, the identity permutation in S_n.

EXERCISE F.1

1 Simplify the following compositions of permutations:

a $\begin{pmatrix} 1 & 2 & 3 & 4 \\ 1 & 4 & 2 & 3 \end{pmatrix}\begin{pmatrix} 1 & 2 & 3 & 4 \\ 4 & 2 & 3 & 1 \end{pmatrix}$ **b** $\begin{pmatrix} 1 & 2 & 3 & 4 \\ 2 & 3 & 1 & 4 \end{pmatrix}\begin{pmatrix} 1 & 2 & 3 & 4 \\ 4 & 3 & 1 & 2 \end{pmatrix}$

c $\begin{pmatrix} 1 & 2 & 3 & 4 \\ 2 & 1 & 4 & 3 \end{pmatrix}\begin{pmatrix} 1 & 2 & 3 & 4 \\ 2 & 1 & 4 & 3 \end{pmatrix}$ **d** $\begin{pmatrix} 1 & 2 & 3 & 4 \\ 3 & 4 & 1 & 2 \end{pmatrix}\begin{pmatrix} 1 & 2 & 3 & 4 \\ 2 & 3 & 1 & 4 \end{pmatrix}\begin{pmatrix} 1 & 2 & 3 & 4 \\ 4 & 1 & 2 & 3 \end{pmatrix}$

2 Find:

a $\begin{pmatrix} 1 & 2 & 3 & 4 \\ 3 & 1 & 4 & 2 \end{pmatrix}^{-1}$ **b** $\begin{pmatrix} 1 & 2 & 3 & 4 \\ 2 & 1 & 4 & 3 \end{pmatrix}^{-1}$

c $\left[\begin{pmatrix} 1 & 2 & 3 & 4 \\ 3 & 4 & 2 & 1 \end{pmatrix}\begin{pmatrix} 1 & 2 & 3 & 4 \\ 2 & 4 & 1 & 3 \end{pmatrix}\right]^{-1}$

3 Prove that for all permutations $p, q \in S_n$, $n \in \mathbb{Z}^+$, $(qp)^{-1} = p^{-1}q^{-1}$.

4 Find permutation p on $\{1, 2, 3, 4\}$ such that:

a $p\begin{pmatrix} 1 & 2 & 3 & 4 \\ 3 & 1 & 2 & 4 \end{pmatrix} = \begin{pmatrix} 1 & 2 & 3 & 4 \\ 2 & 1 & 4 & 3 \end{pmatrix}$ **b** $p\begin{pmatrix} 1 & 2 & 3 & 4 \\ 2 & 3 & 1 & 4 \end{pmatrix} = \begin{pmatrix} 1 & 2 & 3 & 4 \\ 2 & 4 & 1 & 3 \end{pmatrix}$

5 For each of the following, construct a Cayley table and determine whether the set of permutations is a group under composition of permutations.

a $\{A, B, C, D\}$ where $A = \begin{pmatrix} 1 & 2 & 3 & 4 \\ 1 & 2 & 3 & 4 \end{pmatrix}$, $B = \begin{pmatrix} 1 & 2 & 3 & 4 \\ 2 & 3 & 4 & 1 \end{pmatrix}$,
$C = \begin{pmatrix} 1 & 2 & 3 & 4 \\ 3 & 4 & 1 & 2 \end{pmatrix}$, $D = \begin{pmatrix} 1 & 2 & 3 & 4 \\ 4 & 1 & 2 & 3 \end{pmatrix}$

b $\{A, B, C, D\}$ where $A = \begin{pmatrix} 1 & 2 & 3 & 4 \\ 1 & 2 & 3 & 4 \end{pmatrix}$, $B = \begin{pmatrix} 1 & 2 & 3 & 4 \\ 2 & 1 & 4 & 3 \end{pmatrix}$,
$C = \begin{pmatrix} 1 & 2 & 3 & 4 \\ 3 & 4 & 1 & 2 \end{pmatrix}$, $D = \begin{pmatrix} 1 & 2 & 3 & 4 \\ 4 & 3 & 2 & 1 \end{pmatrix}$.

6 Find the order of each permutation:

a $p = \begin{pmatrix} 1 & 2 & 3 & 4 \\ 4 & 3 & 2 & 1 \end{pmatrix}$ **b** $q = \begin{pmatrix} 1 & 2 & 3 & 4 \\ 4 & 1 & 3 & 2 \end{pmatrix}$ **c** $r = \begin{pmatrix} 1 & 2 & 3 & 4 & 5 \\ 4 & 5 & 1 & 3 & 2 \end{pmatrix}$

d $e = \begin{pmatrix} 1 & 2 & 3 & 4 \\ 1 & 2 & 3 & 4 \end{pmatrix}$ **e** $s = \begin{pmatrix} 1 & 2 & 3 & 4 \\ 4 & 1 & 2 & 3 \end{pmatrix}$

7 Consider the symmetric group of degree 4, which is the set S_4 of all possible permutations on $S = \{1, 2, 3, 4\}$. Find all elements of S_4.

8 Consider the symmetric group of degree 2, which is the set S_2 of all possible permutations on $S = \{1, 2\}$.

a Find all elements of S_2.

b Show, by direct verification of the group axioms, that S_2 is a group under composition of permutations.

c Is S_2 Abelian? Explain your answer.

CYCLE NOTATION FOR PERMUTATIONS

If $p = \begin{pmatrix} 1 & 2 & 3 & 4 \\ 3 & 1 & 4 & 2 \end{pmatrix}$ then permutation p maps 1 to 3, 3 to 4, 4 to 2, and 2 to 1

$$p : 1 \mapsto 3 \mapsto 4 \mapsto 2$$

The **cycle notation** for this permutation is $p = (1\ 3\ 4\ 2)$.

Note that the cycle notation $(3\ 4\ 2\ 1)$ represents the same permutation p.

Similarly, $q = \begin{pmatrix} 1 & 2 & 3 & 4 & 5 \\ 4 & 3 & 2 & 5 & 1 \end{pmatrix}$ in S_5 maps 1 to 4, 4 to 5, 5 to 1, 2 to 3, and 3 to 2.

$$q : 1 \mapsto 4 \mapsto 5 \text{ and } 2 \mapsto 3$$

$\therefore$ we write $q = (1\ 4\ 5)(2\ 3)$ in cycle notation.

If $r = \begin{pmatrix} 1 & 2 & 3 & 4 & 5 \\ 1 & 4 & 3 & 5 & 2 \end{pmatrix}$ then we write $r = (1)(2\ 4\ 5)(3)$
or $r = (2\ 4\ 5)$ {we omit the terms (1) and (3)}

in cycle notation.

> A **cycle** $(a_1 a_2 .... a_r)$ has r distinct elements and is said to have **length** r, where $r \in \mathbb{Z}^+$.
>
> If two cycles have no elements in common, they are called **disjoint cycles**.

For the three permutations p, q, r given above, we have:

p is a cycle of length 4.

q is the product (or composition) of two disjoint cycles, one of length 3 and one of length 2.

r is a cycle of length 3.

Clearly every cycle is a permutation, but not every permutation is a cycle. For example, q is not a cycle.

COMPOSITION OF PERMUTATIONS (USING CYCLE NOTATION)

Theorem 10

> Every permutation $p \in S_n$, $n \in \mathbb{Z}^+$, can be written as the composition of disjoint cycles.

For $q = (1\ 4\ 5)(2\ 3)$ and $r = (2\ 4\ 5)$, the permutation qr is the result of performing permutation r followed by permutation q.

$qr = (1\ 4\ 5)(2\ 3)(2\ 4\ 5)$ but this is not yet in simplest form.

Since the first permutation is the one furthest to the right, we work from right to left to trace the mapping of each element, starting with 1.

$(1\ 4\ 5)(2\ 3)(2\ 4\ 5)$ so we write $qr = (1\ 4\ ....$

1 maps to 4

Again working from right to left, we trace the mapping of element 4:

$$(1\ 4\ 5)(2\ 3)(2\ 4\ 5)$$

4 maps to 5

then 5 is mapped to 1

Thus 4 is mapped back to 1, so we have $qr = (1\ 4)\ .....$

We consider the remaining elements in the same way, and obtain:

2 maps to 4 maps to 5, so $2 \mapsto 5$

5 maps to 2 maps to 3, so $5 \mapsto 3$

3 maps to 2, so $3 \mapsto 2$

$\therefore\ \ qr = (1\ 4)(2\ 5\ 3)$ in cycle notation, in simplest form.

We can check the result using permutation notation:

$$\begin{aligned} qr &= \begin{pmatrix} 1 & 2 & 3 & 4 & 5 \\ 4 & 3 & 2 & 5 & 1 \end{pmatrix} \begin{pmatrix} 1 & 2 & 3 & 4 & 5 \\ 1 & 4 & 3 & 5 & 2 \end{pmatrix} \\ &= \begin{pmatrix} 1 & 2 & 3 & 4 & 5 \\ 4 & 5 & 2 & 1 & 3 \end{pmatrix} \\ &= (1\ 4)(2\ 5\ 3) \quad \text{in cycle notation} \end{aligned}$$

Theorem 11

Disjoint cycles commute, so for disjoint cycles $c_1, c_2, ..., c_m$, $(c_1c_2....c_m)^n = c_1^n c_2^n .... c_m^n$.

For example, $((1\ 4)(2\ 5\ 3))^2 = (1\ 4)^2(2\ 5\ 3)^2$.

Theorem 12

1 A cycle $(a_1a_2....a_r)$ of length r is a permutation of order r.

2 The inverse permutation $(a_1a_2....a_r)^{-1}$ has order r, and $(a_1a_2....a_r)^{-1} = (a_ra_{r-1}....a_1)$.

Proof:

1 If $p = (a_1a_2....a_r)$ then $p(a_1) = a_2$ in function notation, since p is a bijection.

$\therefore\ \ p^2(a_1) = p(p(a_1)) = p(a_2) = a_3$

$\vdots$

$\therefore\ \ p^{r-1}(a_1) = a_r$

$\therefore\ \ p^r(a_1) = p(p^{r-1}(a_1)) = p(a_r) = a_1$

Since the elements of a cycle are distinct, r is the least positive integer for which $p^r(a_1) = a_1$.
On checking, $p^r(a_i) = a_i$ for all $i = 1, ..., r$ also.

$\therefore\ \ p^r = \underbrace{ppp....p}_{r \text{ times}} = e$, the identity permutation, and $r \in \mathbb{Z}^+$ is the least positive integer for which this is true.

$\therefore\ \ p$ has order r.

2 Consider $(a_1a_2....a_r)(a_r....a_2a_1)$
$= (a_1)(a_2)(a_3)\,....\,(a_r)$
$= e$

Similarly, $(a_r....a_2a_1)(a_1a_2....a_r)$
$= (a_1)(a_2)\,....\,(a_r)$
$= e$

$\therefore\ (a_1a_2....a_r)^{-1} = (a_ra_{r-1}....a_1)$, which has length r and $\therefore$ by **1** has order r.

Example 53

Find, by direct calculation of the powers of each permutation, the order of each permutation.

a $p = (1\ 2\ 3)$ **b** $q = (2\ 3)(1\ 4\ 5)$ **c** $r = (1\ 2\ 3)(1\ 4\ 5)$

a Let $p = (1\ 2\ 3)$
$\therefore\ p^2 = (1\ 2\ 3)(1\ 2\ 3) = (1\ 3\ 2)$
and $p^3 = p^2p = (1\ 3\ 2)(1\ 2\ 3) = (1)(2)(3) = e$
$\therefore\ p$ has order 3, as expected since p is a cycle of length 3.

b Let $q = (2\ 3)(1\ 4\ 5)$
$\therefore\ q^2 = (2\ 3)(1\ 4\ 5)(2\ 3)(1\ 4\ 5)$
$= (1\ 5\ 4)(2)(3)$
$= (1\ 5\ 4)$
$q^3 = q^2q = (1\ 5\ 4)(2\ 3)(1\ 4\ 5)$
$= (1)(2\ 3)(4)(5)$
$= (2\ 3)$
$q^4 = q^3q = (2\ 3)(2\ 3)(1\ 4\ 5)$
$= (1\ 4\ 5)(2)(3)$
$= (1\ 4\ 5)$
$q^5 = q^4q = (1\ 4\ 5)(2\ 3)(1\ 4\ 5)$
$= (1\ 5\ 4)(2\ 3)$
and $q^6 = q^5q = (1\ 5\ 4)(2\ 3)(2\ 3)(1\ 4\ 5)$
$= (1)(2)(3)(4)(5)$
$= e$
$\therefore\ q$ has order 6.

c $r = (1\ 2\ 3)(1\ 4\ 5)$ is not a product of disjoint cycles, and in fact $r = (1\ 4\ 5\ 2\ 3)$ in simplest form.
$\therefore\ r$ is a cycle of length 5.
$r^2 = (1\ 4\ 5\ 2\ 3)(1\ 4\ 5\ 2\ 3)$
$= (1\ 5\ 3\ 4\ 2)$
$r^3 = r^2r$
$= (1\ 5\ 3\ 4\ 2)(1\ 4\ 5\ 2\ 3)$
$= (1\ 2\ 4\ 3\ 5)$
$r^4 = r^3r$
$= (1\ 2\ 4\ 3\ 5)(1\ 4\ 5\ 2\ 3)$
$= (1\ 3\ 2\ 5\ 4)$
$r^5 = r^4r$
$= (1\ 3\ 2\ 5\ 4)(1\ 4\ 5\ 2\ 3)$
$= (1)(2)(3)(4)(5)$
$= e$
$\therefore\ r$ has order 5, as expected.

Theorem 13

If a permutation $p = c_1c_2....c_m$ is the composition of m disjoint cycles $c_1, c_2, ..., c_m$ of lengths $r_1, r_2, ..., r_m$ respectively, then the order of p equals the lowest common multiple of $r_1, r_2, ...,$ and r_m.

Proof:

Each cycle c_i has length r_i

$\therefore$ c_i has order r_i.

$\therefore$ r_i is the smallest positive integer for which $c_i^{r_i} = e$, and $c_i^n = e$ if and only if n is a multiple of r_i.

It follows that $p^n = e$

$\Leftrightarrow$ $(c_1 c_2 .... c_m)^n = e$

$\Leftrightarrow$ $c_1^n c_2^n .... c_m^n = e$ {since disjoint cycles commute}

$\Leftrightarrow$ $c_1^n = e$, $c_2^n = e$, ..., and $c_m^n = e$ {since the cycles are disjoint, none can be the inverse of any other}

$\Leftrightarrow$ n is a multiple of r_1, r_2, ..., and r_m.

Thus the order of p is the lowest common multiple of r_1, r_2, ..., and r_m.

For example, $q = (2\ 3)(1\ 4\ 5)$ is the product of disjoint cycles of length 2 and 3

$\therefore$ the order of q is the lowest common multiple of 2 and 3, which is 6.

EXERCISE F.2

1 Write each permutation in cycle notation:

a $\begin{pmatrix} 1 & 2 & 3 & 4 \\ 4 & 1 & 2 & 3 \end{pmatrix}$ **b** $\begin{pmatrix} 1 & 2 & 3 & 4 & 5 & 6 \\ 6 & 5 & 2 & 1 & 3 & 4 \end{pmatrix}$ **c** $\begin{pmatrix} 1 & 2 & 3 & 4 & 5 \\ 5 & 3 & 2 & 4 & 1 \end{pmatrix}$

d $\begin{pmatrix} 1 & 2 & 3 & 4 & 5 & 6 \\ 6 & 3 & 1 & 5 & 4 & 2 \end{pmatrix}$ **e** $\begin{pmatrix} 1 & 2 & 3 & 4 & 5 \\ 3 & 2 & 1 & 4 & 5 \end{pmatrix}$ **f** $\begin{pmatrix} 1 & 2 & 3 & 4 & 5 & 6 \\ 1 & 3 & 5 & 6 & 2 & 4 \end{pmatrix}$

g $\begin{pmatrix} 1 & 2 & 3 & 4 & 5 & 6 \\ 6 & 4 & 5 & 2 & 3 & 1 \end{pmatrix}$

2 Write each permutation in cycle notation in simplest form.

a $(1\,2\,3\,4)(1\,5)$ **b** $(1\,3)(1\,2)(1\,5)$ **c** $(1\,2\,3)(1\,4\,2\,3)$

d $(1\,6)(1\,5)(1\,4)(1\,3)(1\,2)$

3 In cycle notation, find the inverse of each permutation.

a $(1\,3\,2\,4\,5)$ **b** $(1\,3\,2)(4\,5)$ **c** $(1\,3)(2\,4\,5)$

d $(1\,2\,3)(1\,4\,5)$ **e** $(1\,3)(1\,4)(1\,5)$ **f** $(1\,2\,3)(1\,5)$

4 Let $p = (1\,3\,2\,4)$ and $q = (1\,2\,3\,5)$. Find, in simplest form:

a p^{-1} **b** q^{-1} **c** $(pq)^{-1}$ **d** $q^{-1}pq$ **e** $p^{-2}q^{-1}$

f the permutation r such that $pr^{-1} = q$

5 Find, by direct calculation of the powers of each permutation, the order of each permutation:

a $(1\,4\,3\,2)$ **b** $(1\,2)(1\,3\,4)$ **c** $(1\,2\,3)(2\,3\,4)$

6 Find the order of each permutation given in **2**.

7 Let $p = (1\,3\,2\,4)$ and let $G = \{e, p, p^2, p^3\}$, where e is the permutation $\begin{pmatrix} 1 & 2 & 3 & 4 \\ 1 & 2 & 3 & 4 \end{pmatrix}$.

a Prove that G forms a group under composition of permutations.

b Let $q = p^2$. Find the permutations:

i q **ii** p^{159} **iii** q^{159} **iv** p^{508}

8 State the order of each permutation.

a (1 2 3)(4 5 6 7)(8 9 10 11 12)

b (1 2)(3 4 5 6)(7 8 9 10 11)

c (1 2)(3 4 5 6)(7 8 9 10 11 12 13 14)

d (1 2 3 4 5 6 7)(8 9 10)(11 12 13 14 15 16)

SYMMETRIES OF PLANE FIGURES

Certain plane figures have lines of symmetry and rotational symmetries. By labelling the vertices of a plane figure $\{1, 2, 3, ..., n\}$, $n \in \mathbb{Z}^+$, and then performing a (symmetric) reflection or rotation, we obtain a permutation p on $\{1, 2, ..., n\}$ where $p(i) = j$ if and only if vertex i is mapped to the original position of vertex j.

Sets of symmetries obtained in this way give rise to important examples of groups.

For example, we now consider the **symmetries of an equilateral triangle**, which give rise to the **dihedral group of degree 3**.

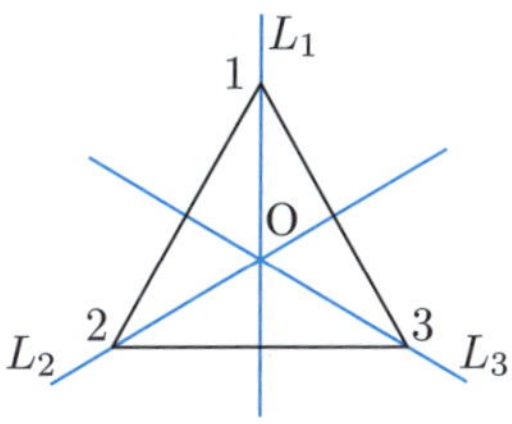

The equilateral triangle shown has centroid O. Lines L_1, L_2, and L_3 contain the three medians of the triangle through the vertices labelled 1, 2, and 3 respectively.

There are six transformations in the plane which map the equilateral triangle onto itself.

There are three **rotations**:

- e an anti-clockwise rotation through $0°$ about O. This is the identity or "do nothing" transformation.
- r an anti-clockwise rotation through $120°$ about O.

  r corresponds to the permutation $\begin{pmatrix} 1 & 2 & 3 \\ 2 & 3 & 1 \end{pmatrix}$ or (1 2 3) in cycle notation.

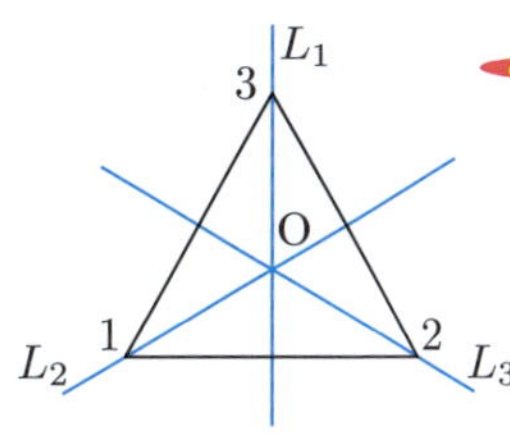

- r^2 an anti-clockwise rotation through $240°$ about O. This is equivalent to two successive applications of r, which is, $r * r$ or r^2.

  r^2 corresponds to the permutation $\begin{pmatrix} 1 & 2 & 3 \\ 3 & 1 & 2 \end{pmatrix}$ or (1 3 2) in cycle notation.

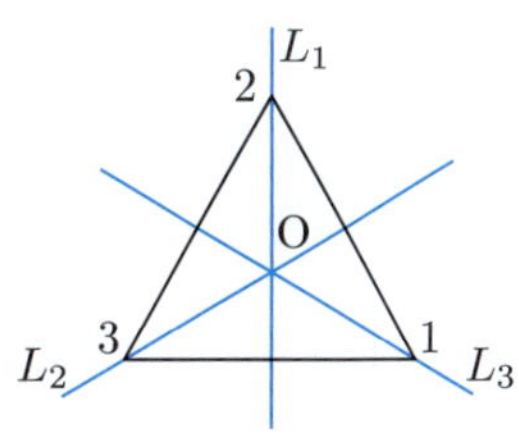

Note that $r^3 = e$ is a rotation through $360°$ which maps every point to itself.

There are also three **reflections**:

- x a reflection in the line L_1 and corresponds to $\begin{pmatrix} 1 & 2 & 3 \\ 1 & 3 & 2 \end{pmatrix}$ or (2 3).

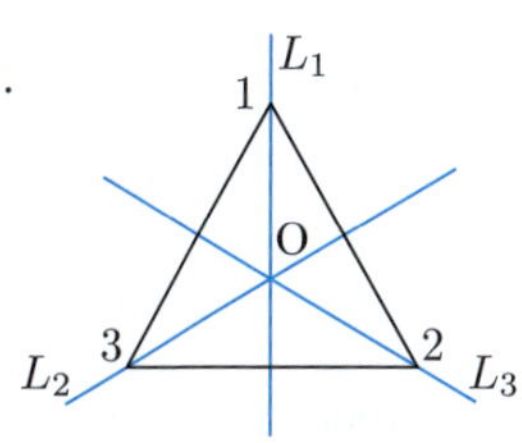

- y a reflection in the line L_2 and corresponds to $\begin{pmatrix} 1 & 2 & 3 \\ 3 & 2 & 1 \end{pmatrix}$ or $(1\ 3)$.

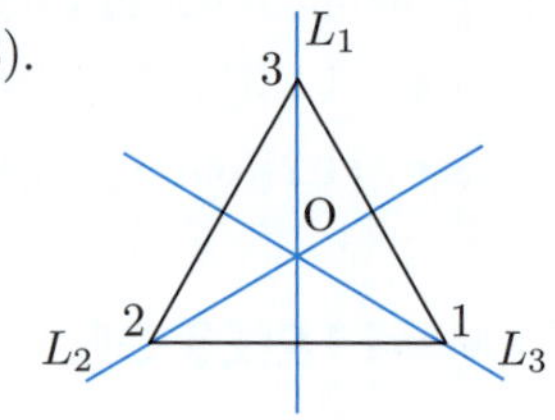

- z a reflection in the line L_3 and corresponds to $\begin{pmatrix} 1 & 2 & 3 \\ 2 & 1 & 3 \end{pmatrix}$ or $(1\ 2)$.

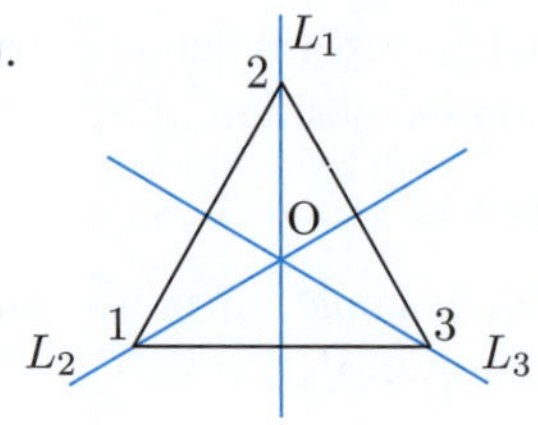

Since x, y, and z are reflections, $x^2 = y^2 = z^2 = e$.

Let $D_3 = \{e, r, r^2, x, y, z\}$.

The Cayley table for D_3 under the operation $*$ of combination of transformations, or equivalently the composition of the corresponding permutations, is:

$*$	e	r	r^2	x	y	z
e	e	r	r^2	x	y	z
r	r	r^2	e	z	x	y
r^2	r^2	e	r	y	z	x
x	x	y	z	e	r	r^2
y	y	z	x	r^2	e	r
z	z	x	y	r	r^2	e

For example, $r * x$ is a reflection in L_1 followed by an anti-clockwise rotation through $120°$. The result is z.

Using a cut-out copy of the triangle may help with recognition of geometric transformations.

Closure: The Cayley table shows that $a * b \in D_3$ for all $a, b \in D_3$.
Therefore D_3 is closed under $*$.

Associative: Transformations in the plane can be considered as bijections on $\mathbb{R}^2$.
Therefore, since composition of bijective functions on a set is associative, composition of transformations is also associative.

Identity: It can be seen from the table that $a * e = e * a = a$ for all $a \in D_3$.
Therefore, since $e \in D_3$, it is the identity element for $*$ in D_3.

Inverse: Since e appears once in every row and column, and $*$ is associative, every element has a unique inverse.

Therefore $\{D_3, *\}$ forms a group.

This group is referred to as the **dihedral group of degree 3**, $\{D_3, *\}$ or just D_3. We notice by comparison with **Example 52** that S_3 and D_3 have the same structure.

The set D_n of all **symmetries** (symmetry transformations) of a regular n-sided polygon in the plane forms a group called the **dihedral group of degree n**.

For example, the dihedral group D_4 is the group of symmetries of a square.

EXERCISE F.3

1 Label the vertices of a square 1, 2, 3, and 4.

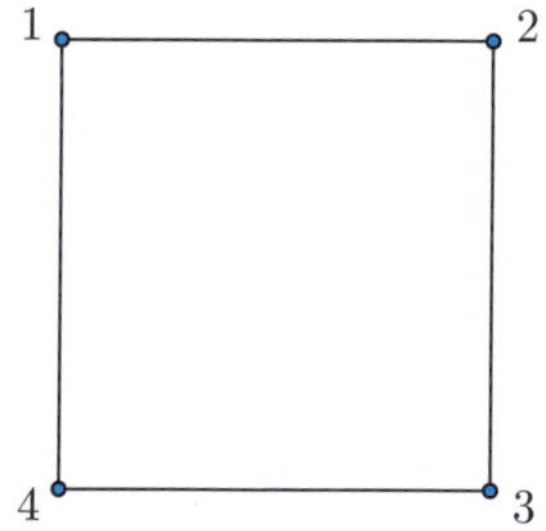

a State the order of rotational symmetry of the square.

b How many lines of symmetry does the square have?

c Write down each symmetry of the square as a permutation in cycle notation.

d Calculate the order of each permutation in c.

e What is the order of the dihedral group D_4?

f What is the order of S_4?

2 Label the vertices of a rectangle (which is **not** a square) 1, 2, 3, and 4.

a List the four symmetries of the rectangle as permutations in cycle notation.

b Calculate the order of each permutation in a.

c Verify that the set of permutations in a forms a group under composition of permutations.

d Is this group Abelian? Explain your answer.

3 Label the vertices of a regular pentagon 1, 2, 3, 4, and 5.

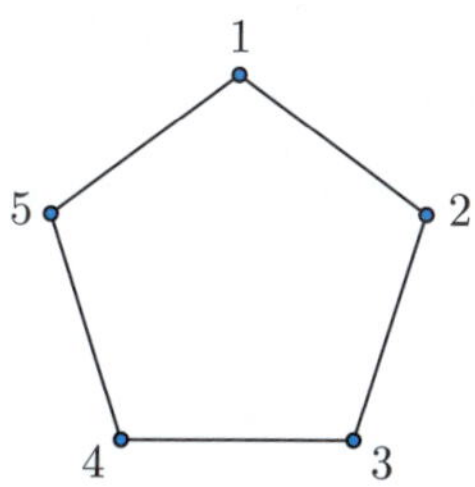

a Write down each symmetry of the regular pentagon as a permutation in cycle notation.

b Calculate the order of each permutation in a.

c What is the order of D_5?

d What is the order of S_5?

4 Prove or disprove (as appropriate) the statement:
"For $n \in \mathbb{Z}^+$, $n > 3$, the symmetric group S_n contains more elements than the dihedral group D_n".

G SUBGROUPS

Consider the group $\{G, *\} = \{\mathbb{Z}_6, +_6\}$ of integers modulo 6 under the operation $+_6$, addition modulo 6.

The Cayley table for G is:

$+_6$	0	1	2	3	4	5
0	0	1	2	3	4	5
1	1	2	3	4	5	0
2	2	3	4	5	0	1
3	3	4	5	0	1	2
4	4	5	0	1	2	3
5	5	0	1	2	3	4

Consider the subsets $H_1 = \{3, 0\}$, $H_2 = \{0, 2, 4\}$ of $\mathbb{Z}_6$.

It can be shown $\{H_1, +_6\}$ and $\{H_2, +_6\}$ are groups. Their Cayley tables are:

H_1:

$+_6$	0	3
0	0	3
3	3	0

and H_2:

$+_6$	0	2	4
0	0	2	4
2	2	4	0
4	4	0	2

We say that $\{H_1, +_6\}$ and $\{H_2, +_6\}$ are **subgroups** of $\{\mathbb{Z}_6, +_6\}$.

> Given a group $\{G, *\}$, if $H \subseteq G$ is
>
> **1** a **non-empty** subset of G and
>
> **2** $\{H, *\}$ is a group,
>
> then $\{H, *\}$ is called a **subgroup** of $\{G, *\}$.
> We write $H < G$ to denote that H is a subgroup of G.

Any group $\{G, *\}$ is a subgroup of itself, and if e is the identity in G, then $\{\{e\}, *\}$ is a subgroup of G. These two groups are called the **trivial subgroups** of the group G.

Any subgroup $\{H, *\}$ of a group $\{G, *\}$ which is not a trivial subgroup is called a **proper subgroup** of G.

Some examples of subgroups are:

- the **nested** subgroups $\{\mathbb{Z}, +\} < \{\mathbb{Q}, +\} < \{\mathbb{R}, +\} < \{\mathbb{C}, +\}$
- $\{\mathbb{Q}^+, \times\} < \{\mathbb{R}^+, \times\}$, both of which are infinite groups
- $\{\{1, i, -i, -1\}, \times\} < \{\mathbb{C} \setminus \{0\}, \times\}$ which is a finite subgroup of an infinite group.

Consider again $\{\mathbb{Z}_6, +_6\}$ and subgroups $\{H_1, +_6\}$ and $\{H_2, +_6\}$ defined above.

Notice that $3 +_6 3 = 0$, therefore in **group notation**, $3 +_6 3 = 3^2 = 0$.

Also, $2 +_6 2 = 4$ and $2 +_6 2 +_6 2 = 0$
$\therefore\ 2^2 = 4$ $\qquad \therefore\ 2^3 = 0$.

Hence we can write $H_1 = \{3, 3^2 = 0\} = \{3^n \mid n \in \mathbb{Z}\}$
and $H_2 = \{2, 2^2, 2^3 = 0\} = \{2^n \mid n \in \mathbb{Z}\}$

where 0 is the identity in $\{\mathbb{Z}_6, +_6\}$.

Theorem 14

Let $\{G, *\}$ be any group with identity e. For any element $a \in G$, the subset $H = \{a^n \mid n \in \mathbb{Z}\}$ of G, where $a^0 = e$ and $a^{-n} = (a^{-1})^n$ for $n \in \mathbb{Z}^+$, forms a subgroup $\{H, *\}$ of $\{G, *\}$.

Proof:

Firstly, $a^1 = a \in H \quad \therefore \quad H$ is non-empty.

For $a^p, a^q \in H$, $a^p * a^q = a^{p+q} \in H \quad \{\text{since } p+q \in \mathbb{Z}\}$

$\therefore \quad *$ is closed in H.

Since $*$ is closed in G, $a^n = a * a * .... * a \in H \cap G$ for all $n \in \mathbb{Z}^+$.

For $n = 0$, $a^0 = e \in H \cap G$, so H contains the identity.

For $n = -1$, $a^{-1} \in H \cap G$ and $a^{-n} = (a^{-1})^n \in H \cap G$ for $n \in \mathbb{Z}^+$.

Thus all elements of H are elements of G.

$\therefore \quad H \subseteq G$, $e \in H$, and each element $a^n \in H$ has inverse $(a^{-1})^n = a^{-n} \in H$.

Since $*$ is associative in G, $*$ is associative in $H \subseteq G$.

Thus H is a group, and so $H < G$.

Corollary: If $\{G, *\}$ is a group with identity e, and $a \in G$, the subgroup

$$H = \{a^n \mid n \in \mathbb{Z}\} = \{a^0 = e, a, a^2, a^3, ..., a^{-1}, a^{-2}, a^{-3}, ...\}$$
$$= \{e, a, a^2, a^3, ..., a^{-1}, (a^{-1})^2, (a^{-1})^3, ...\}$$

of G is the smallest subgroup of G which contains element a.

Proof:

By **Theorem 14**, H contains a and $H < G$.

Suppose H' is a subgroup of G, and that $a \in H'$.

$\therefore \quad \{H', *\}$ is a group

$\therefore \quad H'$ is closed under $*$, and $a^{-1} \in H'$.

$\therefore \quad (a^{-1})^n = a^{-n}$, $a^n \in H'$ for all $n \in \mathbb{Z}^+ \quad \{H' \text{ is closed under } *\}$
and $aa^{-1} = e \in H' \quad \{H' \text{ is closed under } *\}$

$\therefore \quad H \subseteq H'$.

But $\{H, +\}$ is a group, so $H < H'$.

Consider a subset H of a group $\{G, *\}$. From the definition of a subgroup, to determine whether H is a subgroup of G we need to verify if

1. H is non-empty, and
2. $\{H, *\}$ is a group.

The following theorems concerning subgroups provide quicker methods, in some cases, for determining whether or not a subset of a group is in fact a subgroup.

Theorem 15 (The subgroup test)

Consider a non-empty subset H of a group $\{G, *\}$. If $a * b^{-1} \in H$ for all $a, b \in H$, then $\{H, *\}$ is a subgroup of $\{G, *\}$.

Proof:

If we prove that H is a group, then since $H \subseteq G$, this will give us $H < G$.

Associative: $*$ is associative in G

$\therefore$ $*$ is associative in $H \subseteq G$.

Identity: Let e denote the identity in G.

Since H is non-empty, it contains an element a.

If $a = e$, then $e \in H$.

If $a \neq e$, then letting $b = a$ we have, by the given property of H, $a * a^{-1} = e \in H$.

Since e is the identity in G, and $H \subseteq G$, e is an identity in H.

Inverse: For all $a, b \in H$, $a * b^{-1} \in H$.

Since $e \in H$, we can let $a = e$.

$\therefore$ $e * b^{-1} = b^{-1} \in H$ for all $b \in H$.

Hence each element in H has an inverse in H.

Closure: For all $a, b \in H$, $a * b^{-1} \in H$.

For any $c \in H$, we know that $c^{-1} \in H$ {from above} and $(c^{-1})^{-1} = c$.

Letting $b = c^{-1}$, then for all $a, c^{-1} \in H$, $a * (c^{-1})^{-1} = a * c \in H$.

Thus for all $a, c \in H$, $a * c \in H$

$\therefore$ H is closed under $*$.

Hence H is a group, and $\therefore$ $H < G$.

So, given a non-empty subset H of a group $\{G, *\}$, to show H is a subgroup of G it is sufficient to show that $a * b^{-1} \in H$ for all $a, b \in H$. Conversely, if $a * b^{-1} \notin H$ for any pair of elements $a, b \in H$ then H is not a subgroup of $\{G, *\}$.

We can improve the result of **Theorem 15** for $\{G, *\}$ a **finite** group:

Theorem 16 (The subgroup test for finite groups)

Suppose $\{G, *\}$ is a *finite* group and H is a non-empty subset of G. $\{H, *\}$ is a subgroup of $\{G, *\}$ if $a * b \in H$ for all $a, b \in H$.

Proof:

Associative: The associativity of $*$ applies to all elements of G and it therefore must apply to all elements of H, a subset of G.

Closure: The property $a * b \in H$ for all $a, b \in H$ means $\{H, *\}$ is closed {by definition}.

Identity: As $\{G, *\}$ is a finite group, the order of any $x \in H$ is finite, m say, where $m \in \mathbb{Z}^+$.

$\therefore$ $x^m = e$, but $x^m \in H$ by closure, so $e \in H$.

$\therefore$ the identity element is in H.

Inverse: Firstly, we note that e is its own inverse.

For all other $x \in H$, suppose x has order m in G.

$\therefore$ $x^m = e$ where $m \in \mathbb{Z}^+$, $m \geqslant 2$.

Now $x^m = x^{(m-1)+1} = x^{1+(m-1)}$ where $m - 1 \in \mathbb{Z}^+$

$\therefore$ $e = x^{m-1} * x = x * x^{m-1}$

$\therefore$ x^{m-1} is the inverse of x, and since $x \in H$, $x * x \in H$, ..., $x^{m-1} \in H$.

Since we can do this for all $x \in H$ other than e, and we already know that e is its own inverse, every element $x \in H$ has an inverse in H.

Therefore $\{H, *\}$ is a group, and in particular $\{H, *\}$ is a subgroup of $\{G, *\}$.

Example 54

Let $H_2 = \{0, 2, 4\}$. Prove that $\{H_2, +_6\}$ is a subgroup of $\{\mathbb{Z}_6, +_6\}$.

$\mathbb{Z}_6 = \{0, 1, 2, 3, 4, 5\}$

$\therefore$ H_2 is a non-empty subset of $\mathbb{Z}_6$.

$\{\mathbb{Z}_6, +_6\}$ is a finite group of order 6.

$\therefore$ to check $\{H_2, +_6\}$ is a subgroup we need only check H_2 is closed under $+_6$ {by **Theorem 16**}

The Cayley table of H_2 is:

$+_6$	0	2	4
0	0	2	4
2	2	4	0
4	4	0	2

Each element in the table is an element in H_2, so H_2 is closed under $+_6$

$\therefore$ $\{H_2, +_6\} < \{\mathbb{Z}_6, +_6\}$ as required.

Example 55

Find all subgroups of the Klein 4-group $V_4 = \{e, a, b, c\}$ with Cayley table

$*$	e	a	b	c
e	e	a	b	c
a	a	e	c	b
b	b	c	e	a
c	c	b	a	e

$\{\{e\}, *\}$ and $\{V_4, *\}$ are the trivial subgroups of $\{V_4, *\}$.

$\{e, a\}$ is a non-empty subset of V_4, and under $*$ has Cayley table

$*$	e	a
e	e	a
a	a	e

$\therefore$ $\{e, a\}$ is closed under $*$, and therefore $\{\{e, a\}, *\}$ is a subgroup of V_4.
{subgroup test for finite groups}

We can similarly show that $\{\{e, b\}, *\}$ and $\{\{e, c\}, *\}$ are subgroups of V_4.

Any other subgroup must contain e.

If it contains a and b, then since $a * b = c$, it must also contain c, by closure.

$\therefore$ this subgroup is V_4.

Similarly, $a * c = b$
and $b * c = a$.

Thus we have found all subgroups of V_4.

Using the above example, we can draw a **lattice diagram** of the subgroups of the Klein 4-group $\{V_4, *\}$:

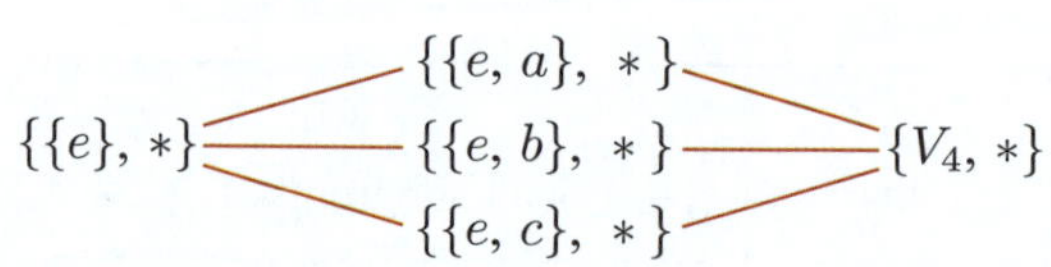

where two groups are connected (left to right) by branch(es) if and only if the first is a subgroup of the second.

Example 56

Consider the subset $\mathbb{Z}\setminus\{0\}$ of $\mathbb{R}\setminus\{0\}$.

a Show that $\mathbb{Z}\setminus\{0\}$ is closed under multiplication.

b Is $\{\mathbb{Z}\setminus\{0\}, \times\}$ a subgroup of the group $\{\mathbb{R}\setminus\{0\}, \times\}$?

c Does your answer in **b** contradict **Theorem 16**?

a For $a, b \in \mathbb{Z}\setminus\{0\}$, $ab \in \mathbb{Z}$.
Since $a \neq 0$ and $b \neq 0$, $ab \neq 0$.
$\therefore$ $ab \in \mathbb{Z}\setminus\{0\}$.
Hence $\mathbb{Z}\setminus\{0\}$ is closed under multiplication.

b $\{\mathbb{R}\setminus\{0\}, \times\}$ is a group, but $\{\mathbb{Z}\setminus\{0\}, \times\}$ is *not* a subgroup of $\{\mathbb{R}\setminus\{0\}, \times\}$ since $\{\mathbb{Z}\setminus\{0\}, \times\}$ is not a group!
For example, the identity in $\{\mathbb{Z}\setminus\{0\}, \times\}$ is 1, and $5 \in \mathbb{Z}\setminus\{0\}$ has no inverse in $\mathbb{Z}\setminus\{0\}$, since $5a \neq 1$ for all $a \in \mathbb{Z}\setminus\{0\}$.

c This does not contradict **Theorem 16**, since $\{\mathbb{R}\setminus\{0\}, \times\}$ is an infinite group, not a finite group.

EXERCISE G

1 For the following sets H and G and given operation $*$, determine whether $\{H, *\}$ is a subgroup of the group $\{G, *\}$.

- a $H = \mathbb{R}^+,\ G = \mathbb{R}\setminus\{0\},\ * = \times$
- b $H = \mathbb{Q}^+,\ G = \mathbb{R}^+,\ * = \times$
- c $H = \mathbb{Z}^+,\ G = \mathbb{Z}\setminus\{0\},\ * = \times$
- d $H = \{0, 2, 4\},\ G = \mathbb{Z}_6,\ * = +_6$
- e $H = \{4n \mid n \in \mathbb{Z}\},\ G = \mathbb{Z},\ * = +$
- f $H = \{8n \mid n \in \mathbb{Z}\},\ G = \{4n \mid n \in \mathbb{Z}\},\ * = +$
- g $H = \{e, r, r^2\}$ where r is a clockwise rotation of $\frac{2\pi}{3}$ radians, $G = D_3$, the dihedral group of degree 3, and $*$ = composition of transformations
- h $H = \{e, a, b, c\}$ where $\begin{cases} a = (1\,2)(3\,4) \\ b = (1\,3)(2\,4) \\ c = (1\,4)(2\,3) \end{cases}$, $G = S_4$, $*$ = composition of permutations
- i For $n \in \mathbb{Z}^+$, $H = U_n = \{z \mid z^n = 1,\ z \in \mathbb{C}\}$, $G = \mathbb{C}\setminus\{0\}$, $* = \times$
- j $H = \{a + ib\sqrt{5} \mid a, b \in \mathbb{R}\}$, $G = \mathbb{C}$, $* = +$

2 Let $S = \{(x, y) \mid x, y \in \mathbb{Z}\}$. Define the operation $*$ to be the composition of points such that $(a, b) * (c, d) = (a + c, (-1)^c b + d)$.

- a Prove that S is a group with respect to the operation $*$.
- b Is the group $\{S, *\}$ Abelian?
- c Do the following sets with the operation $*$ form subgroups of S?
  - i $H_1 = \{(a, 0) \mid a \in \mathbb{Z}\}$
  - ii $H_2 = \{(0, b) \mid b \in \mathbb{Z}\}$
  - iii $H_3 = \{(1, a) \mid a \in \mathbb{Z}\}$

3 Let $\{G, *\}$ be a group and let $\{H_1, *\}$ and $\{H_2, *\}$ be subgroups of $\{G, *\}$. Prove that $\{H_1 \cap H_2, *\}$ is a subgroup of $\{G, *\}$.

4 Let $\{G, *\}$ be a group with identity e and let $a \in G$ be a fixed element in G. Show that $H = \{x \mid x \in G \text{ and } x * a = a * x\}$ is a subgroup of G.

5 Let G be an Abelian group with subgroup $H < G$. Let $S = \{x \mid x \in G \text{ and } x^2 \in H\}$. Show that S is a subgroup of G.

6 a Show that $\alpha = \dfrac{1}{\sqrt{2}} + \dfrac{i}{\sqrt{2}}$ generates a group H under multiplication.

- b State the order of this group.
- c State a group G such that H is a proper subgroup of $\{G, \times\}$, where
  - i G is an infinite group
  - ii G is a finite group.

7 Suppose $\{G, *\}$ is a finite group with identity e and with an element $x \neq e$. Let $H = \{x^n \mid n \in \mathbb{Z}^+\}$. Prove that $\{H, *\}$ is a group.

H CYCLIC GROUPS

In **Theorem 14** we showed that if $a \in G$ is any element of a group $\{G, *\}$, the set $H = \{a^n \mid n \in \mathbb{Z}\} = \{e, a, a^2, a^3, ..., a^{-1}, a^{-2}, a^{-3}, ...\}$ is a subgroup of G. We say the group H is **generated** by element a, and write $H = \langle a \rangle$.

For example: $\{\mathbb{Z}, +\} = \{0, 1, 1+1, 1+1+1, ..., 1-1-1, 1-1-1-1, ...\}$
$= \langle 1 \rangle$

The group $\{\mathbb{Z}, +\}$ is generated by the element 1 since for each element $n \in \mathbb{Z}^+$ $n = \underbrace{1+1+....+1}_{n \text{ times}} = \underbrace{1*1*....*1}_{n \text{ times}} = 1^n$ in group notation, and for each element $n \in \mathbb{Z},\ n < 0,\quad n = \underbrace{-1-1-....-1}_{-n \text{ times}}$ where -1 is the inverse of 1.

We see that $1^0 = e = 0$ by definition.

A group $\{G, *\}$ is called **cyclic** if there exists an element $g \in G$ such that for all $x \in G$, $x = g^m$ for some $m \in \mathbb{Z}$. In this case $G = \{e, g, g^2, g^3, ..., g^{-1}, g^{-2}, g^{-3}, ...\}$.

The element g is called a **generator** of the cyclic group, and we write $G = \langle g \rangle$ to indicate g is a generator of G.

Theorem 17

All cyclic groups are Abelian.

Proof:

Let $\{G, *\}$ be a cyclic group and let $g \in G$ be a generator of the group.

Let $x, y \in G$.

Since the group is cyclic, there exists $p, q \in \mathbb{Z}$ such that $x = g^p$ and $y = g^q$

$\therefore\ x * y = g^p * g^q$
$= g^{p+q}$
$= g^{q+p}$ {addition of integers is commutative}
$= g^q * g^p$
$= y * x$ Therefore all cyclic groups are Abelian.

Examples of cyclic groups are:

- $\{\mathbb{Z}, +\} = \langle 1 \rangle$ is an **infinite cyclic group**, which is a cyclic group of infinite order, with generator 1.
- $\{\{1, i, -1, -i\}, \times\} = \langle i \rangle$ is a finite cyclic group with generator i, since $i^2 = -1$, $i^3 = -i$, $i^4 = 1$
  $\therefore\ \{1, i, -1, -i\} = \{1, i, i^2, i^3\} = \langle i \rangle$.
  We also see that $(-i)^2 = -1$, $(-i)^3 = i$, $(-i)^4 = 1$,
  so $\{1, -i, -1, i\} = \{1, -i, (-i)^2, (-i)^3\} = \langle -i \rangle$
  $\therefore\ -i$ is also a generator of this cyclic group.

- $\{2\mathbb{Z}, +\}$, the group consisting of the even integers under addition, is an infinite cyclic group.
  0 is the identity, and by definition $2^0 = e = 0$ is an element of the group.
  For all positive elements $2n \in 2\mathbb{Z}$, $n \in \mathbb{Z}$, and $n > 0$:
  $$2n = \underbrace{2 + 2 + ... + 2}_{n \text{ times}} = 2^n \quad \text{in group notation}$$
  For all negative elements $2n \in 2\mathbb{Z}$, $n \in \mathbb{Z}$, and $n < 0$:
  $$\begin{aligned} 2n &= \underbrace{(-2) + (-2) + ... + (-2)}_{-n \text{ times}} \\ &= (-n) \times (-2) \\ &= (2^{-1})^{-n} \\ &= 2^n \quad \text{in group notation} \end{aligned}$$
  Hence every element can be written as 2^n where $n \in \mathbb{Z}$, and so 2 is the generator of this group:
  $$\{2\mathbb{Z}, +\} = \langle 2 \rangle = \{0, 2, 4, 6, ..., -2, -4, -6, ...\}.$$

By contrast, the Klein 4-group is not cyclic.

$$\begin{aligned} \langle a \rangle &= \{a, a^2 = e\} \neq V_4 \\ \langle b \rangle &= \{b, b^2 = e\} \neq V_4 \\ \langle c \rangle &= \{c, c^2 = e\} \neq V_4 \\ \langle e \rangle &= \{e\} \qquad\quad\; \neq V_4 \end{aligned}$$

Since each element of the Klein 4-group has order 1 or 2, no element generates the 4 distinct elements of the group.

Example 57

Let $H_2 = \{0, 2, 4\}$. Show that $\{H_2, +_6\}$ is a cyclic group and find all the generators of H_2.

0 is the identity.

$0 +_6 0 = 0$ so 0 cannot be a generator.

$$2^2 = 2 +_6 2 = 4 \quad \text{and} \quad 2^3 = 4 +_6 2 = 0$$

$\therefore \; H_2 = \{0, 2, 2^2\}$

$\therefore \; H_2 = \langle 2 \rangle$ and 2 is a generator of this cyclic group.

$$4^2 = 4 +_6 4 = 2 \quad \text{and} \quad 4^3 = 2 +_6 4 = 0$$

$\therefore \; H_2 = \{0, 4, 4^2\}$

$\therefore \; H_2 = \langle 4 \rangle$, and 4 is also a generator of this cyclic group.

Hence H_2 has two generators, 2 and 4.

Note that in the above example elements 2 and 4 each have order 3, which is the order of the group H_2.

We now investigate further the orders of elements in finite cyclic groups, and compare them with the order of the group.

Consider the group $\{\mathbb{Z}_7 \setminus \{0\}, \times_7\}$ where $\times_7$ is multiplication modulo 7.

The Cayley table is shown alongside:

$\times_7$	1	2	3	4	5	6
1	1	2	3	4	5	6
2	2	4	6	1	3	5
3	3	6	2	5	1	4
4	4	1	5	2	6	3
5	5	3	1	6	4	2
6	6	5	4	3	2	1

Clearly, the identity element is 1.

We determine the order of the other elements of the group:

$2^1 = 2,\ 2^2 = 4,\ 2^3 = 1$ so the element 2 has order 3.

$3^1 = 3,\ 3^2 = 2,\ 3^3 = 6,\ 3^4 = 4,\ 3^5 = 5,\ 3^6 = 1$ so the element 3 has order 6.

$4^1 = 4,\ 4^2 = 2,\ 4^3 = 1$ so the element 4 has order 3.

$5^1 = 5,\ 5^2 = 4,\ 5^3 = 6,\ 5^4 = 2,\ 5^5 = 3,\ 5^6 = 1$ so the element 5 has order 6.

$6^1 = 6,\ 6^2 = 1$ so the element 6 has order 2.

Note also that the elements 3 and 5 each have order 6, the same as the order of the group. Every element of $\{\mathbb{Z}_7 \setminus \{0\}, \times_7\}$ can therefore be written as powers of 3 or 5. The group is therefore **cyclic**, and 3 and 5 are the **generators** of the group.

The cyclic nature of $\{\mathbb{Z}_7 \setminus \{0\}, \times_7\}$ can be seen in a rearrangement of the Cayley table. We let $a = 3$ and replace 2 by a^2, 6 by a^3, 4 by a^4, and 5 by a^5.

		1	3	2	6	4	5
	$\times_7$	1	a	a^2	a^3	a^4	a^5
1	1	1	a	a^2	a^3	a^4	a^5
3	a	a	a^2	a^3	a^4	a^5	1
2	a^2	a^2	a^3	a^4	a^5	1	a
6	a^3	a^3	a^4	a^5	1	a	a^2
4	a^4	a^4	a^5	1	a	a^2	a^3
5	a^5	a^5	1	a	a^2	a^3	a^4

Theorem 18

A finite cyclic group $\{G, *\}$ of order n has a generator g of order n, and $G = \langle g \rangle = \{e, g, g^2, ..., g^{n-1}\}$.

Proof:

By the definition of a cyclic group, G has a generator g, and $G = \langle g \rangle = \{e, g, g^2, ..., g^{-1}, g^{-2}, ...\}$.

We must prove that g has order n equal to the order of the group. Since G has finite order n, g has finite order $m \leqslant n$, by **Theorem 7**.

Suppose the order m of g is $m < n$.

$\therefore$ $\langle g \rangle = \{e, g, g^2, ..., g^{m-1}\}$, and since $g^m = e$, $g^{m+1} = g$ and so on are already listed as elements of the group.

Also, since $g^m = e$

$$gg^{m-1} = g^{m-1}g = e$$

$$\therefore\ g^{-1} = g^{m-1}$$

Similarly $g^{-2} = (g^2)^{-1} = g^{m-2}$ since $g^2 g^{m-2} = g^{m-2} g^2 = g^m = e$, and so on, are already listed as elements in $\langle g \rangle$.

$\therefore$ there are no further elements.

Thus $\langle g \rangle = \{e, g, g^2, ..., g^{m-1}\} = \{g^n \mid n \in \mathbb{Z}\}$.

This is a contradiction, since $G = \langle g \rangle$ contains n distinct elements.

Therefore the element g has order $m = n$, and $G = \langle g \rangle = \{e, g, g^2, ..., g^{n-1}\}$.

Theorem 19

For all $n \in \mathbb{Z}^+$, there exists a cyclic group of order n.

Proof:

A group of order 1 must contain the identity e, and $\{\{e\}, *\}$ is cyclic.

Let $G = \{a, a^2, a^3, ..., a^n\}$ where element a has order n.

$\therefore$ $G = \{e, a, a^2, ..., a^{n-1}\}$ has n distinct elements, and $a^n = e$.

For example, when $n = 1$, $G = \{a\} = \{e\}$; when $n = 2$, $G = \{a, a^2\} = \{a, e\}$.

Closure: Let $a^p, a^q \in G$ where $p, q \in \mathbb{Z}^+$ and $1 \leqslant p, q \leqslant n$

Then $a^p * a^q = a^{p+q}$

Now *either* $2 \leqslant p + q \leqslant n$ in which case $a^{p+q} \in G$

or $p + q = n + r$ where $1 \leqslant r \leqslant n$

$\Rightarrow$ $a^{p+q} = a^{n+r} = a^n * a^r = e * a^r = a^r$

$\Rightarrow$ as $1 \leqslant r \leqslant n$, $a^r \in G$, and so $a^{p+q} \in G$

Hence G is closed under $*$.

Associative: For all $x, y, z \in G$,

$$\begin{aligned} x * (y * z) &= a^p * (a^q * a^r) \\ &= a^p * a^{q+r} \\ &= a^{p+q+r} \\ &= a^{p+q} * a^r \\ &= (a^p * a^q) * a^r \\ &= (x * y) * z \end{aligned}$$

$\Rightarrow$ $*$ is an associative operation on G.

Identity: $a^n = e$ is the identity.

Inverse: If $1 \leqslant p \leqslant n - 1$ is an integer then $1 \leqslant n - p \leqslant n - 1$ and therefore $a^p, a^{n-p} \in G$.

Then

$$\begin{aligned} a^p * a^{n-p} &= a^{n-p} * a^p \\ &= e \end{aligned}$$

$\therefore$ $(a^p)^{-1} = a^{n-p}$ and $(a^{n-p})^{-1} = a^p$.

$a^n = e$ is its own inverse.

Hence each element has an inverse.

Therefore $\{G, *\}$ is a group.

For each $n \in \mathbb{Z}^+$, the cyclic group $\{\mathbb{Z}_n, +_n\} = \langle 1 \rangle$ is a classic example of a cyclic group of order n. The generators of this group are the elements 1, and any element in $\{2, 3, ..., n-1\}$ which is *coprime* to n, which means they have no common factors with n apart from 1.

Theorem 20

A subgroup of a cyclic group is cyclic.

The proof of this theorem is omitted.

EXERCISE H

1 Consider the group $\{\mathbb{Z}_{12}, +_{12}\}$ of integers modulo 12 under addition modulo 12.

- **a** State the order of each element in the group.
- **b** Find a subgroup:
  - **i** of order 3
  - **ii** of order 4.
- **c** Show that $\{\{0, 2, 4, 6, 8, 10\}, +_{12}\}$ is a subgroup of $\{\mathbb{Z}_{12}, +_{12}\}$.
- **d** Find, if possible, all generators of
  - **i** $\{\mathbb{Z}_{12}, +_{12}\}$
  - **ii** the subgroup in **c**.

2 Consider each of the following groups.

- **i** Determine if the group is cyclic.
- **ii** If the group is cyclic, find all of its generators. If the group is not cyclic, list all of its subgroups.

a $\{\{2, 4, 6, 8\}, \times_{10}\}$ **b** $\{\{1, 3, 5, 7\}, \times_8\}$

3 Consider the group $\{G, \times_n\}$ where G is the set containing the $n-1$ residue classes modulo n excluding 0, and $\times_n$ is multiplication modulo n. Which elements are generators of $\{G, \times_n\}$ when:

a $n = 3$ **b** $n = 5$ **c** $n = 7$ **d** $n = 11$?

4 Let $\{G, *\}$ be a finite cyclic group of order n. Prove that if g is a generator of G, then g^{-1} is also a generator of G.

5 Let R be the set of all rotations in D_n, the dihedral group of degree n, including the identity transformation e. Show that R is a cyclic subgroup of D_n and find a generator for R.

6 Let $\{G, *\}$ be an Abelian group of order 6 with identity e. Suppose G contains an element α of order 2, and an element β of order 3.

- **a** Write down all elements of G in terms of e, α, and β.
- **b** Construct a Cayley table for G.
- **c** Determine the order of each element in G.
- **d** Show that G is a cyclic group, and find all generators of G.
- **e** Write down the unique subgroup of order 3 of G.
- **f** Determine whether or not $\{e, \alpha, \beta, \alpha\beta\}$ is a subgroup of G.

7 Find the order of each of the following cyclic groups, where the operation is multiplication in $\mathbb{C}$:

a $\langle i \rangle$ **b** $\left\langle \frac{1}{2} + i\frac{\sqrt{3}}{2} \right\rangle$ **c** $\left\langle \frac{\sqrt{3}}{2} + \frac{i}{2} \right\rangle$ **d** $\langle \sqrt{3} + i \rangle$

8 Suppose $G = \langle g \rangle$ is a cyclic group of order 12 with generator g.

a Show that $\langle g^4 \rangle$ is a subgroup of G, and find its order.

b Find the order of the group: **i** $\langle g^2 \rangle$ **ii** $\langle g^3 \rangle$ **iii** $\langle g^6 \rangle$.

HISTORICAL NOTE — ÉVARISTE GALOIS

The young French mathematician **Évariste Galois** (1811 - 1832) was the first to use the term *group*. At age 14, Galois studied his first mathematics course, and by 15 was already studying the works of **Legendre** and **Lagrange**.

Galois studied polynomial equations and under what conditions they were solvable by radicals. This means the polynomial must by solvable in a finite number of steps, using only the coefficients and simple formulae with the operations addition, subtraction, multiplication, division, and taking nth roots. For example, quadratic equations are solvable in one step using the quadratic formula, and cubic and quartic equations can be solved using similar formulae in several steps.

Working with another mathematician **Abel**, Galois proved that a general polynomial equation of degree greater than or equal to 5 is not solvable by radicals. The original methods used by Galois in this proof became foundations for **Group theory**, and in particular **Galois theory**.

Applications of Group theory can be found in many other fields of study including Physics and Chemistry. Groups are used when describing symmetries in physical systems, symmetries in molecules, and in classifying crystalline structures. Even the famous puzzle the Rubik's cube can be solved with the help of Group theory.

In his short life, Galois failed examinations, spent time in jail for political activities, and sent his work to other mathematicians only for the manuscripts to be lost on at least two occasions. Tragically, he died from wounds from a duel at age 20.

Some of Galois' great contributions to mathematics may have been lost, if not for a letter written to his friend **Chevalier** on the night before the duel. In this letter he made annotations and corrections to some of his previous work. Famously, he wrote next to one theorem, "There are a few things left to be completed in this proof. I have not the time."

I HOMOMORPHISM

Let $\{G, *\}$ and $\{H, \circ\}$ be two groups with identities e_G and e_H respectively.

A function $f : G \to H$ with domain G is a **homomorphism from G to H** if $f(a * b) = f(a) \circ f(b)$ for all $a, b \in G$.

The **kernel of f**, denoted $\text{Ker}(f)$, is the set $\text{Ker}(f) = \{a \in G \mid f(a) = e_H\}$ of elements in G which are mapped by f to the identity in H.

The **range of f**, denoted $R(f)$, is the set $R(f) = \{f(a) \mid a \in G\}$ of elements in H to which the elements in G are mapped by f.

If f is also a bijection, then f is an **isomorphism** from G to H and we say the groups G and H are **isomorphic**. We denote this by $G \cong H$.

Example 58

For the two given groups, determine whether the function f is a homomorphism. If it is, find $\text{Ker}(f)$ and $R(f)$, and determine whether f is an isomorphism.

- **a** $f : \{\mathbb{R}, +\} \to \{\mathbb{R}, +\}$ where $f(x) = 3x$
- **b** $f : \{\mathbb{R}, +\} \to \{\mathbb{R}^+, \times\}$ where $f(x) = 2^x$
- **c** $f : \{\mathbb{R}^2, +\} \to \{\mathbb{R}^2, +\}$ where $f(x, y) = (x, 0)$
- **d** $f : \{\mathbb{R}^+, \times\} \to \{\mathbb{R}^+, \times\}$ where $f(x) = \sqrt{x}$
- **e** $f : \{G, *\} \to \{H, \circ\}$ where $f(x) = e_H$.

a $f(x + y) = 3(x + y)$
$= 3x + 3y$
$= f(x) + f(y)$ for all $x, y \in \mathbb{R}$

$\therefore$ f is a homomorphism.

$\text{Ker}(f) = \{0\}$ since $f(x) = 3x = 0$, the identity in $\{\mathbb{R}, +\}$, if and only if $x = 0$.

$R(f) = \mathbb{R}$

f is one-to-one and onto

$\therefore$ f is an isomorphism.

b $f(x + y) = 2^{x+y}$
$= 2^x \times 2^y$
$= f(x) \times f(y)$ for all $x, y \in \mathbb{R}$

$\therefore$ f is a homomorphism.

$\text{Ker}(f) = \{0\}$ since $f(x) = 1$, the identity in $\{\mathbb{R}^+, \times\}$, if and only if $x = 0$.

$R(f) = \mathbb{R}^+$

f is one-to-one and onto

$\therefore$ f is an isomorphism.

c $f((x_1, y_1) + (x_2, y_2)) = f(x_1 + x_2, y_1 + y_2)$
$= (x_1 + x_2, 0)$
$= (x_1, 0) + (x_2, 0)$
$= f(x_1, y_1) + f(x_2, y_2)$ for all $(x_1, y_1), (x_2, y_2) \in \mathbb{R}^2$

$\therefore$ f is a homomorphism.

$\text{Ker}(f) = \{(0, y) \mid y \in \mathbb{R}\}$ since $f(x, y) = (0, 0)$, the identity in $\{\mathbb{R}^2, +\}$, if and only if $x = 0$

$= \{\text{all points on the } y\text{-axis}\}$

$R(f) = \{(x, 0) \mid x \in \mathbb{R}\}$

$= \{\text{all points on the } x\text{-axis}\}$

f is not one-to-one, and f is not onto

$\therefore$ f is not an isomorphism.

d $f(x + y) = \sqrt{x + y}$

$f(x) + f(y) = \sqrt{x} + \sqrt{y}$

$\therefore$ in general, $f(x + y) \neq f(x) + f(y)$

$\therefore$ f is not a homomorphism.

e $f(x * y) = e_H$ and $f(x) \circ f(y) = e_H \circ e_H = e_H$

$\therefore$ $f(x * y) = f(x) \circ f(y)$

$\therefore$ f is a homomorphism.

$\text{Ker}(f) = G$ $\therefore$ f is not one-to-one unless $|G| = |H| = 1$.

$R(f) = \{e_H\}$ $\therefore$ f is not onto unless $|H| = 1$.

$\therefore$ in general, f is not an isomorphism.

Theorem 21

If $f : G \to H$ is a homomorphism between groups $\{G, *\}$ and $\{H, \circ\}$ with identities e_G and e_H respectively, then:

1 $f(e_G) = e_H$, or in other words $e_G \in \text{Ker}(f)$, and

2 $f(a^{-1}) = f(a)^{-1}$ for all $a \in G$.

Proof:

1 Let $a \in G$ be any element in G.

Since f is a homomorphism, $f(a * e_G) = f(a) \circ f(e_G)$

$\therefore$ $f(a) = f(a) \circ f(e_G)$

Since $f(a)$ is an element of H, let $f(a) = h \in H$.

Since H is a group, h^{-1} exists in H.

We have $h = h \circ f(e_G)$

$\therefore$ $h^{-1} \circ h = h^{-1} \circ (h \circ f(e_G))$ {multiply on the left by h^{-1}}

$= (h^{-1} \circ h) \circ f(e_G)$ {associativity in H}

$\therefore$ $e_H = e_H \circ f(e_G)$

$\therefore$ $e_H = f(e_G)$ {$f(e_G)$ is an element of H, so $e_H \circ f(e_G) = f(e_G)$}

2 Let $a \in G$ be any element in G.

$e_G = a * a^{-1} = a^{-1} * a$

$\therefore$ $f(e_G) = f(a * a^{-1}) = f(a^{-1} * a)$

$e_H = f(a) \circ f(a^{-1}) = f(a^{-1}) \circ f(a)$ {by part **1** and since f is a homomorphism}

$\therefore$ by the definition of inverse, $f(a)$ and $f(a^{-1})$ are inverses of each other in group H.

$\therefore$ $f(a^{-1}) = f(a)^{-1}$ as required.

Theorem 22

A homomorphism $f : G \to H$ of groups $\{G, *\}$ and $\{H, \circ\}$ is an injection if and only if $\text{Ker}(f) = \{e_G\}$, where e_G is the identity of G.

Proof:

Let e_H be the identity of H.

$(\Rightarrow)$ If f is an injection, since $f(e_G) = e_H$ from **Theorem 21**, no further element of G can map to e_H in H.

$\therefore \quad \text{Ker}(f) = \{e_G\}$

$(\Leftarrow)$ Suppose $\text{Ker}(f) = \{e_G\}$.

Suppose $f(a) = f(b)$ for $a, b \in G,\ a \neq b$.

Consider
$$\begin{aligned} f(a * b^{-1}) &= f(a) \circ f(b^{-1}) && \{\text{since } f \text{ is a homomorphism}\} \\ &= f(b) \circ f(b^{-1}) && \{\text{since } f(a) = f(b)\} \\ &= f(b) \circ f(b)^{-1} && \{\text{by } \textbf{Theorem 21}\} \\ &= e_H \end{aligned}$$

$\therefore \quad a * b^{-1} \in \text{Ker}(f)$

$\therefore \quad a * b^{-1} = e_G$

$\therefore \quad a * b^{-1} * b = e_G * b \qquad \{\text{multiply on the right by } b\}$

$\therefore \quad a = b$

Hence $f(a) = f(b)$ implies $a = b$

$\therefore \quad f$ is an injection.

Theorem 23

Suppose $f : \{G, *\} \to \{H, \circ\}$ is a homomorphism of groups.

1. The kernel of f is a subgroup of G. $\text{Ker}(f) < G$.
2. The range of f is a subgroup of H. $R(f) < H$.

Proof:

1. $\text{Ker}(f) = \{a \in G \mid f(a) = e_H\}$ is clearly a subset of G.

$e_G \in \text{Ker}(f)$ by **Theorem 21**

$\therefore \quad \text{Ker}(f)$ is non-empty.

Suppose a, b are two elements in $\text{Ker}(f)$, so $f(a) = f(b) = e_H$.

Consider
$$\begin{aligned} f(a * b^{-1}) &= f(a) \circ f(b^{-1}) && \{f \text{ is a homomorphism}\} \\ &= e_H \circ (f(b))^{-1} && \{\textbf{Theorem 21}\} \\ &= e_H \circ (e_H)^{-1} \\ &= e_H \circ e_H \\ &= e_H \end{aligned}$$

$\therefore \quad a * b^{-1} \in \text{Ker}(f)$.

Since $\text{Ker}(f)$ is a non-empty subset of G, and since $a * b^{-1} \in \text{Ker}(f)$ for all $a, b \in \text{Ker}(f)$, by **Theorem 15**, the subgroup test, $\text{Ker}(f) < G$.

2 $R(f) = \{f(a) \mid a \in G\}$ is clearly a subset of H.

$f(e_G) = e_H$ {**Theorem 21**}

$\therefore$ $e_H \in R(f)$

$\therefore$ $R(f)$ is non-empty.

Suppose h_1, h_2 are two elements in $R(f)$, so $h_1 = f(a_1)$ and $h_2 = f(a_2)$ for some elements $a_1, a_2 \in G$.

Consider $h_1 \circ h_2^{-1} = f(a_1) \circ f(a_2)^{-1}$

$= f(a_1) \circ f(a_2^{-1})$ {**Theorem 21**}

$= f(a_1 * a_2^{-1})$

$= f(a)$ for some element $a = a_1 * a_2^{-1} \in G$ since $a_1, a_2 \in G$ and G is a group

$\therefore$ $h_1 \circ h_2^{-1} \in R(f)$.

Since $R(f)$ is a non-empty subset of H, and since $h_1 \circ h_2^{-1} \in R(f)$ for all $h_1, h_2 \in R(f)$, by **Theorem 15**, the subgroup test, $R(f) < H$.

EXERCISE I

1 For the two given groups, determine whether the function f is a homomorphism. If it is:

i find $\text{Ker}(f)$ and $R(f)$ **ii** determine whether f is an isomorphism.

- **a** $f : \{\mathbb{R}, +\} \to \{\mathbb{R}, +\}$ where $f(x) = x^2$
- **b** $f : \{\mathbb{R}, +\} \to \{\mathbb{R}, +\}$ where $f(x) = 7x$
- **c** $f : \{\mathbb{R} \setminus \{0\}, \times\} \to \{\mathbb{R} \setminus \{0\}, \times\}$ where $f(x) = x^2$
- **d** $f : \{\mathbb{R}^+, \times\} \to \{\mathbb{R}^+, \times\}$ where $f(x) = x^2$
- **e** $f : \{\mathbb{R}, +\} \to \{\mathbb{R}^+, \times\}$ where $f(x) = e^x$, where e is the exponential constant
- **f** $f : \{\mathbb{Z}, +\} \to \{5\mathbb{Z}, +\}$ the additive group of integer multiples of 5, where $f(x) = 5x$
- **g** $f : \{\mathbb{Z}, +\} \to \{\mathbb{Z}_5, +_5\}$ where $f(x) = x \pmod 5$
- **h** $f : S_3 \to S_3$ where $f(x) = x^2$ for $x \in S_3$, the symmetric group of degree 3.

2 Let $\{P, +\}$ be the additive group of polynomials of degree $n \in \mathbb{Z}^+ \cup \{0\}$, in x, $x \in \mathbb{R}$. Define $f : \{P, +\} \to \{P, +\}$ by $f(p(x)) = p'(x)$, the derivative of p.

- **a** Show that f is a homomorphism.
- **b** Determine: **i** $\text{Ker}(f)$ **ii** $R(f)$
- **c** Write down a proper subgroup of $\{P, +\}$.
- **d** Is f an isomorphism? Explain your answer.

3 Suppose $f : \{G, *\} \to \{H, \circ\}$ is a homomorphism of groups, and let $f(G) = \{f(g) \mid g \in G\}$.

- **a** Prove that $f(G)$ is a group.
- **b** Show that if G is Abelian, then $f(G)$ is Abelian.
- **c** Show that if G is cyclic, then $f(G)$ is cyclic.
- **d** Determine whether the following statement is true or false, giving reasons for your answer: "If $g \in G$ has finite order $|g| = m$, then $f(g)$ has order m in H."

J ISOMORPHISM

We have seen that a bijective homomorphism is called an isomorphism.
Specifically, we can define:

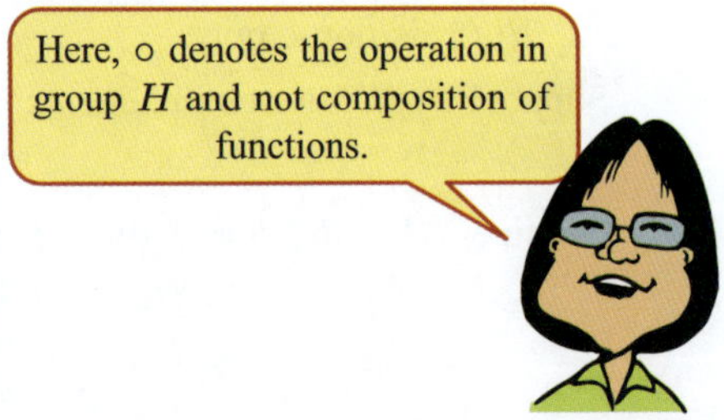

> Two groups $\{G, *\}$ and $\{H, \circ\}$ are **isomorphic** if:
>
> - there is a bijection $f : G \mapsto H$ and
> - $f(a * b) = f(a) \circ f(b)$ for all $a, b \in G$.
>
> Such a function f is called an **isomorphism** from G to H.

We can sometimes use Cayley tables to help establish isomorphism. It requires that for every p and q in G, if $f(p) = p' \in H$ and $f(q) = q' \in H$ then the element in the p' row and q' column of the Cayley table of $\{H, \circ\}$ is $f(p * q) = p' \circ q' = (p * q)'$.

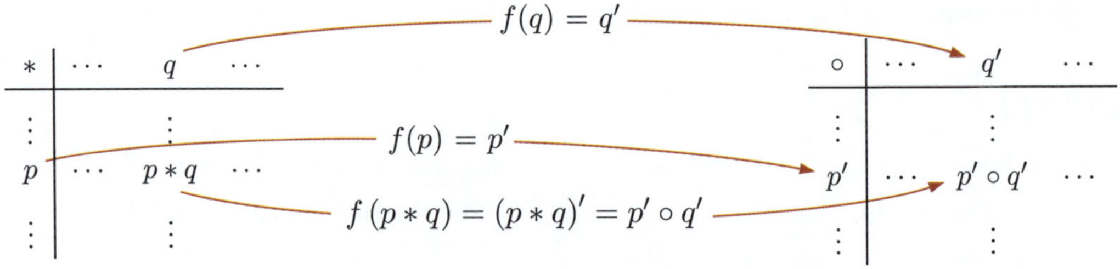

For two isomorphic groups, their elements are in one-to-one correspondence.

It follows that:

> - If two finite groups are isomorphic, then they have the same order.
> - If two groups do not have the same order, then they cannot be isomorphic.

Example 59

Show that groups $\{\mathbb{Z}_4, +_4\}$ and $\{\mathbb{Z}_5 \setminus \{0\}, \times_5\}$ are isomorphic.

We must construct an isomorphism f from one group to the other.

The Cayley table for $\{\mathbb{Z}_5 \setminus \{0\}, \times_5\}$ is:

$\times_5$	1	2	3	4
1	1	2	3	4
2	2	4	1	3
3	3	1	4	2
4	4	3	2	1

The Cayley table for the group $\{\mathbb{Z}_4, +_4\}$ is:

$+_4$	0	1	2	3
0	0	1	2	3
1	1	2	3	0
2	2	3	0	1
3	3	0	1	2

The two groups $\{\mathbb{Z}_5 \setminus \{0\}, \times_5\}$ and $\{\mathbb{Z}_4, +_4\}$ have the same structure.

We define a mapping $f : \{\mathbb{Z}_4, +_4\} \to \{\mathbb{Z}_5 \setminus \{0\}, \times_5\}$ by
$$\begin{aligned}0 &\mapsto 1\\ 1 &\mapsto 2\\ 2 &\mapsto 4\\ 3 &\mapsto 3.\end{aligned}$$

f is one-to-one and onto, so f is a bijection.

The similarity of the Cayley tables ensures that for all $a, b \in \mathbb{Z}_4$, $f(a +_4 b) = f(a) \times_5 f(b)$.

$\therefore$ f is a homomorphism, and since f is also a bijection, it is an isomorphism.

$\therefore$ by definition, the groups $\{\mathbb{Z}_4, +_4\}$ and $\{\mathbb{Z}_5 \setminus \{0\}, \times_5\}$ are isomorphic.

Example 60

Prove that the group of integers $\mathbb{Z}$ under addition is isomorphic to the group of even integers, $2\mathbb{Z}$, under addition.

We must construct an isomorphism f from one group to the other.

Let $f : \mathbb{Z} \to 2\mathbb{Z}$ be defined by $f(x) = 2x$

First, we establish that f is a bijection.

Suppose $f(a) = f(b)$, where $a, b \in \mathbb{Z}$
$$\therefore\ 2a = 2b$$
$$\therefore\ a = b.$$

No two distinct elements $a, b \in \mathbb{Z}$ can map to the same value in $2\mathbb{Z}$, so f is an injection.

Suppose $q \in 2\mathbb{Z}$, then $q = 2a$ for some $a \in \mathbb{Z}$
$$\therefore\ f(a) = q.$$

Thus each element in $2\mathbb{Z}$ is the image of an element in $\mathbb{Z}$ under f, and so f is a surjection.

Since f is both an injection and a surjection, f is a bijection.

Now $f(a + b) = 2(a + b) = 2a + 2b = f(a) + f(b)$ for all $a, b \in \mathbb{Z}$.

$\therefore$ f is a homomorphism.

Since f is both a bijection and a homomorphism, it is an isomorphism, and the two groups $\{\mathbb{Z}, +\}$ and $\{2\mathbb{Z}, +\}$ are isomorphic.

The above example shows us that $2\mathbb{Z} < \mathbb{Z}$, $2\mathbb{Z} \neq \mathbb{Z}$, and yet $\{2\mathbb{Z}, +\} \cong \{\mathbb{Z}, +\}$. So, for the *proper* subgroup $\{2\mathbb{Z}, +\}$ of $\{\mathbb{Z}, +\}$, the groups $2\mathbb{Z}$ and $\mathbb{Z}$ are in fact isomorphic! It is crucial here that both groups have infinite order. This could not occur for finite groups.

PROPERTIES OF ISOMORPHISM

Determining whether two groups are isomorphic is not always easy, so it is useful to know some properties of isomorphism between groups. If any one of these fails in a particular instance, then the two given groups cannot be isomorphic.

Since every isomorphism is also a homomorphism, **Theorem 21** gives us our first two properties. For completeness we include direct proofs of *Property 1* and *Property 2* for this section.

Property 1: If $\{G, *\}$ and $\{H, \circ\}$ are isomorphic then the identity of $\{G, *\}$ is mapped to the identity of $\{H, \circ\}$.

Proof:

Let e_G be the identity element of $\{G, *\}$ and let $f : G \to H$ be the isomorphism.

$\therefore$ for all $a, b \in G$, $f(a * b) = f(a) \circ f(b)$.

Now $e_G \in G$ and $a * e_G = e_G * a = a$ for all $a \in G$.

$\therefore$ $f(a * e_G) = f(a) \circ f(e_G) = f(a)$ and $f(e_G * a) = f(e_G) \circ f(a) = f(a)$

$\therefore$ $f(a) = f(a) \circ f(e_G) = f(e_G) \circ f(a)$ for all $f(a) \in H$

Since f is an isomorphism, $f(e_G) \in H$. Furthermore, each $h \in H$ is the image of an element in G, since f is onto.

$\therefore$ letting $h = f(a)$, we find $h = h \circ f(e_G) = f(e_G) \circ h$ for all $h \in H$

$\therefore$ $f(e_G)$ is the identity element e_H of $\{H, \circ\}$.

Property 2: If $\{G, *\}$ and $\{H, \circ\}$ are isomorphic then the inverse of an element of $\{G, *\}$ is mapped to the inverse of the corresponding element in $\{H, \circ\}$.
So, for f an isomorphism from G to H, $(f(a))^{-1} = f(a^{-1})$ for all $a \in G$.

Proof:

For f an isomorphism from G to H, $f(a * b) = f(a) \circ f(b)$ for all $a, b \in G$.

Now $a^{-1} \in G$ and $a * a^{-1} = a^{-1} * a = e_G$ the identity of G

$\therefore$ $f(a * a^{-1}) = f(a) \circ f(a^{-1}) = f(e_G) = e_H$ the identity in H {by *Property 1*}

and $f(a^{-1} * a) = f(a^{-1}) \circ f(a) = f(e_G) = e_H$

$\therefore$ $e_H = f(a) \circ f(a^{-1}) = f(a^{-1}) \circ f(a)$

$\therefore$ $f(a^{-1})$ is an inverse of $f(a)$ in H, written $(f(a))^{-1} = f(a^{-1})$.

Property 3: If $\{G, *\}$ and $\{H, \circ\}$ are isomorphic then for all $a \in G$, a and $f(a)$ will have the same order.

Property 4: If $\{G, *\}$ and $\{H, \circ\}$ are isomorphic, $\{G, *\}$ is Abelian if and only if $\{H, \circ\}$ is Abelian.

Property 5: If $\{G, *\}$ and $\{H, \circ\}$ are isomorphic, $\{G, *\}$ is cyclic if and only if $\{H, \circ\}$ is cyclic.

Proof:

Suppose G is cyclic, so $G = \langle g \rangle$ where $g \in G$ is a generator of G.

Let $h = f(g) \in H$, where $f : \{G, *\} \to \{H, \circ\}$ is an isomorphism from G to H.

By *Property 3*, g and $h = f(g)$ have the same order.

By **Theorem 14**, where $\langle h \rangle = \{h^n \mid n \in \mathbb{Z}\}$, $\langle h \rangle < H$.

Thus if G and H are **finite** groups,

$$\begin{aligned} |\langle g \rangle| &= |G| && \{g \text{ is a generator of } G\} \\ &= |H| && \{G \cong H\} \\ &= |\langle h \rangle| && \{\text{since } |g| = |h|\} \end{aligned}$$

$\therefore$ since $\langle h \rangle < H$ and $|H| = |\langle h \rangle|$, $H = \langle h \rangle$ and $\therefore$ H is cyclic.

If G and H are **infinite** groups, we prove $\langle h \rangle = H$ by showing $\langle h \rangle \subseteq H$ and $H \subseteq \langle h \rangle$:

By **Theorem 14**, $\langle h \rangle < H$ and so $\langle h \rangle \subseteq H$.

For any $y \in H$, $y = f(x)$ for some x in G. $\{$the isomorphism f is a surjection$\}$

But $x \in G$, so $x = g^m$ for some $m \in \mathbb{Z}$. $\{G = \langle g \rangle\}$

$$\begin{aligned} \therefore\ y &= f(g^m) \\ &= f(\underbrace{g * \dots * g}_{m \text{ times}}) \\ &= \underbrace{f(g) \circ f(g) \circ \dots \circ f(g)}_{m \text{ times}} && \{f \text{ is an isomorphism}\} \\ &= (f(g))^m \\ &= h^m \end{aligned}$$

$\therefore$ $y \in \langle h \rangle$, and y was any element of H.

$\therefore$ $H \subseteq \langle h \rangle$.

It follows that $H = \langle h \rangle$ and therefore H is a cyclic group.

Similarly, if H is cyclic then G is cyclic.

Property 6: If $\{G, *\}$ and $\{H, \circ\}$ are isomorphic then any subgroup of $\{G, *\}$ will be isomorphic to some subgroup of $\{H, \circ\}$.

Proof:

Let $M < G$ be a subgroup of G.

Let $f(M) = \{f(x) \mid x \in M\}$ where $f : \{G, *\} \to \{H, \circ\}$ is an isomorphism.

$\therefore$ $f(M) \subseteq H$.

(continued next page)

For $f(x), f(y) \in f(M)$, $(f(y))^{-1} = f(y^{-1})$ by *Property 2*, and since $y \in M$, $y^{-1} \in M$ since M is a group.

$\therefore$ $f(y^{-1}) \in f(M)$.

Consider $f(x) \circ (f(y))^{-1} = f(x) \circ f(y^{-1}) = f(x * y^{-1})$. $\{f$ is an isomorphism$\}$

Now $x * y^{-1} \in M$, since $x, y \in M$ and M is a subgroup of G

$\therefore$ $f(x * y^{-1}) \in f(M)$

We have shown $f(M) \subseteq H$, and that $f(x) \circ (f(y))^{-1} \in f(M)$ for all $f(x), f(y) \in f(M)$.

$\therefore$ $f(M)$ is a subgroup of H, by the subgroup test **Theorem 15**.

Theorem 24

For any $n \in \mathbb{Z}^+$, all cyclic groups of finite order n are isomorphic to each other.

Proof:

Let $\{G, *\}$ and $\{H, \circ\}$ be cyclic groups of order n. Suppose $G = \langle g \rangle = \{e_G, g, g^2, ..., g^{n-1}\}$ where $e_G = g^0 = g^n$ and $H = \langle h \rangle = \{e_H, h, h^2, ..., h^{n-1}\}$ where $e_H = h^0 = h^n$.

Let $f : G \to H$ be defined by $f(g^i) = h^i$, $i = 0, 1, ..., n-1$.

f is clearly one-to-one and onto, so f is a bijection.

For $g^i, g^j \in G$ where $0 \leqslant i, j \leqslant n-1$ we find $f(g^i * g^j) = f(g^{i+j})$ where $0 \leqslant i+j \leqslant 2n-2$.

$\therefore$ $i+j = r$ or $i+j = n+r$ for some integer r such that $0 \leqslant r \leqslant n-1$.

$\therefore$ $g^{i+j} = g^r$ or $g^{i+j} = g^{n+r} = g^n * g^r = e_G * g^r = g^r$.

$\therefore$ $g^{i+j} = g^r$ for some integer $0 \leqslant r \leqslant n-1$.

$\therefore$ $f(g^i * g^j) = f(g^{i+j}) = f(g^r) = h^r$ for some integer $0 \leqslant r \leqslant n-1$.

Similarly, $f(g^i) \circ f(g^j) = h^i \circ h^j$
$= h^{i+j}$

$\therefore$ $h^{i+j} = h^r$ or h^{n+r} for the same integer $0 \leqslant r \leqslant n-1$.
$= h^r$ or $h^n \circ h^r$
$= h^r$ or $e_H \circ h^r$
$= h^r$ or h^r
$\therefore$ $h^{i+j} = h^r$.

$\therefore$ $f(g^i * g^j) = f(g^i) \circ f(g^j)$ for all $g^i, g^j \in G$.

Hence f is an isomorphism, and $\therefore$ $G \cong H$.

Since $\{\mathbb{Z}_n, +_n\}$, $n \in \mathbb{Z}^+$, is a cyclic group of order n (generated by element 1), the above theorem tells us that, up to isomorphism, this is the unique cyclic group of order n, $n \in \mathbb{Z}^+$.

Earlier we investigated the groups of order 1, 2, 3, and 4.

The groups of orders 1, 2, and 3 were all shown to be cyclic. Therefore, up to isomorphism, there is a unique group of order 1, a unique group of order 2, and a unique group of order 3.

We can of course find examples of *different* groups of order 2, but the above theorem tells us they are isomorphic.

For $n = 4$, we found the cyclic group $\{\mathbb{Z}_4, +_4\}$ and the Klein 4-group. These groups are not isomorphic, since for example $\mathbb{Z}_4$ contains element 1 of order 4 and the Klein 4-group contains no element of order 4. {*Property 3* of isomorphism fails}

Thus, up to isomorphism, there are two distinct groups of order 4.

EXERCISE J

1 Show that the group $\{0, 1, 2\}$ under $+_3$ addition modulo 3 is isomorphic to the group $\{1, 2, 4\}$ under $\times_7$ multiplication modulo 7.

2 Show that the group $\{1, -\frac{1}{2} + i\frac{\sqrt{3}}{2}, -\frac{1}{2} - i\frac{\sqrt{3}}{2}\}$ under multiplication is isomorphic to the group $\{1, 2, 4\}$, where 1, 2, and 4 are residue classes modulo 7 under $\times_7$ multiplication modulo 7.

3 Show that the group $\{0, 1, 2, 3, 4\}$ under $+_5$ addition modulo 5 is isomorphic to the group of the five fifth roots of unity under multiplication.

4 Determine, with reasons, whether or not the two given groups are isomorphic.

   a $G = \{\{2, 4, 6, 8\}, \times_{10}\}$ and $H = \{\{1, 3, 5, 7\}, \times_8\}$

   b $G = \{\mathbb{Z}, +\}$ and $H = \{n\mathbb{Z}, +\}$, where $n \in \mathbb{Z}^+$ is a constant

   c $G = \{\mathbb{Z}_6, +_6\}$ and $H = S_6$, the symmetric group of degree 6

   d $G = \{\mathbb{Z}_6, +_6\}$ and $H = S_3$, the symmetric group of degree 3

   e $G = \{\langle i \rangle, \times\}$ and $H = \{\langle 1+i \rangle, \times\}$

5 Prove that the multiplicative group of positive real numbers is isomorphic to the additive group of real numbers.
Hint: Use $f(x) = \ln x$.

6 Prove *Property 3*:
If $\{G, *\}$ and $\{H, \circ\}$ are isomorphic then for all $a \in G$, a and $f(a)$ will have the same order.

7 Prove *Property 4*:
If $\{G, *\}$ and $\{H, \circ\}$ are isomorphic, $\{G, *\}$ is Abelian if and only if $\{H, \circ\}$ is Abelian.

8 For the dihedral groups and symmetric groups:

   a Show that $D_3 \cong S_3$.

   b Explain why $D_n \not\cong S_n$ for $n \in \mathbb{Z}^+$ and $n > 3$.

9 Suppose groups $\{G_1, *\}$ and $\{G_2, \triangle\}$ are isomorphic, so $G_1 \cong G_2$.
Suppose also groups $\{G_2, \triangle\}$ and $\{G_3, \square\}$ are isomorphic, so $G_2 \cong G_3$.
Prove that $G_1 \cong G_3$
Hint: You need to construct an isomorphism between G_1 and G_3.

10 Let R be the group of symmetries of a rectangle which is *not* a square.

a Show that R is isomorphic to the group of permutations $G = \{e, (1\,2)(3\,4), (1\,3)(2\,4), (1\,4)(2\,3)\}$ under composition of permutations.

b Show that G is isomorphic to the Klein 4-group.

c What can you deduce about R and the Klein 4-group?

K COSETS AND LAGRANGE'S THEOREM

If $\{H, *\}$ is a subgroup of a group $\{G, *\}$ and $g \in G$ is any fixed element in G, then $gH = \{g * h \mid h \in H\}$ is called a **left coset** of H in G, and $Hg = \{h * g \mid h \in H\}$ is called a **right coset** of H in G.

Theorem 25

If $\{G, *\}$ is an Abelian group and $\{H, *\}$ is a subgroup of G, then $gH = Hg$ for any $g \in G$. In other words, the left coset of H defined by g equals the right coset of H defined by g.

Example 61

For $\{H_2 = \{0, 2, 4\}, +_6\} < \{\mathbb{Z}_6, +_6\}$, find the left cosets of H_2 in $\mathbb{Z}_6$.

The left cosets of H_2 in $\mathbb{Z}_6$ are:

$$\begin{aligned} 0H_2 &= \{0 +_6 h \mid h \in H_2\} \\ &= \{0 +_6 0, 0 +_6 2, 0 +_6 4\} \\ &= \{0, 2, 4\} \\ &= H_2 \\ 1H_2 &= \{1 +_6 0, 1 +_6 2, 1 +_6 4\} = \{1, 3, 5\} \\ 2H_2 &= \{2 +_6 0, 2 +_6 2, 2 +_6 4\} = \{2, 4, 0\} = H_2 \\ 3H_2 &= \{3 +_6 0, 3 +_6 2, 3 +_6 4\} = \{3, 5, 1\} \\ 4H_2 &= \{4 +_6 0, 4 +_6 2, 4 +_6 4\} = \{4, 0, 2\} = H_2 \\ 5H_2 &= \{5 +_6 0, 5 +_6 2, 5 +_6 4\} = \{5, 1, 3\} \end{aligned}$$

Hence H_2 has only two distinct cosets in $\mathbb{Z}_6$:
$\{0, 2, 4\} = H_2$ (itself) and $\{1, 3, 5\}$.

We proceed using left cosets, but all statements and results hold equivalently for right cosets of a subgroup of a (not necessarily Abelian) group.

In the above example, note that:

- the only coset which is a subgroup of the group is the coset equal to the original subgroup itself
- $gH = H$ for all $g \in H$
- two left cosets of the subgroup are either identical or are disjoint
- the set union of the left cosets of the subgroup equals the given group.

Theorem 26

Suppose $\{H, *\}$ is a subgroup of group $\{G, *\}$, and $x, y \in G$ are any elements in G.

1 The cardinality of a left coset of H equals the cardinality of H.

2 $xH = H$ if and only if $x \in H$.

3 $xH = yH$ if and only if $x^{-1}y \in H$.

4 $xH \cap yH = xH$ or $\varnothing$.

5 The only left coset of H which is a subgroup of G is H itself.

6 G is the disjoint union of the left cosets of H in G.

Proof:

1 Consider the left coset xH of H in G.
For $h_1, h_2 \in H$, $x * h_1 = x * h_2$
$\Leftrightarrow$ $x^{-1} * x * h_1 = x^{-1} * x * h_2$ {multiplying on the left by x^{-1}, where x has inverse x^{-1} in G}
$\Leftrightarrow$ $e * h_1 = e * h_2$ {for e the identity in G}
$\Leftrightarrow$ $h_1 = h_2$
Hence the elements of H and xH are in one-to-one correspondence
$\therefore$ the cardinality of H equals the cardinality of xH.
If H is a finite group, then $|H| = |gH|$, which means H and any coset of H have the same number of elements.

2 $(\Rightarrow)$ Suppose $xH = H$
$\therefore$ $xh \in H$ for all $h \in H$
$\therefore$ $xe \in H$ since $e \in H$ {where e is the identity of G}
$\therefore$ $x \in H$

$(\Leftarrow)$ If $x \in H$, then $xH = \{x * h \mid h \in H\}$
But H is closed under $*$
$\therefore$ $x * h \in H$ for all $h \in H$
$\therefore$ $xH \subseteq H$ .... (1)
Also, for $h \in H$ consider $x^{-1} * h \in H$ {since $x \in H$, $x^{-1} \in H$}
$\therefore$ $x * (x^{-1} * h) \in xH$
$\therefore$ $(x * x^{-1}) * h \in xH$ {since $*$ is associative}
$\therefore$ $e * h \in xH$
$\therefore$ $h \in xH$
This result is true for all $h \in H$, so $H \subseteq xH$ .... (2)
(1) and (2) $\Rightarrow$ $H = xH$.

3 For $x, y \in G$:
$xH = yH$
$\Leftrightarrow$ $x^{-1} * xH = x^{-1} * yH$ {multiplying on the left by x^{-1} and since $*$ is associative}
$\Leftrightarrow$ $H = (x^{-1} * y)H$ {since $x^{-1} * x = e$ and $eH = H$}
$\Leftrightarrow$ $x^{-1} * y \in H$ by part **2**

4 Consider $x, y \in G$ and suppose $xH \cap yH \neq \varnothing$

$\therefore\ x * h_1 = y * h_2$ for some $h_1, h_2 \in H$

$\therefore\ h_1 = x^{-1}yh_2$ {multiplying on the left by x^{-1} and simplifying}

$\therefore\ h_1h_2^{-1} = x^{-1}y$ {multiplying on the right by h_2^{-1} and simplifying}

Now $h_1h_2^{-1} \in H$, so $x^{-1}y \in H$

$\therefore\ xH = yH$ {by part **3**}

$\therefore\ xH \cap yH = xH$

$\therefore$ either $xH \cap yH = xH$, or $xH \cap yH = \varnothing$.

5 Since $e \in H$, $e \notin gH$ for all $g \in G$ and $g \notin H$ by properties **2** and **4**.

$\therefore$ no coset of H besides H contains e, the unique identity in G.

$\therefore$ H is the only coset of H which is a subgroup of G.

6 For $g \in G$, $g = g * e \in gH$, so each element of G lies in a coset of H.
The result now follows from **4**.

Theorem 27 (Lagrange's theorem)

If $\{H, *\}$ is a subgroup of a finite group $\{G, *\}$ then the order of H is a factor of the order of G.

This fundamental result of Group theory is attributed to **Joseph Lagrange** (1736 - 1813).

Proof:

Suppose $|H| = m$ and $|G| = n$, where $m, n \in \mathbb{Z}$.

By **Theorem 26**, $|gH| = |H| = m$ for any left coset gH of H in G.

Since G is the disjoint union of the left cosets of H in G, it follows that $|G| = n = rm = r\,|H|$ for some $r \in \mathbb{Z}^+$.

$\therefore\ n = rm$ and hence the result.

Corollary 1:

The order of an element of a finite group is a factor of the order of the group.

Proof:

Let $\{G, *\}$ be a finite group of order n.

Suppose $h \in G$ has order m, so $H = \langle h \rangle = \{e, h, h^2, ..., h^{m-1}\}$ is a subgroup of G by **Theorem 14**, and $|H| = |h| = m$.

By Lagrange's theorem, $|G| = n = r\,|H|$ for some $r \in \mathbb{Z}^+$.

$\therefore\ n = rm$

$\therefore\ |G| = r\,|h|$

$\therefore$ the order of h is a factor of the order of G.

Corollary 2:

For $n \in \mathbb{Z}^+$ and n a prime, up to isomorphism, there is a unique group of order n.

Proof:

Existence: By previous results, for each $n \in \mathbb{Z}^+$, $\{\mathbb{Z}_n, +_n\}$ is a (cyclic) group of order n.

Uniqueness: If $\{G, *\}$ is a group and $|G| = n$ is prime, then 1 and n are the only factors of n.

$\therefore$ by Corollary 1 to Lagrange's theorem, each non-identity element in G has order n.

$\therefore$ for $g \in G$, $g \neq e$, $|g| = |\langle g \rangle| = n = |G|$

$\therefore$ $G = \langle g \rangle = \{e, g^1, g^2, ..., g^{n-1}\}$ and $\therefore$ G is a cyclic group.

Hence by **Theorem 24**, up to isomorphism, G is unique.

GROUPS OF ORDER n

We have shown that, up to isomorphism, there is a unique group of order $n = 1, 2, 3$, or n prime, and these are all cyclic groups.

For $n = 4$, we have shown there are two non-isomorphic groups of order 4, a cyclic group and the Klein 4-group.

For $n \in \mathbb{Z}^+$, $n > 4$ and n not prime, current results show there exist p pairwise non-isomorphic groups of order n, according to the following table:

n	4	6	8	9	10	12	14	15	16	18	20	21	22	24	25
p	2	2	5	2	2	5	2	1	14	5	5	2	2	15	2

EXERCISE K

1 Consider $\{H = \{0, 3\}, +_6\} < \{\mathbb{Z}_6, +_6\}$.

a Find the left cosets of H in $\{\mathbb{Z}_6, +_6\}$.

b Hence find the right cosets of H in $\{\mathbb{Z}_6, +_6\}$.

2 The group $H = \{0, 4, 8\}$ under $+_{12}$ addition modulo 12 is a subgroup of $\{\mathbb{Z}_{12}, +_{12}\}$.

a Find the left cosets of H in $\{\mathbb{Z}_{12}, +_{12}\}$.

b Explain why each left coset of H is also a right coset of H.

3 Consider $R = \{e, (1\ 2)(3\ 4), (1\ 3)(2\ 4), (1\ 4)(2\ 3)\}$, a subgroup of S_4 under composition of permutations.

a How many distinct left cosets does R have in S_4?

b How many left cosets of R in S_4 are subgroups of S_4?

c Consider the cosets $(1\ 2\ 3)R$ and $(1\ 2\ 3\ 4)R$. How many elements do they have in common?

d Show that $(1\ 2)R$ and $R(1\ 2)$ are equal sets, even though S_4 is not an Abelian group.

4 Suppose H is a subgroup of a group G. Prove that if $gH = Hg$ for $g \in G$, then $g^{-1}H = Hg^{-1}$.

5 Explain why S_5 has no subgroup of order 7.

6 Consider the symmetric group S_n of degree n, $n \in \mathbb{Z}^+$. Find a subgroup of order $m!$ for each $m \in \mathbb{Z}^+$ and $m \leqslant n$.

7 Suppose G is a finite group of order 36 with identity e. Show that there is a unique element $g \in G$ such that $g^7 = e$.

8 Prove that any group of prime order is cyclic.

REVIEW SET A

1 Suppose $A = \{a, b, c, d, e, f\}$ and $B = \{c, e, g, h\}$. Find:

a $A \cup B$ **b** $A \backslash B$ **c** $A \Delta B$

2 If $A = \{1, 2, 3\}$ and $B = \{2, 4\}$, find $A \times B$.

3 Prove that $(A \cap B) \times (C \cap D) = (A \times C) \cap (B \times D)$.

4 A relation R in $\{0, 1, 2, 3, 4, 5\}$ is such that xRy if and only if $|x - y| < 3$.

a Write R as a set of ordered pairs.

b Is R **i** reflexive **ii** symmetric **iii** transitive?

c Is R an equivalence relation? Explain your answer.

5 Determine whether each of the following functions is: **i** an injection **ii** a surjection.

a $f : \mathbb{R} \to \mathbb{R},\ f(x) = 2x^3 + 3x - 1$ **b** $f : \mathbb{Z} \to \mathbb{Z}^+,\ f(x) = x^2$

c $f : \mathbb{C} \to \mathbb{R}^+ \cup \{0\},\ f(z) = |z|$ **d** $f : \mathbb{Z}^+ \to \mathbb{R}^+,\ f(x) = \sqrt{x}$

6 A Cayley table for the binary operation $*$ is shown alongside.

$*$	1	2	3	4
1	2	1	3	1
2	3	2	4	2
3	4	1	3	2
4	1	4	2	1

a Find:

i $3 * 4$ **ii** $2 * (1 * 3)$ **iii** $(2 * 1) * 3$

b Is the Cayley table a Latin square? Explain the significance of your answer.

7 Let $f = \begin{pmatrix} 1 & 2 & 3 & 4 \\ 1 & 3 & 4 & 2 \end{pmatrix}$ and $g = \begin{pmatrix} 1 & 2 & 3 & 4 \\ 2 & 3 & 1 & 4 \end{pmatrix}$ be permutations.

a Find: **i** gf **ii** fg **iii** f^{-1} **iv** g^{-1}

b Find n such that $f^n = \begin{pmatrix} 1 & 2 & 3 & 4 \\ 1 & 2 & 3 & 4 \end{pmatrix}$.

8 Consider the group $\{G, +_n\}$ where G is the set containing the n residue classes modulo n. Which elements are generators of $\{G, +_n\}$ when:

a $n = 3$ **b** $n = 5$ **c** $n = 6$?

9 Let $p = c_1 c_2 .... c_r$ be a permutation in S_n, $n \in \mathbb{Z}^+$, where $c_1, c_2, ..., c_r$ are disjoint cycles. Define a relation R on $\{1, 2, ..., n\}$ by

$xRy \Leftrightarrow p^a(x) = y$ for some $a \in \mathbb{Z}^+$

$\Leftrightarrow$ it is possible to map x to y using permutation p (more than once if necessary).

a Prove that R is an equivalence relation. **b** Describe the equivalence classes of R.

c For $p = (1\ 2)(3\ 6\ 8)(4\ 5) \in S_8$, write down explicitly the equivalence classes defined by R.

10 A system of elements with binary operation $*$ is called a **semigroup** if and only if the system is closed under the operation and $*$ is associative.

Show that the following are all semigroups, and indicate which are also groups.

a

$*$	1	2
1	1	2
2	1	2

b

$*$	1	2
1	1	2
2	2	1

c

$*$	1	2	3
1	1	2	3
2	2	3	1
3	3	1	2

d

$*$	1	2	3
1	1	2	3
2	3	2	3
3	3	2	3

11 Suppose $\{G, *\}$ is a group with identity element e, and $\{G', \circ\}$ is a group with identity element e'. Let $S = G \times G'$. Define the "product" of pairs of elements $(a, a'), (b, b') \in S$ by $(a, a')(b, b') = (a * b, a' \circ b')$.

a Prove that S is a group under the "product" operation.

b Show that $S_1 = \{(g, e') \mid g \in G\}$ is a subgroup of S under the "product" operation.

12 Explain why a non-Abelian group must have at least six elements.

13 Let $G = \{(x, y) \mid x \in \mathbb{Z},\ y \in \mathbb{Q}\}$ and define the composition of points by $(a, b) * (c, d) = (a + c, 2^c b + d)$.

a Prove that G forms a group under $*$.

b Is $\{G, *\}$ Abelian?

c Do the following sets with the operation $*$ form subgroups of G?

i $H_1 = \{(a, 0) \mid a \in \mathbb{Z}\}$ **ii** $H_2 = \{(0, b) \mid b \in \mathbb{Q}\}$

14 Consider groups $G_1 : \{\mathbb{R} \setminus \{1\}, *\}$ where $a * b = a + b - ab$, and $G_2 : \{\mathbb{R}^+, \times\}$. Define $f : G_1 \to G_2$ by $f(a) = |a - 1|$.

a Show that f is a homomorphism.

b Determine whether f is:

i an injection **ii** a surjection **iii** an isomorphism.

15 Let $f : S_4 \to S_4$ be defined by $f(p) = p^2$ for all permutations $p \in S_4$, the symmetric group of degree 4.

a Find two permutations $p, q \in S_4$ such that $pq \neq qp$.

b Explain clearly why f is not a homomorphism.

REVIEW SET B

1 Consider the sets $A = \{0, 3, 6, 9, 12\}$, $B = \{1, 2, 3, 4, 5, 6\}$, $C = \{2, 4, 6, 8, 10\}$, and $\mathbb{U} = \{0, 1, 2, 3, 4, 5, 6, 7, 8, 9, 10, 11, 12, 13\}$. Find:

a $A \cap (B \cup C)$ **b** $A \Delta (B \setminus C)$ **c** $B' \cup C'$

d $A \cup (B \Delta C)$ **e** $A' \cap (B' \Delta C')$

In each case, illustrate the set on a Venn diagram.

2 Prove De Morgan's rule $(A \cap B)' = A' \cup B'$.

3 Let R be a relation on $\mathbb{Z}$ such that xRy if and only if $x - y$ is divisible by 6.

a Show that R is an equivalence relation. **b** Describe the equivalence classes.

4 Let R be a relation on $\mathbb{R} \times \mathbb{R}$ such that for $(a, b), (x, y) \in \mathbb{R} \times \mathbb{R}$,

$(a, b)R(x, y)$ if and only if $|x| + |y| = |a| + |b|$.

a Show that R is an equivalence relation.

b Describe how R partitions $\mathbb{R} \times \mathbb{R}$, and state the equivalence classes.

5 Determine which of the following is a bijection. If it is a bijection, state $f^{-1}(x)$.

a $f\colon\ \mathbb{R} \to \mathbb{R},\ f(x) = x^3 + 5$ **b** $f\colon\ \mathbb{R}^+ \to \mathbb{R},\ f(x) = \ln x$

c $f\colon\ \mathbb{Z} \to \mathbb{Z},\ f(x) = 2x$ **d** $f\colon\ \mathbb{R} \to \mathbb{R},\ f(x) = 2x$

e $f\colon\ \mathbb{R} \to [-1, 1],\ f(x) = \sin x$

6 For each of the given operations on $\mathbb{R}$, answer the following:

i Is the operation associative? **ii** Is the operation commutative?

iii If possible, find the identity element.

iv If possible, find the inverse of a, for $a \in \mathbb{R}$.

a $a * b = ab + 2$ **b** $a * b = |a + b|$ **c** $a * b = |ab|$

7 Find the order of each of the following permutations in S_4, and rewrite each permutation in cycle notation:

a $\begin{pmatrix} 1 & 2 & 3 & 4 \\ 3 & 1 & 2 & 4 \end{pmatrix}$ **b** $\begin{pmatrix} 1 & 2 & 3 & 4 \\ 1 & 2 & 4 & 3 \end{pmatrix}$ **c** $\begin{pmatrix} 1 & 2 & 3 & 4 \\ 2 & 1 & 4 & 3 \end{pmatrix}$

8 The Cayley table for a set $S = \{I, A, B, C, D\}$ under the operation $*$ is shown. Determine, with proof, which of the group axioms hold.

$*$	I	A	B	C	D
I	I	A	B	C	D
A	A	I	D	B	C
B	B	C	I	D	A
C	C	D	A	I	B
D	D	B	C	A	I

9 **a** Show that $\{1, 3, 5, 9, 11, 13\}$ under $\times_{14}$ multiplication modulo 14, is a group.

b State the order of each element of the group in **a**.

c Is the group in **a** cyclic? If so find the generators of the group.

10 Suppose $\{G, *\}$ is a group with identity e. If each element $a \in G,\ a \neq e$, has order 2, prove that $\{G, *\}$ is an Abelian group.

11 Let $\{A, +_m\}$ be a group where $A = \{0, 1, 2, ..., (m-1)\}$.
Let $\{B, +_{m^2}\}$ be a group where $B = \{0, 1, 2, ..., (m^2-1)\}$.
Prove that $G = \{(a, b) \mid a \in A,\ b \in B\}$ is a non-Abelian group of order m^3 under the operation $*$ defined by $(a, b) * (x, y) = (a + x \pmod m,\ b + y + mxb \pmod{m^2})$.

12 Consider a group $\{G, *\}$ for which $|G|$ is an odd prime number. Prove that there is only one element which is its own inverse.

13 Let $\{G, *\}$ be a group, and let $\{H_1, *\}$ and $\{H_2, *\}$ be subgroups of $\{G, *\}$.
Determine whether $\{H_1 \cup H_2, *\}$ is a subgroup of $\{G, *\}$.

14 Consider the two isomorphisms f and g, where
$f : \{\mathbb{R}^+, \times\} \to \{\mathbb{R}^+, \times\}$ is defined by $f(x) = x^2$, and
$g : \{\mathbb{R}^+, \times\} \to \{\mathbb{R}, +\}$ is defined by $g(x) = \ln x$.

a Find: **i** $f \circ g$ **ii** $g \circ f$

b Is $f \circ g$ an isomorphism from $\{\mathbb{R}^+, \times\}$ to $\{\mathbb{R}^+, \times\}$?

c Is $g \circ f$ an isomorphism from $\{\mathbb{R}^+, \times\}$ to $\{\mathbb{R}, +\}$?

15 For each of the given groups, determine whether the function f is a homomorphism. If it is:

i find $\text{Ker}(f)$ and $R(f)$ **ii** determine whether f is an isomorphism.

a $f : \{\mathbb{R}, +\} \to \{\mathbb{R}^+, \times\}$, where $f(x) = 3^x$.

b Let $G = \langle p \rangle$ for permutation $p = (1\ 4\ 3\ 2)$.
Define $f : G \to \{\mathbb{C} \setminus \{0\}, \times\}$ by $f(p^m) = i^m$, $m \in \mathbb{Z}^+ \cup \{0\}$.

REVIEW SET C

1 Find the power set $P(A)$ if $A = \{1, 2\}$. Determine whether $P(A)$ forms a group under:

a $\cap$ **b** $\cup$

2 Prove that $(A \backslash B) \times C = (A \times C) \setminus (B \times C)$.

3 Let R be a relation on $(\mathbb{R} \setminus \{0\}) \times \mathbb{R}^+$ such that for $(a, b), (x, y) \in (\mathbb{R} \setminus \{0\}) \times \mathbb{R}^+$,
$(a, b)R(x, y)$ if and only if $bx^2 = a^2y$.

a Show that R is an equivalence relation.

b Describe how R partitions $(\mathbb{R} \setminus \{0\}) \times \mathbb{R}^+$, and state the equivalence classes.

4 Is the following argument valid? Explain your answer.

Consider a symmetric and transitive relation R on a set S.
If xRy then yRx for all $x, y \in S$ {symmetry}
If xRy and yRx then xRx for all $x, y \in S$ {transitivity}
Since xRx, R must be reflexive.
Therefore a symmetric and transitive relation on a set is always an equivalence relation.

5 Let $f_1(x) = x$, $f_2(x) = -x$, $f_3(x) = \dfrac{1}{x}$, and $f_4(x) = -\dfrac{1}{x}$.

a Show that $\{f_1, f_2, f_3, f_4\}$ is a group under the composition of functions. Denote this group G.

b Find the order of each element in G.

c Is G a cyclic group? Explain your answer.

d State a group of order 4, distinct from G, but which is isomorphic to G.

6 For each of the given operations, answer the following:

i Is the operation associative?

ii Is the operation commutative?

iii If possible, find the identity element.

iv If possible, find the inverse of an element a.

a $a \circ b = \dfrac{1}{ab}$ on $\mathbb{R} \setminus \{0\}$ **b** $a \circ b = (a+2)(b+3)$ on $\mathbb{R}$

c $a \circ b = a + b + 3ab$ on $\mathbb{R}$

7 **a** Write down all the permutations in S_3 in cycle notation.

b Find p^{-1} for **i** $p = (2\ 3)$ **ii** $p = (1\ 3\ 2)$

c Find the order of all permutations in S_3.

8 **a** Show that the set $\{1, 7, 9, 15\}$ forms a group under $\times_{16}$ multiplication modulo 16.

b State the order of each element of the group in **a**.

c Is the group in **a** cyclic? Explain your answer.

9 Suppose $S = \{(a, b) \mid a, b \in \mathbb{R}\}$, and define the operation $*$ by $(a, b) * (c, d) = (ac, bc + d)$.

a Is $*$ associative?

b Is $*$ commutative?

c Is there an identity element for $*$ in S?

d Does each element have an inverse?

e Is $\{S, *\}$ a group?

10 Let $\{G, *\}$ be a finite group of order n with identity e. Explain why $a^n = e$ for all $a \in G$.

11 Let $\{G, *\}$ be a finite group of order n with identity e. If $H \neq \varnothing$ is a subset of G, and $a * b \in H$ for all $a, b \in H$, prove that $\{H, *\}$ is a group.

12 Consider the group $\{G, \times\}$ where $G = \{1, -1, i, -i\}$. $S = \{1, -1\}$ and $T = \{i, -1\}$ are subsets of G.
Under multiplication, determine whether each of S or T is a subgroup of $\{G, \times\}$.

13 Prove that a cyclic group of order m, $m \in \mathbb{Z}^+$, is isomorphic to the additive group of residue classes modulo m.

14 Suppose $f : \{G, *\} \to \{H, \circ\}$ is a homomorphism of finite groups.
Prove that f is an isomorphism if and only if $\text{Ker}(f) = \{e_G\}$, the identity in G.

15 Let $p = (1\,5\,3\,4\,2)$, $q = (1\,3\,2)$, $r = (1\,4\,5)$ be permutations in S_5, the symmetric group of degree 5.
Let $G = \langle p \rangle$ be the cyclic subgroup of S_5 generated by p.

a List the elements of the only left coset of G in S_5 which is a subgroup of S_5.

b How many distinct left cosets does G have in S_5?

c Find the permutation $q^{-1}r$.

d Hence or otherwise determine whether the cosets qG and rG are disjoint.

16 Suppose $f : \{G, *\} \to \{H, \circ\}$ is a homomorphism of groups, and that G and H have identities e_G and e_H respectively.
We have a theorem that states that $e_G \in \text{Ker}\{f\}$ and that $f(a^{-1}) = f(a)^{-1}$ for all $a \in G$.
Using this theorem and the subgroup test, prove that:

a $\text{Ker}(f)$ is a subgroup of G

b $R(f)$ is a subgroup of H.

REVIEW SET D

1 Use Venn diagrams to illustrate the following distributive laws:

a $A \cap (B \cup C) = (A \cap B) \cup (A \cap C)$

b $A \cup (B \cap C) = (A \cup B) \cap (A \cup C)$

2 **a** Find the power set $P(A)$ if $A = \{1, 2, 3\}$.

b Determine whether $P(A)$ forms a group under:

i $\cap$

ii $\cup$

3 Let R be a relation on $\mathbb{R}\times\mathbb{R}$ such that for $(a, b), (x, y) \in \mathbb{R}\times\mathbb{R}$,

$(a, b)R(x, y)$ if and only if $x^2+y^2=a^2+b^2$.

a Show that R is an equivalence relation.

b Describe how R partitions $\mathbb{R}\times\mathbb{R}$, and state the equivalence classes.

4 Let R be a relation on $\mathbb{Z}\times\mathbb{Z}$ such that for $(a, b), (x, y) \in \mathbb{Z}\times\mathbb{Z}$,

$(a, b)R(x, y)$ if and only if $y=b$.

a Show that R is an equivalence relation.

b Describe how R partitions $\mathbb{Z}\times\mathbb{Z}$, and state the equivalence classes.

5 Show that the set $\{f_1, f_2, f_3, f_4, f_5, f_6\}$ is a group under composition of functions where $f_1(x)=x$, $f_2(x)=\dfrac{1}{1-x}$, $f_3(x)=\dfrac{x-1}{x}$, $f_4(x)=\dfrac{1}{x}$, $f_5(x)=1-x$, and $f_6(x)=\dfrac{x}{x-1}$.

6 Determine whether the binary operation $*$ on $\mathbb{R}$ is associative, where $*$ is defined by:

a $a*b=\dfrac{a+b}{a^2}$ **b** $a*b=2^{a+b}$ **c** $a*b=a+b-3ab$

7 If $p=\begin{pmatrix}1 & 2 & 3 & 4 & 5 & 6\\ 2 & 6 & 1 & 4 & 3 & 5\end{pmatrix}$, find in cycle notation: **a** p^{-1} **b** p^6

8 Determine whether each of the following Cayley tables defines a group:

a

$*$	a	b	c	d	e
a	a	b	c	d	e
b	b	c	d	e	a
c	c	d	e	a	b
d	d	e	a	b	c
e	e	a	b	c	d

b

$*$	a	b	c	d	e
a	a	b	c	d	e
b	b	e	d	a	c
c	c	a	b	e	d
d	d	c	e	b	a
e	e	d	a	c	b

9 An operation $*$ on $S=\{0, 1, 2, 3, 4, 5\}$ is defined by $a*b=a\times_6(a+b)$, where $\times_6$ is multiplication modulo 6.

a Construct a Cayley table for this operation on the given set.

b Is the table a Latin square? **c** Is $\{S, *\}$ a group?

10 Consider the group $\{A, +_m\}$ where $A=\{0, 1, 2, ..., (m-1)\}$ and $+_m$ is addition modulo m.

a Prove that $\{G, *\}$ is a group where $G=\{(a, b, c) \mid a, b, c\in A\}$ and $*$ is defined by $(a, b, c)*(x, y, z)=(a+x(\text{mod}\, m),\ b+y(\text{mod}\, m),\ c+z-xb(\text{mod}\, m))$.

b Is the group G Abelian?

c State the order of the group G.

11 Consider a set G under the associative operation $*$, which is closed on G. Suppose there are unique solutions $x, y\in G$ for the equations $x*a=b$ and $a*y=b$ for each chosen $a, b\in G$. Prove that $\{G, *\}$ is a group.

12 Show that the rational numbers of the form $\dfrac{2a+1}{2b+1}$ where $a, b\in\mathbb{Z}$, form a subgroup of the group $\{\mathbb{Q}\backslash\{0\}, \times\}$.

13 **a** For each of the following sets, construct a Cayley table under the given operation.

i $\{1, 9, 11, 19\}$ under multiplication modulo 20

ii $\{1, 3, 7, 9\}$ under multiplication modulo 20

iii $\{1, 9, 13, 17\}$ under multiplication modulo 20

b Show that each set in **a** forms a group under the given operation.

c Are any pairs of the groups in **a** isomorphic?

14 **a** Consider the subgroup $\{2\mathbb{Z}, +\}$ of the group $\{\mathbb{Z}, +\}$, where $G = 2\mathbb{Z} = \{...., -4, -2, 0, 2, 4, ...\}$ is the set of all even integers.

i Find the left cosets $0G$ and $1G$.

ii How many distinct left cosets does G have in $\{\mathbb{Z}, +\}$?

iii Hence or otherwise, explain why the set of odd integers does not form a subgroup of $\{\mathbb{Z}, +\}$.

b Let $n \in \mathbb{Z}^+$ be any fixed integer.
Let G be the subgroup $\{n\mathbb{Z}, +\}$ of $\{\mathbb{Z}, +\}$, where $n\mathbb{Z}$ is the set of all integer multiples of n.

i Find the left cosets $0G$, $1G$, ..., and $(n-1)G$.

ii How many distinct left cosets does G have in $\{\mathbb{Z}, +\}$?

iii Hence or otherwise, explain why the residue classes modulo n, $[0], [1], ..., [n-1]$, partition the set of integers.

THEORY OF KNOWLEDGE — MATHEMATICAL PARADOX

In mathematics, a **naive** theory is one which is described using natural language rather than rigidly defined symbols. Such theories are often used when a mathematical principle is first investigated, before the boundaries of the principle are fully investigated and the theory formalised. Naive theories are often sufficient for everyday usage of the principles, but may fail under certain specific conditions.

At the end of the 19th century, German mathematician **Georg Cantor** (1845 - 1918) described a naive set theory, and indeed is considered the inventor of the set theory we use today. However, the English mathematician **Bertrand Russell** showed in 1901 that the naive set theory leads to a contradiction known as **Russell's paradox**:

> Let R be the set of all sets that are not members of themselves.
>
> If R is a member of itself, it would contradict its own definition as a set containing all sets that are not members of themselves.
>
> If R is not a member of itself, it would qualify as a member of itself by the same definition.

1 Does Russell's paradox present a problem for the everyday application of set theory?

2 What does Russell's paradox teach us about naive theories in mathematics?

3 In what ways is an axiomatic theory an advancement from naive theory?

Kurt Friedrich Gödel was born in 1906 in Brno, in what is now the Czech Republic. In 1931 Gödel published two famous "incompleteness" theorems, the first of which was summarised by Kleene[1] as:

> Any effectively generated theory capable of expressing elementary arithmetic cannot be both consistent and complete. In particular, for any consistent, effectively generated formal theory that proves certain basic arithmetic truths, there is an arithmetical statement that is true, but not provable in the theory.

4 Is Russell's paradox an example of Gödel's first incompleteness theorem?

5 What does Gödel's first incompleteness theorem tell us about the nature of mathematics? How is mathematics believable if cannot be both consistent and complete?

In 1989, **George Boolos** proved Gödel's first incompleteness theorem using a formalised version of the **Berry paradox**. This paradox is self-referential, and arises from an expression like "the smallest positive integer not definable in under eleven words." Since this expression contains only 10 words, the smallest positive integer not definable by any other expression of less than eleven words, would be defined by this expression, thus leaving the expression self-contradictory.

6 How can a paradox be used in a mathematical proof?

[1] Stephen Cole Kleene, 1967, *Mathematical Logic.*

APPENDIX: METHODS OF PROOF

Greek mathematicians more than 2000 years ago realised that progress in mathematical thinking could be brought about by conscious formulation of the methods of **abstraction** and **proof**.

By considering a few examples, one might notice a certain common quality or pattern from which one could predict a rule or formula for the general case. In mathematics this prediction is known as a **conjecture**. Mathematicians love to find patterns, and try to understand why they occur.

Experiments and further examples might help to convince you that the conjecture is true. However, problems will often contain extra information which can sometimes obscure the essential detail, particularly in applied mathematics. Stripping this away is the process of **abstraction**.

For example, by considering the given table of values one may conjecture:

a	b	a^2	b^2
1	2	1	4
3	5	9	25
4	5	16	25
5	7	25	49
6	9	36	81

"If a and b are real numbers then $a < b$ implies that $a^2 < b^2$."

However, on observing that $-2 < 1$ but $(-2)^2 \not< 1^2$ we have a **counter-example**.

In the light of this we reformulate and refine our conjecture:

"If a and b are *positive* real numbers then $a < b$ implies $a^2 < b^2$."

The difficulty is that this process might continue with reformulations, counter-examples, and revised conjectures indefinitely. At what point are we certain that the conjecture is true? A **proof** is a flawless logical argument which leaves no doubt that the conjecture is indeed a truth. If we have a proof then the conjecture can be called a **theorem**.

Mathematics has evolved to accept certain types of arguments as valid proofs. They include a mixture of both logic and calculation. Generally mathematicians like elegant, efficient proofs. It is common not to write every minute detail. However, when you write a proof you should be prepared to expand and justify every step if asked to do so.

We have already examined in the HL Core text, proof by **the principle of mathematical induction**. Now we consider other methods.

DIRECT PROOF

In a **direct proof** we start with a known truth and by a succession of correct deductions finish with the required result.

Example 1: Prove that if $a, b \in \mathbb{R}$ then $a < b \;\Rightarrow\; a < \dfrac{a+b}{2}$

Proof: $a < b \;\Rightarrow\; \dfrac{a}{2} < \dfrac{b}{2}$ {as we are dividing by 2 which is > 0}

$\Rightarrow\; \dfrac{a}{2} + \dfrac{a}{2} < \dfrac{a}{2} + \dfrac{b}{2}$ {adding $\dfrac{a}{2}$ to both sides}

$\Rightarrow\; a < \dfrac{a+b}{2}$

Sometimes it is not possible to give a direct proof of the full result and so the different possible cases (called **exhaustive cases**) need to be considered and proved separately.

Example 2: Prove the **geometric progression**: For $n \in \mathbb{Z},\ n \geqslant 0,$

$$1 + r^1 + r^2 + ... + r^n = \begin{cases} \dfrac{r^{n+1} - 1}{r - 1}, & r \neq 1 \\ n + 1, & r = 1 \end{cases}$$

Proof: **Case $r = 1$:** $1 + r^1 + r^2 + ... + r^n$

$= 1 + 1 + 1 + ... + 1 \quad \{n + 1 \text{ times}\}$

$= n + 1$

Case $r \neq 1$: Let $S_n = 1 + r^1 + r^2 + ... + r^n$.

Then $rS_n = r^1 + r^2 + r^3 + ... + r^{n+1}$

$\therefore\ rS_n - S_n = r^{n+1} - 1$ {after cancellation of terms}

$\therefore\ (r - 1)S_n = r^{n+1} - 1$

$\therefore\ S_n = \dfrac{r^{n+1} - 1}{r - 1}$ {dividing by $r - 1$ since $r \neq 1$}

Example 3: Alice looks at Bob and Bob looks at Clare. Alice is married, but Clare is not. Prove that a married person looks at an unmarried person.

Proof: We do not know whether Bob is married or not, so we consider the different (exhaustive) cases:

Case: Bob is married. If Bob is married, then a married person (Bob) looks at an unmarried person (Clare).

Case: Bob is unmarried. If Bob is unmarried, then a married person (Alice) looks at an unmarried person (Bob).

Since we have considered all possible cases, the full result is proved.

EXERCISE

1 Let $I = \sqrt{2}$, which is irrational. Consider I^I and I^{I^I}, and hence prove that an irrational number to the power of an irrational number can be rational.

PROOF BY CONTRADICTION (AN INDIRECT PROOF)

In **proof by contradiction** we deliberately assume the opposite to what we are trying to prove. By a series of correct steps we show that this is impossible, our assumption is false, and hence its opposite is true.

Example 4: Consider **Example 1** again but this time use proof by contradiction:

Prove that if $a,\ b \in \mathbb{R}$ then $a < b \Rightarrow a < \dfrac{a+b}{2}$.

Proof (by contradiction):

For $a < b$, suppose that $a \geqslant \dfrac{a+b}{2}$.

$$\Rightarrow \ 2a \geqslant 2\left(\frac{a+b}{2}\right) \qquad \{\text{multiplying both sides by 2}\}$$
$$\Rightarrow \ 2a \geqslant a+b$$
$$\Rightarrow \ a \geqslant b \qquad \{\text{subtracting } a \text{ from both sides}\}$$

which is false.

Since the steps of the argument are correct, the supposition must be false and the alternative, $a < \dfrac{a+b}{2}$ must be true.

Example 5: Prove that the solution of $3^x = 8$ is irrational.

Proof (by contradiction):

Suppose the solution of $3^x = 8$ is rational, or in other words, that x is rational. Notice that $x > 0$.

$$\Rightarrow \quad x = \frac{p}{q} \quad \text{where } p,\ q \in \mathbb{Z},\ q \neq 0 \quad \{\text{and since } x > 0, \text{ integers } p,\ q > 0\}$$
$$\Rightarrow \quad 3^{\frac{p}{q}} = 8$$
$$\Rightarrow \quad \left(3^{\frac{p}{q}}\right)^q = 8^q$$
$$\Rightarrow \quad 3^p = 8^q$$

which is impossible since for the given possible values of p and q, 3^p is always odd and 8^q is always even. Thus, the assumption is false and its opposite must be true. Hence x is irrational.

Example 6: Prove that no positive integers x and y exist such that $x^2 - y^2 = 1$.

Proof (by contradiction):

Suppose $x,\ y \in \mathbb{Z}^+$ exist such that $x^2 - y^2 = 1$.

$$\Rightarrow \ (x+y)(x-y) = 1$$
$$\Rightarrow \ \underbrace{x+y = 1 \text{ and } x-y = 1}_{\text{case 1}} \quad \textbf{or} \quad \underbrace{x+y = -1 \text{ and } x-y = -1}_{\text{case 2}}$$
$$\Rightarrow \ x = 1,\ y = 0 \ \text{(from case 1)} \quad \textbf{or} \quad x = -1,\ y = 0 \ \text{(from case 2)}$$

Both cases provide a contradiction to $x,\ y > 0$.
Thus, the supposition is false and its opposite is true.
There do *not* exist positive integers x and y such that $x^2 - y^2 = 1$.

Indirect proof often seems cleverly contrived, especially if no direct proof is forthcoming. It is perhaps more natural to seek a direct proof for the first attempt to prove a conjecture.

ERRORS IN PROOF

One must be careful not to make errors in algebra or reasoning. Examine carefully the following examples.

Example 7: Consider **Example 5** again: Prove that the solution of $3^x = 8$ is irrational.

Invalid argument:

$$\begin{aligned} 3^x &= 8 \\ \Rightarrow \quad \log 3^x &= \log 8 \\ \Rightarrow \quad x \log 3 &= \log 8 \\ \Rightarrow \quad x &= \frac{\log 8}{\log 3} \quad \text{where both } \log 8 \text{ and } \log 3 \text{ are irrational.} \\ \Rightarrow \quad & x \text{ is irrational.} \end{aligned}$$

The last step is not valid. The argument that an irrational divided by an irrational is always irrational is not correct. For example, $\frac{\sqrt{2}}{\sqrt{2}} = 1$, and 1 is rational.

Dividing by zero is *not* a valid operation. $\dfrac{a}{0}$ is not defined for any $a \in \mathbb{R}$, in particular $\frac{0}{0} \neq 1$.

Example 8: Invalid "proof" that $5 = 2$

$$\begin{aligned} 0 &= 0 \\ \Rightarrow \quad 0 \times 5 &= 0 \times 2 \\ \Rightarrow \quad \frac{0 \times 5}{0} &= \frac{0 \times 2}{0} \qquad \{\text{dividing through by } 0\} \\ \Rightarrow \quad 5 &= 2, \quad \text{which is clearly false.} \end{aligned}$$

This invalid step is not always obvious, as illustrated in the following example.

Example 9: Invalid "proof" that $0 = 1$:

$$\begin{aligned} \text{Suppose} \quad a &= 1 \\ \Rightarrow \quad a^2 &= a \\ \Rightarrow \quad a^2 - 1 &= a - 1 \\ \Rightarrow \quad (a+1)(a-1) &= a - 1 \\ \Rightarrow \quad a + 1 &= 1 \quad .... \ (*) \\ \Rightarrow \quad a &= 0 \\ \text{So,} \quad 0 &= 1 \end{aligned}$$

The invalid step in the argument is $(*)$ where we divide both sides by $a - 1$.
Since $a = 1$, $a - 1 = 0$, and so we are dividing both sides by zero.

Another trap to be avoided is to begin by assuming the result we wish to prove is true. This readily leads to invalid circular arguments.

Example 10: Prove without decimalisation that $\sqrt{3}-1>\frac{1}{\sqrt{2}}$.

Invalid argument:

$$\begin{aligned} & \sqrt{3}-1>\tfrac{1}{\sqrt{2}} \\ \Rightarrow\ & (\sqrt{3}-1)^2>\left(\tfrac{1}{\sqrt{2}}\right)^2 && \{\text{both sides are } >0\text{, so we can square them}\} \\ \Rightarrow\ & 4-2\sqrt{3}>\tfrac{1}{2} \\ \Rightarrow\ & \tfrac{7}{2}>2\sqrt{3} \\ \Rightarrow\ & 7>4\sqrt{3} \\ \Rightarrow\ & 7^2>48 && \{\text{squaring again}\} \\ \Rightarrow\ & 49>48 && \text{which is true.} \end{aligned}$$

Hence $\sqrt{3}-1>\frac{1}{\sqrt{2}}$ is true.

Although $\sqrt{3}-1>\frac{1}{\sqrt{2}}$ is in fact true, the above argument is invalid because we began by assuming the result.

A valid method of proof for $\sqrt{3}-1>\frac{1}{\sqrt{2}}$ can be found by either:

- reversing the steps of the above argument, or by
- using proof by contradiction (supposing $\sqrt{3}-1\leqslant\frac{1}{\sqrt{2}}$).

It is important to distinguish **errors in proof** from a **false conjecture**.

Consider the table alongside, which shows values of n^2-n+41 for various values of $n\in\mathbb{N}$.

n	n^2-n+41
1	41
2	43
3	47
4	53
5	61
6	71
7	83
8	97
9	113
10	131
11	151
12	173
13	197
...	...
30	911
...	...
99	9743
...	...

From the many examples given, one might conjecture:

"For all natural numbers n, n^2-n+41 is prime."

This conjecture is in fact false.

For example, for $n=41$, $n^2-n+41=41^2$ is clearly not prime.

IMPLICATIONS AND THEIR CONVERSE

If .... then ....

Many statements in mathematics take the form of an **implication** "If A then B", where A and B are themselves statements. The statement A is known as the **hypothesis**. The statement B is known as the **conclusion**.

Implications can be written in many forms in addition to "If A then B". For example, the following all have the same meaning:

$$A \left\{ \begin{array}{c} \text{implies} \\ \text{so} \\ \text{hence} \\ \text{thus} \\ \text{therefore} \end{array} \right\} B.$$

Given a statement of the form "If A then B", we can write a **converse** statement "If B then A".

If we know the truth, or otherwise, of a given statement, we can say nothing about the truth of the converse. It could be true or false.

A statement and its converse are said to be (logically) *independent.*

For example, suppose x is an integer.

- The statement "If x is odd, then $2x$ is even" is *true*, but its converse "If $2x$ is even, then x is odd" is *false*.
- The statement "If $2x$ is even, then x is odd" is *false*, but its converse "If x is odd, then $2x$ is even" is *true*.
- The statement "If $x > 1$, then $\ln x > 0$" is *true*, and its converse "If $\ln x > 0$, then $x > 1$" is also *true*.
- The statement "If $x = 5$, then $x^2 = 16$" is *false*, and its converse "If $x^2 = 16$, then $x = 5$" is also *false*.

EXERCISE

Prove or disprove:

1 If x is rational then $2^x \neq 3$.

2 If $2^x \neq 3$ then x is rational.

EQUIVALENCE

Some conjectures with two statements A and B involve **logical equivalence** or simply **equivalence**.

We say A *is equivalent to* B, or A is true *if and only if* B is true.

The phrase "if and only if" is often written as "iff" or $\Leftrightarrow$.

$A \Leftrightarrow B$ means $A \Rightarrow B$ *and* $B \Rightarrow A$

In order to prove an equivalence, we need to prove both implications: $A \Rightarrow B$ *and* $B \Rightarrow A$.

For example: $x^2 = 9 \;\Leftrightarrow\; x = 3$ is a false statement.

$x = 3 \;\Rightarrow\; x^2 = 9$ is true

but $x^2 = 9 \;\not\Leftarrow\; x = 3$ as x may be -3.

Example 11: Prove that $(n+2)^2 - n^2$ is a multiple of 8 $\Leftrightarrow$ n is odd.

Proof: $(\Rightarrow)$ $(n+2)^2 - n^2$ is a multiple of 8

$\Rightarrow\; n^2 + 4n + 4 - n^2 = 8a$ for some integer a

$\Rightarrow\; 4n + 4 = 8a$

$\Rightarrow\; n + 1 = 2a$

$\Rightarrow\; n = 2a - 1$

$\Rightarrow\; n$ is odd.

$(\Leftarrow)$ n is odd

$\Rightarrow\; n = 2a - 1$ for some integer a

$\Rightarrow\; n + 1 = 2a$

$\Rightarrow\; 4n + 4 = 8a$

$\Rightarrow\; (n^2 + 4n + 4) - n^2 = 8a$

$\Rightarrow\; (n+2)^2 - n^2$ is a multiple of 8.

In the above example the $(\Rightarrow)$ argument is clearly reversible to give the $(\Leftarrow)$ argument. However, this is not always the case.

Example 12: Prove that for all $x \in \mathbb{Z}^+$, x is not divisible by 3 $\Leftrightarrow$ $x^2 - 1$ is divisible by 3.

Proof: $(\Rightarrow)$ x is not divisible by 3

$\Rightarrow$ either $x = 3k + 1$ or $x = 3k + 2$ for some $k \in \mathbb{Z}^+ \cup \{0\}$

$\Rightarrow\; x^2 - 1 = 9k^2 + 6k$ or $9k^2 + 12k + 3$

$= 3(3k^2 + 2)$ or $3(3k^2 + 4k + 1)$

$\Rightarrow\; x^2 - 1$ is divisible by 3.

$(\Leftarrow)$ $x^2 - 1$ is divisible by 3

$\Rightarrow\; 3 \mid x^2 - 1$

$\Rightarrow\; 3 \mid (x+1)(x-1)$

$\Rightarrow\; 3 \mid (x+1)$ or $3 \mid (x-1)$ {as 3 is a prime number}

$\Rightarrow\; 3 \nmid x$

or in other words, x is not divisible by 3.

NEGATION

For any given statement A, we write not A or $\neg A$ to represent the negation of the statement A.

For example:

	A	$\neg A$
	$x > 0$	$x \leqslant 0$
	x is prime	x is not prime
	x is an integer	x is not an integer
For $x \in \mathbb{R}$:	x is rational	x is irrational
For $z \in \mathbb{C}$:	z is real	$z = a + bi,\ a, b \in \mathbb{R},\ b \neq 0$
For $x \in \mathbb{Z}^+ \cup \{0\}$:	x is a multiple of 3	x is not a multiple of 3 *or* $x = 3k+1$ or $3k+2$ for $k \in \mathbb{Z}^+ \cup \{0\}$

PROOF OF THE CONTRAPOSITIVE

To prove the statement "If A then B", we can provide a direct proof, or we can prove the logically equivalent **contrapositive** statement "If not B, then not A" which we can also write as "If $\neg B$, then $\neg A$".

For example, the statement "If it is Jon's bicycle, then it is blue"
is logically equivalent to "If that bicycle is not blue, then it is not Jon's".

Example 13: Prove that for $a, b \in \mathbb{R}$, "ab is irrational $\Rightarrow$ either a or b is irrational".

Proof using contrapositive:

a and b are both rational $\Rightarrow$ $a = \frac{p}{q}$ and $b = \frac{r}{s}$ where $p, q, r, s \in \mathbb{Z},\ q \neq 0,\ s \neq 0$

$\Rightarrow$ $ab = \left(\frac{p}{q}\right)\left(\frac{r}{s}\right) = \frac{pr}{qs}$ {where $qs \neq 0$, since $q, s \neq 0$}

$\Rightarrow$ ab is rational {since $pr, qs \in \mathbb{Z}$}

Thus ab is irrational $\Rightarrow$ either a or b is irrational.

Example 14: Prove that if n is a positive integer of the form $3k+2,\ k \geqslant 0,\ k \in \mathbb{Z}$, then n is not a square.

Proof using contrapositive:

If n is a square then
n has one of the forms $(3a)^2$, $(3a+1)^2$ or $(3a+2)^2$, where $a \in \mathbb{Z}^+ \cup \{0\}$.

$\Rightarrow$ $n = 9a^2,\ 9a^2 + 6a + 1$ or $9a^2 + 12a + 4$

$\Rightarrow$ $n = 3(3a^2),\ 3(3a^2 + 2a) + 1$ or $3(3a^2 + 4a + 1) + 1$

$\Rightarrow$ n has the form $3k$ or $3k+1$ only, where $k \in \mathbb{Z}^+ \cup \{0\}$

$\Rightarrow$ n does not have form $3k+2$.

Thus if n is a positive integer of the form $3k+2,\ k \geqslant 0,\ k \in \mathbb{Z}$, then n is not a square.

USING PREVIOUS RESULTS

In mathematics we build up collections of important and useful results, each depending on previously proven statements.

Example 15: Prove the conjecture:

"The recurring decimal $0.\overline{9} = 0.999\,999\,99....$ is exactly equal to 1".

Proof (by contradiction):

Suppose $0.\overline{9} < 1$

$\Rightarrow\ 0.\overline{9} < \dfrac{0.\overline{9}+1}{2}$ {We proved earlier that $a < b \ \Rightarrow\ a < \dfrac{a+b}{2}$}

$\Rightarrow\ 0.\overline{9} < \dfrac{1.\overline{9}}{2}$ {Ordinary division: $2\ \underline{\big|\ 1.999\,999\,99....}$ $\ 0.999\,999\,99....$}

$\Rightarrow\ 0.\overline{9} < 0.\overline{9}$ clearly a contradiction

Therefore the supposition is false, and so $0.\overline{9} \geqslant 1$ is true.

Since, $0.\overline{9} > 1$ is absurd, $0.\overline{9} = 1$.

Proof (Direct Proof):

$$\begin{aligned}
0.\overline{9} &= 0.999\,999\,99.... \\
&= 0.9 + 0.09 + 0.009 + 0.0009 + ... \\
&= 0.9\left(1 + \tfrac{1}{10} + \tfrac{1}{100} + \tfrac{1}{1000} + ...\right) \\
&= \tfrac{9}{10}\left(\sum_{i=0}^{\infty}\left(\tfrac{1}{10}\right)^i\right) \\
&= \tfrac{9}{10}\left(\frac{1}{1-\frac{1}{10}}\right) \qquad \text{\{Using the previously proved Geometric Series with } r = \tfrac{1}{10} \text{ and } \left|\tfrac{1}{10}\right| < 1\text{\}} \\
&= \tfrac{9}{10} \times \tfrac{10}{9} \\
&= 1
\end{aligned}$$

THEORY OF KNOWLEDGE — AXIOMS AND OCCAM'S RAZOR

In order to understand complicated concepts, we often try to break them down into simpler components. But when mathematicians try to understand the foundations of a particular branch of the subject, they consider the question "What is the minimal set of assumptions from which all other results can be deduced or proved?" The assumptions they make are called **axioms**. Whether the axioms accurately reflect properties observed in the physical world is less important to pure mathematicians than the theory which can be developed and deduced from the axioms.

Occam's razor is a principle of economy that among competing hypotheses, the one that makes the fewest assumptions should be selected.

1 What value does Occam's razor have in understanding the long-held belief that the world was flat?

2 Is the simplest explanation to something always true?

3 Is it reasonable to construct a set of mathematical axioms under Occam's razor?

One of the most famous examples of a set of axioms is given by Euclid in his set of 13 books called *Elements*. He gives five axioms, which he calls "postulates", as the basis for his study of Geometry:

1. Any two points can be joined by a straight line.
2. Any straight line segment can be extended indefinitely in a straight line.
3. Given any straight line segment, a circle can be drawn having the segment as radius and one endpoint as centre.
4. All right angles are congruent.
5. **Parallel postulate**: If two lines intersect a third in such a way that the sum of the inner angles on one side is less than two right angles, then the two lines inevitably must intersect each other on that side if extended far enough.

4 Is the parallel postulate genuinely an axiom, or can it be proved from the others?

5 What happens if you change the list of axioms or do not include the parallel postulate?

6 What other areas of mathematics can we reduce to a concise list of axioms?

Worked Solutions

EXERCISE A.1

1 **a** $S = \{a, b, c\}$ and $n(S) = 3$

b $S = \{2, 3, 5, 7\}$ and $n(S) = 4$

c $S = \{3, 4, 5, 6, 7\}$ and $n(S) = 5$

d As $x^2 = -9$ has no real solution $S = \varnothing$ and $n(S) = 0$.

e $S = \{3, 4\}$ on removing repetitions and $n(S) = 2$

f $S = \{\varnothing\}$ is the set containing the symbol $\varnothing$ and $n(S) = 1$

2 **a** $S = \{1, 2, 3, 4, 5, ..., 98, 99\}$ is finite with cardinality $n(S) = 99$.

b There are infinitely many rational numbers in S.
That is, S is an infinite set.

3 **a** 7 is an integer, $\therefore$ $7 \in \mathbb{Z}$ is true.

b $\sqrt{13}$ is not rational, $\therefore$ $\sqrt{13} \in \mathbb{Q}$ is false.

c If e is exponential e, then $e \in \mathbb{R}$ is true.

d $-3.5 = \frac{-7}{2} \in \mathbb{Q}$ is true.

e $4.\overline{1} = 4\frac{1}{9}$ is not a positive integer
$\therefore$ $4.\overline{1} \in \mathbb{Z}^+$ is false.

f $\sqrt{-2} = i\sqrt{2}$ is of the form $a + ib$ where $a = 0$, $b = \sqrt{2}$ are both real
$\therefore$ $\sqrt{-2} \in \mathbb{C}$ is true.

g $(\sqrt{3})^2 = 3$ which is an integer
$\therefore$ $(\sqrt{3})^2 \in \mathbb{Z}$ is true.

h $\pi^2 \approx 9.869\,604....$ can be placed on the real number line
$\therefore$ $\pi^2 \in \mathbb{R}$ is true.

4 **a** Sets are *equal* as repetitions are ignored.

b Sets are *equal* as the order of listing elements is not important.

c Both sets contain -2 and 2 only.
$\therefore$ the sets are *equal*.

d Both sets are empty sets.
$\therefore$ the sets are *equal*.

e These sets are *not equal* as the first set does not contain 2 and 5 which are in the second set.

EXERCISE A.2

1 **a** $P(A) = \{\varnothing, \{p\}, \{q\}, \{p, q\}\}$

b $P(A) = \{\varnothing, \{1\}, \{2\}, \{3\}, \{1, 2\}, \{1, 3\}, \{2, 3\}, \{1, 2, 3\}\}$

c $P(A) = \{\varnothing, \{0\}\}$

2 **a** $A = \{\text{a, e, i, o, u}\}$ and $B = \{\text{s, e, q, u, o, i, a}\}$
$\therefore$ the elements of A are also in B
$\therefore$ $A \subseteq B$ is true.

b $A = \{0\}$ and $B = \{\ \}$.
$\therefore$ $0 \in A$, but $0 \notin B$
$\therefore$ $A \subseteq B$ is false.

c $A = \{3, 5, 9\}$ and $B = \{2, 3, 5, 7, 11, ...\}$
$\therefore$ $9 \in A$, but $9 \notin B$
$\therefore$ $A \subseteq B$ is false.

d $2 \in A$ {when $a = 2$ and $b = 0$} and 2 is rational.
Thus $2 \notin B$
$\therefore$ $A \subseteq B$ is false.

3 **a** <u>If $x \in A$</u> $\Rightarrow x = a + \frac{1}{2}$ where $a \in \mathbb{Z}$
$\Rightarrow x = (a+1) - \frac{1}{2}$ where $a \in \mathbb{Z}$
$\Rightarrow x = b - \frac{1}{2}$, $b = a + 1 \in \mathbb{Z}$
{if $a \in \mathbb{Z}$ then $a + 1 \in \mathbb{Z}$}
$\Rightarrow x \in B$

<u>If $x \in B$</u> $\Rightarrow x = b - \frac{1}{2}$ where $b \in \mathbb{Z}$
$\Rightarrow x = (b-1) + \frac{1}{2}$ where $b \in \mathbb{Z}$
$\Rightarrow x = a + \frac{1}{2}$, $a = b - 1 \in \mathbb{Z}$
{if $b \in \mathbb{Z}$ then $b - 1 \in \mathbb{Z}$}
$\Rightarrow x \in A$

Thus $A \subseteq B$ and $B \subseteq A \Rightarrow A = B$.

b <u>If $x \in A$</u> $\Rightarrow x = \sqrt{y}$ where $y > 0$
$\Rightarrow x > 0$
{$\sqrt{y} \geqslant 0$ by definition, but $y > 0$ given}
$\Rightarrow x \in B$

<u>If $x \in B$</u> $\Rightarrow x > 0$
$\Rightarrow x = \sqrt{x^2}$
$\Rightarrow x = \sqrt{y}$ for $y = x^2 \in \mathbb{R}^+$
$\Rightarrow x \in A$

Thus $A \subseteq B$ and $B \subseteq A \Rightarrow A = B$.

4 **a** In A, x is even
$\therefore$ $x + 1$ is odd
$\therefore$ $(x+1)^2 = 1, 9, 25, 49, ...$
$\therefore$ $y = 1, 9, 25, 49, ...$
$\therefore$ $A = \{1, 9, 25, 49, ...\}$
Thus $A \subseteq B$ as $B = \{1, 3, 5, 7, 9, 11, ...\}$
However $A \neq B$.

b In A, $y \in \mathbb{C}$
$\therefore$ $y = a + ib$ where $a, b \in \mathbb{R}$
$\therefore$ $yy^* = (a+ib)(a-ib) = a^2 + b^2$
$\therefore$ $x = yy^*$ is real and $\geqslant 0$
$\therefore$ $A \subseteq B$
However $A \neq B$ {B includes negatives}

5 P_m is: $n(P(A)) = 2^m$ where $n(A) = m$ for all $m \in \mathbb{Z}^+ \cup \{0\}$.

Proof by induction (on m)

(1) If $m = 0$ $\therefore$ $n(A) = 0$ then $A = \varnothing$
$\therefore$ $P(A) = \{\varnothing\}$
$\therefore$ $n(P(A)) = 1$
$\therefore$ $n(P(A)) = 2^0$
$\therefore$ P_0 is true.

(2) If P_k is true, $n(P(A)) = 2^k$, $n(A) = k$.
Consider set $B = \{\text{elements of } A, b\}$ where $b \notin A$.
Then $n(B) = n(A) + 1 = k + 1$ and
$P(B) = \{\text{elements of } P(A), \{b\} \cup S \text{ for each } S \in P(A)\}$
$\therefore$ $n(P(B)) = 2 \times n(P(A))$
$= 2 \times 2^k$
$= 2^{k+1}$

Thus as P_0 is true and the truth of $P_k \Rightarrow$ the truth of P_{k+1}, P_m is true. {Principle of mathematical induction}

EXERCISE A.3

1 **a** $\{0, 1, 2, 3, 4, 5, 7\}$ **b** $\{7\}$ **c** $\varnothing$ **d** $\{1, 3, 7\}$
e $\{1, 3, 7\}$ **f** $\{5, 6, 7, 8, 9\}$ **g** $\{6, 8, 9\}$ **h** $\{6, 8, 9\}$

2 **a**

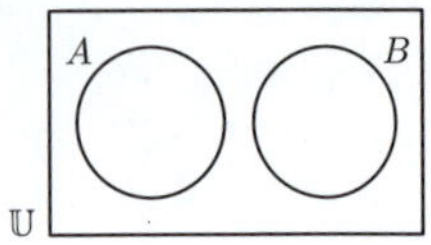

b

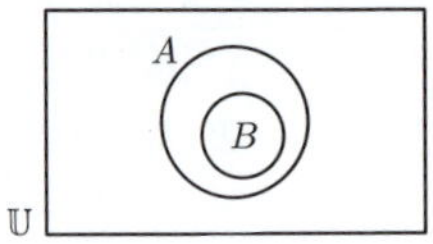

c

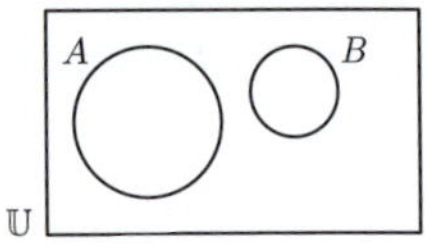

d not possible

e

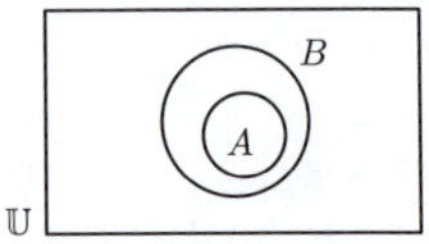

f

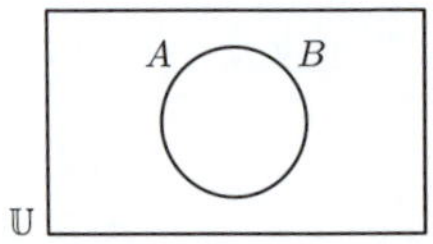

g not possible

3

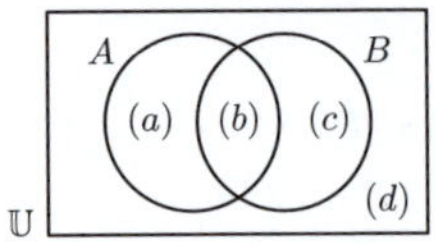

Consider the Venn diagram

$n(A \cup B) = a + b + c$

$n(A' \cap B) = c$

If $A \cup B = A' \cap B$

then $n(A \cup B) = n(A' \cap B)$

$\therefore\ a + b + c = c$

$\therefore\ a + b = 0,\ a \geqslant 0,\ b \geqslant 0$

$\therefore\ a = b = 0$

$\therefore\ n(A) = 0$

a contradiction to 'A is non-empty'

Thus $A \cup B = A' \cap B$ is not possible.

4 **To prove:** $A \cap B = A \Leftrightarrow A \subseteq B$

$(\Rightarrow)$ Suppose $A \cap B = A$

If $A = \varnothing$, we know $A \subseteq B$

If $A \neq \varnothing$ we let $x \in A$, say

$\Rightarrow x \in A \cap B$ $\{A = A \cap B$, given$\}$

$\Rightarrow x \in B$

So, $x \in A \Rightarrow x \in B$

$\therefore\ A \subseteq B$

$(\Leftarrow)$ Suppose $A \subseteq B$

If $A = \varnothing$, $A \cap B = \varnothing \cap B = \varnothing = A$

$\therefore\ A \cap B = A$

If $A \neq \varnothing$ and $x \in A$

As $A \subseteq B$, then $x \in B$ also

$\therefore\ x \in A \cap B$

$\therefore\ x \in A \Rightarrow x \in A \cap B$

Hence $A \subseteq A \cap B$ .... (1)

If $A \neq \varnothing$ and $x \in A \cap B$ then $x \in A$

$\therefore\ x \in A \cap B \Rightarrow x \in A$

$\therefore\ A \cap B \subseteq A$ .... (2)

From (1) and (2), $A \cap B = A$

Thus $A \cap B = A \Leftrightarrow A \subseteq B$.

5 A and B are disjoint, and $A \cup B = \mathbb{U} \Rightarrow A$ and B partition $\mathbb{U}$

We need to show that $B \subseteq A'$ **and** $A' \subseteq B$

Proof: If $x \in A' \Rightarrow x \notin A$

$\Rightarrow x \in B$ $\{A$ and B partition $\mathbb{U}\}$

$\therefore\ A' \subseteq B$ .... (1)

If $x \in B \Rightarrow x \notin A$

$\Rightarrow x \in A'$

$\therefore\ B \subseteq A'$ .... (2)

From (1) and (2), $B = A'$.

6 **a**

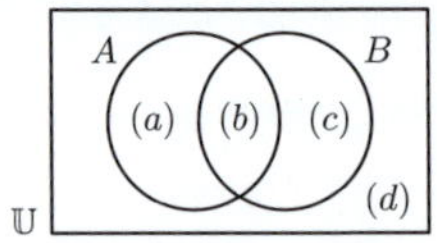

$n(A) + n(B) - n(A \cap B)$

$= (a + b) + (b + c) - b$

$= a + b + c$

$= n(A \cup B)$

b $n(A \cup B) = n(\mathbb{U}) - n((A \cup B)')$

$= 30 - 6$

$= 24$

From **a**, $24 = 16 + 15 - n(A \cap B)$

$\therefore\ n(A \cap B) = 7$

$\therefore$ 7 play both sports.

7 As $A \subseteq B$, if $x \in A$ then $x \in B$

As $B \subseteq C$, if $x \in B$ then $x \in C$

Thus if $A \subseteq B$ **and** $B \subseteq C$ then

$x \in A \Rightarrow x \in B \Rightarrow x \in C$

Thus $A \subseteq C$.

EXERCISE A.4

1 **a** $\{o, n, u, a, c, e\}$ **b** $\{n, a\}$

c $\{c, j, g, t, e\}$ **d** $\{c, o, j, u, g, t, e\}$

e $\{c, o, j, u, g, t, e\}$ **f** $\{o, n, u, a\}$

2 **a**

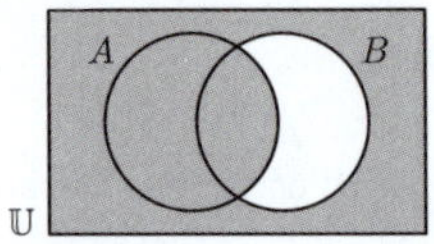

b

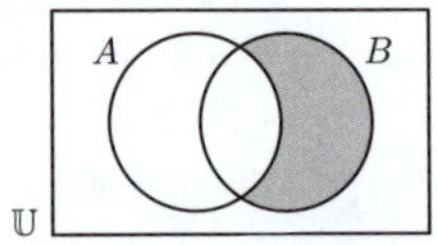

c

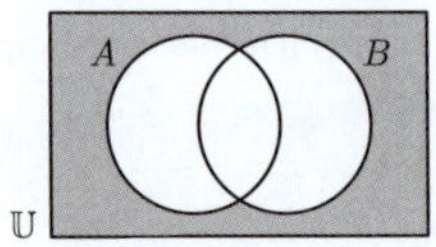

d

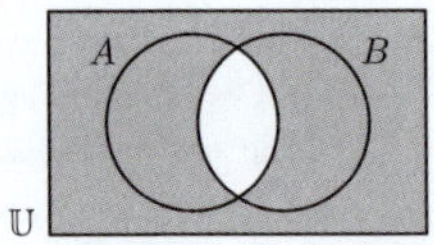

e

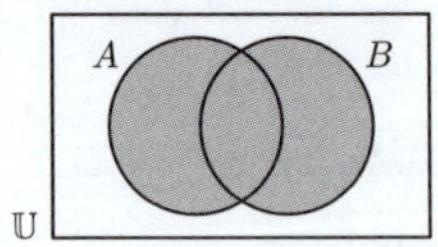

3 **a** $(\Rightarrow)$ Let $x \in (A \cup B) \cup C$

$\therefore\ x \in A \cup B$ or $x \in C$

$\therefore\ x \in A$ or B or C

$\therefore\ x \in A$ or $B \cup C$

$\therefore\ x \in A \cup (B \cup C)$

Thus $(A \cup B) \cup C \subseteq A \cup (B \cup C)$ .... (1)

$(\Leftarrow)$ Let $x \in A \cup (B \cup C)$

$\therefore\ x \in A$ or $x \in (B \cup C)$

$\therefore\ x \in A$ or B or C

$\therefore\ x \in A \cup B$ or $x \in C$

$\therefore\ x \in (A \cup B) \cup C$

Thus $A \cup (B \cup C) \subseteq (A \cup B) \cup C$ .... (2)

From (1) and (2), $(A \cup B) \cup C = A \cup (B \cup C)$

b $(\Rightarrow)$ Let $x \in A \cap (B \cup C)$

$\therefore\ x \in A$ and $x \in B$ or $x \in C$

$\therefore\ x \in A$ and $x \in B$ or $x \in A$ and $x \in C$

$\therefore\ x \in A \cap B$ or $x \in A \cap C$

$\therefore\ x \in (A \cap B) \cup (A \cap C)$

Thus $A \cap (B \cup C) \subseteq (A \cap B) \cup (A \cap C)$ .... (1)

$(\Leftarrow)$ Let $x \in (A \cap B) \cup (A \cap C)$

$\therefore\ x \in A \cap B$ or $x \in A \cap C$

$\therefore\ x \in A$ and B or $x \in A$ and C

$\therefore\ x \in A$ and $x \in B \cup C$

$\therefore\ x \in A \cap (B \cup C)$

Thus $(A \cap B) \cup (A \cap C) \subseteq A \cap (B \cup C)$ .... (2)

From (1) and (2), $A \cap (B \cup C) = (A \cap B) \cup (A \cap C)$

c $(A \cup B) \cap (A' \cup B)$

$= (B \cup A) \cap (B \cup A')$ {Commutative law}

$= B \cup (A \cap A')$ {Distributive law}

$= B \cup \varnothing$ {Complement law}

$= B$ {Identity law}

or

$(\Rightarrow)$ Let $x \in (A \cup B) \cap (A' \cup B)$

$\therefore\ x \in A \cup B$ and $x \in A' \cup B$

Now suppose $x \notin B$ then $x \in A$ and $x \in A'$ which is a contradiction as $A \cap A' = \varnothing$

$\therefore\ x \in B$

Hence $(A \cup B) \cap (A' \cup B) \subseteq B$ .... (1)

$(\Leftarrow)$ Let $x \in B$

$\therefore\ x \in A \cup B$ and $x \in A' \cup B$

$\therefore\ x \in (A \cup B) \cap (A' \cup B)$

$\therefore\ B \subseteq (A \cup B) \cap (A' \cup B)$ .... (2)

From (1) and (2), $(A \cup B) \cap (A' \cup B) = B$.

d $(\Rightarrow)$ Let $x \in (A \cap B)'$

$\therefore\ x \notin A \cap B$

$\therefore\ x \notin A$ and B

$\therefore\ x \in A'$ or $x \in B'$

$\therefore\ x \in A' \cup B'$

Thus $(A \cap B)' \subseteq A' \cup B'$ .... (1)

$(\Leftarrow)$ Let $x \in A' \cup B'$

$\therefore\ x \in A'$ or $x \in B'$

$\therefore\ x \notin A$ and B

$\therefore\ x \notin A \cap B$

$\therefore\ x \in (A \cap B)'$

Thus $A' \cup B' \subseteq (A \cap B)'$ .... (2)

From (1) and (2), $(A \cap B)' = A' \cup B'$.

EXERCISE A.5

1 **a** **i** {2, 4} **ii** $\varnothing$ **b** **i** {irrational numbers} **ii** $\varnothing$

c **i** {0, 1} **ii** {4, 5} **d** **i** {2, 3, 4} **ii** {0, 1, 5}

2 **a** {*b*, *c*, *d*} **b** {1, 2, 5} **c** {1, 2, 3, 4, 5, 6}

d {9, 11, 13}

3 **a** **i**

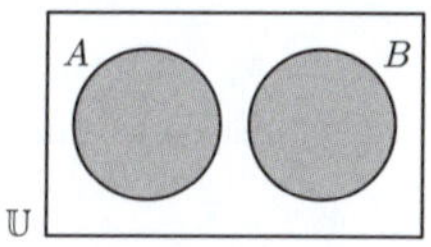

ii

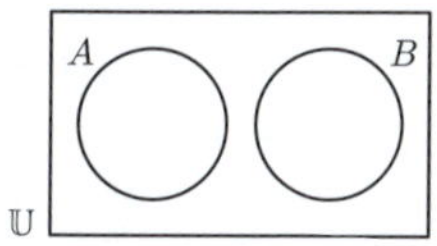

iii

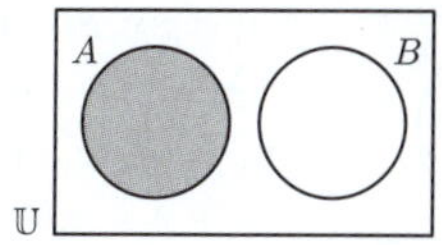

iv

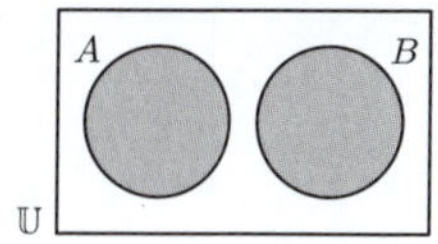

b **i**

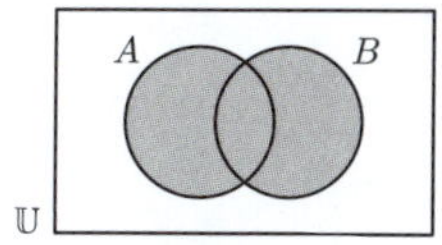

ii

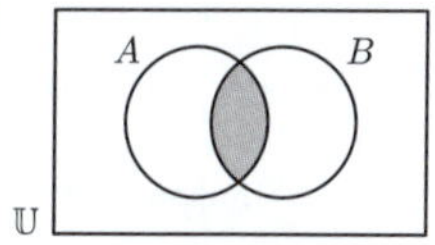

iii

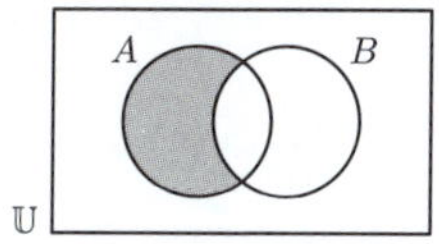

iv

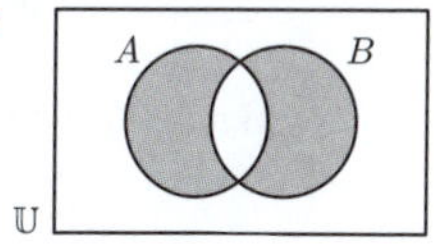

c **i**

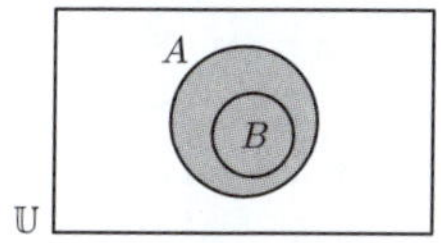

ii

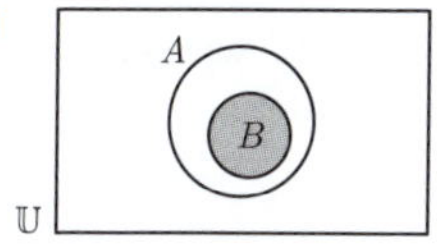

iii

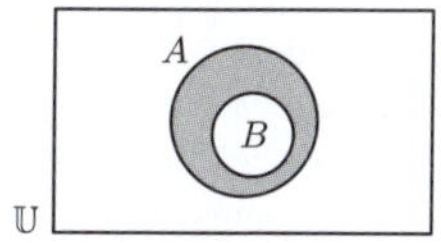

iv

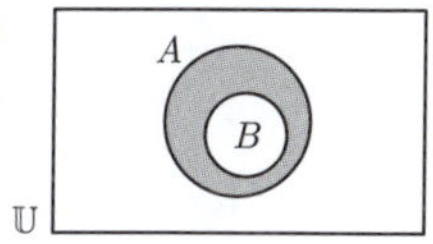

4 **a**

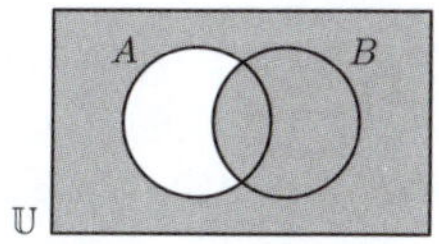

b

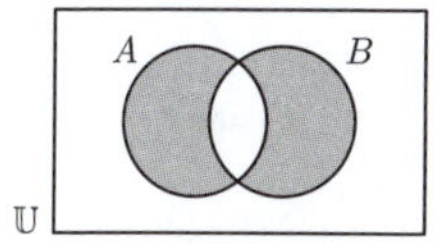

c

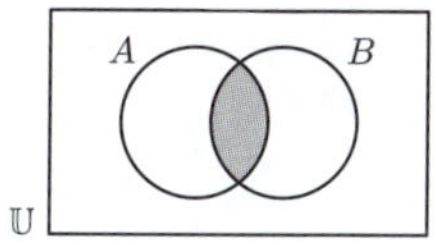

d

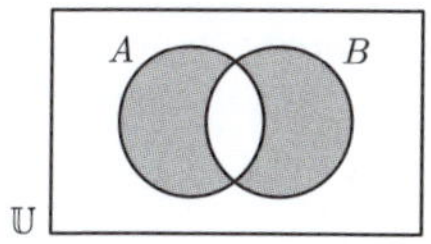

5 **To prove:** $A\Delta B = A \cup B \Leftrightarrow A \cap B = \varnothing$.

Proof:

$(\Leftarrow)$ Suppose $A \cap B = \varnothing$

If $x \in A\Delta B$, then $x \in A$ or $x \in B$ but $x \notin A \cap B$

$\therefore\ x \in A \cup B$

$\therefore\ A\Delta B \subseteq A \cup B$ .... (1)

If $x \in A \cup B$, then $x \in A$ or $x \in B$

$\therefore\ x \in A$ or $x \in B$ but $x \notin A \cap B$

{as $A \cap B = \varnothing$ has no elements}

$\therefore\ x \in A\Delta B$

$\therefore\ A \cup B \subseteq A\Delta B$ .... (2)

From (1) and (2), $A\Delta B = A \cup B$.

$(\Rightarrow)$ Suppose $A\Delta B = A \cup B$

$$A\Delta B \stackrel{\text{def}}{=} \{x \mid x \in A \text{ or } x \in B, \text{ but } x \notin A \cap B\}$$

and $A \cup B \stackrel{\text{def}}{=} \{x \mid x \in A \text{ or } x \in B\}$

So $A\Delta B = A \cup B \Rightarrow$ there are no elements in $A \cap B$

$\Rightarrow A \cap B = \varnothing$

Note:

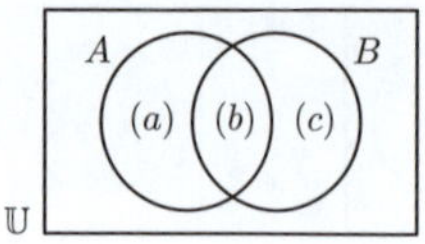

In the general case, if $A\Delta B = A \cup B$ then

$n(A\Delta B) = n(A \cup B)$

$\therefore \; a + c = a + b + c$

$\therefore \; b = 0$

$\therefore \; A \cap B = \varnothing$

6 **To prove:** $A \cap (B \backslash C) = (A \cap B) \backslash (A \cap C)$.

Proof:

$(\Rightarrow)$ Suppose $x \in A \cap (B \backslash C)$

$\therefore \; x \in A$ and $x \in B \backslash C$

$\therefore \; x \in A$ and $x \in B$ and $x \notin C$
(1) (2) (3)

From (1) and (2), $x \in A \cap B$

From (1) and (3), $x \notin A \cap C$

Thus $x \in (A \cap B) \backslash (A \cap C)$

So, $x \in A \cap (B \backslash C) \subseteq (A \cap B) \backslash (A \cap C)$ .... (4)

$(\Leftarrow)$ Suppose $x \in (A \cap B) \backslash (A \cap C)$

$\therefore \; x \in A \cap B$ and $x \notin A \cap C$

$\therefore \; x \in A \cap B$ and $x \in (A \cap C)'$

$\therefore \; x \in A \cap B$ and $x \in A' \cup C'$ {De Morgan's law}

$\therefore \; x \in A$ and $x \in B$ and $x \in A'$ or $x \in C'$

$\therefore \; x \in A$ and $x \in B$ and $x \in A'$
or $x \in A$ and $x \in B$ and $x \in C'$

As $x \in A$ and $x \in A'$ is not possible, $x \in A$ and $x \in B$ and $x \notin C$

$\therefore \; x \in A \cap (B \backslash C)$

So, $(A \cap B) \backslash (A \cap C) \subseteq A \cap (B \backslash C)$ .... (5)

From (4) and (5), $A \cap (B \backslash C) = (A \cap B) \backslash (A \cap C)$

7 Firstly if $A = B$ then both $A\Delta B$ and $A'\Delta B'$ are both $\varnothing$.

$\therefore$ the result is trivially true.

We now prove the result if $A \neq B$.

We need to prove (1) $A\Delta B \subseteq A'\Delta B'$
and (2) $A'\Delta B' \subseteq A\Delta B$.

(1) Suppose $x \in A\Delta B$

$\therefore \; x \in (A \backslash B) \cup (B \backslash A)$

$\therefore \; x \in A \backslash B$ *or* $x \in B \backslash A$

$\therefore \; x \in A$ and $x \notin B$	$\therefore \; x \in B$ and $x \notin A$
$\therefore \; x \notin A'$ and $x \in B'$	$\therefore \; x \notin B'$ and $x \in A'$
$\therefore \; x \in B' \backslash A'$	$\therefore \; x \in A' \backslash B'$

$\therefore \; x \in (A' \backslash B') \cup (B' \backslash A')$

$\therefore \; x \in A'\Delta B'$

So, $x \in A\Delta B \Rightarrow x \in A'\Delta B'$

Thus $A\Delta B \subseteq A'\Delta B'$

(2) Suppose $x \in A'\Delta B'$

$\therefore \; x \in (A' \backslash B') \cup (B' \backslash A')$

$\therefore \; x \in A' \backslash B'$ *or* $x \in B' \backslash A'$

$\therefore \; x \in A'$ and $x \notin B'$	$\therefore \; x \in B'$ and $x \notin A'$
$\therefore \; x \notin A$ and $x \in B$	$\therefore \; x \notin B$ and $x \in A$
$\therefore \; x \in B \backslash A$	$\therefore \; x \in A \backslash B$

$\therefore \; x \in (A \backslash B) \cup (B \backslash A)$

$\therefore \; x \in A\Delta B$

So, $x \in A'\Delta B' \Rightarrow x \in A\Delta B$

Thus $A'\Delta B' \subseteq A\Delta B$

Hence from (1) and (2), $A\Delta B = A'\Delta B'$.

EXERCISE B.1

1 **a** **i** $\{(1, 3), (1, 4), (1, 5), (2, 3), (2, 4), (2, 5)\}$

ii $\{(3, 1), (3, 2), (4, 1), (4, 2), (5, 1), (5, 2)\}$

b **i** $\{(a, a), (a, b)\}$ **ii** $\{(a, a), (b, a)\}$

c **i** $\varnothing$ **ii** $\varnothing$

2 **a**

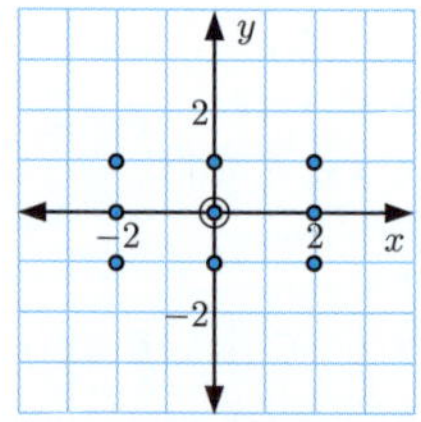

b

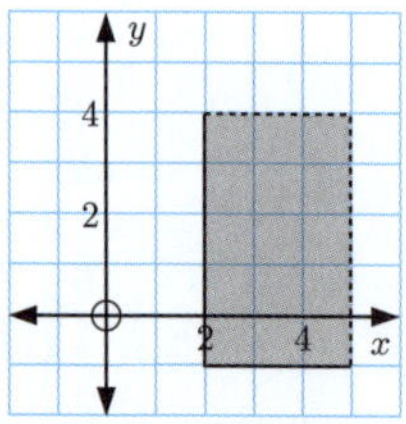

3 **To prove:** $A \times (B \cup C) = (A \times B) \cup (A \times C)$

Proof:

$(\Rightarrow)$ Suppose $(x, y) \in A \times (B \cup C)$

$\therefore \; x \in A$ and $y \in B \cup C$

$\therefore \; x \in A$ and $y \in B$ or $y \in C$

$\therefore \; (x, y) \in A \times B$ or $(x, y) \in A \times C$

$\therefore \; (x, y) \in (A \times B) \cup (A \times C)$

Thus $A \times (B \cup C) \subseteq (A \times B) \cup (A \times C)$ .... (1)

$(\Leftarrow)$ Suppose $(x, y) \in (A \times B) \cup (A \times C)$

$\therefore \; (x, y) \in A \times B$ or $(x, y) \in A \times C$

$\therefore \; x \in A$ and $y \in B$ or $x \in A$ and $y \in C$

$\therefore \; x \in A$ and $y \in B \cup C$

$\therefore \; (x, y) \in A \times (B \cup C)$

Thus $(A \times B) \cup (A \times C) \subseteq A \times (B \cup C)$ .... (2)

From (1) and (2), $A \times (B \cup C) = (A \times B) \cup (A \times C)$

EXERCISE B.2

1 **a** domain $= \{0, 1, 2\}$, range $= \{2, 3, 5\}$

b domain $= \{-3, -2, -1, 0, 1, 2, 3\}$,
range $= \{-3, -2\sqrt{2}, -\sqrt{5}, 0, \sqrt{5}, 2\sqrt{2}, 3\}$

c domain $= \{x \mid x \in \mathbb{R}\}$,
range $= \{y \mid y \in \mathbb{R}, \; -1 \leqslant y \leqslant 1\}$

d domain $= \{5, 10\}$,
range $= \{(3, 4), (4, 3), (6, 8), (8, 6)\}$

2 **a** $\{(2, 6), (2, 8), (3, 6), (4, 8), (5, 5)\}$

b $\{(2, 5), (3, 6), (4, 7), (5, 8)\}$

c $\{(3, 6), (4, 8)\}$

d $\{(2, 5), (2, 6), (2, 7), (2, 8), (3, 7), (3, 8)\}$

3 **a** **i** If x is female then $x \not{R} x$

$\therefore$ R is not reflexive.

ii If x is female and y is a brother of x then x *is not* a brother of y.

$\therefore \; xRy \not\Rightarrow yRx$

$\therefore$ R is not symmetric.

iii Since xRy and yRz, z is a brother of y who is a brother of x.

$\therefore$ z and x are siblings and z is a 'brother'.

$\therefore \; xRz$

$\therefore$ R is transitive.

b **i** xRx is *not possible*, as a person cannot be older than oneself.

$\therefore \; x \not{R} x$

Thus R is not reflexive.

ii If xRy then y is older than x.
$\therefore$ x is not older than y
$\therefore$ $y \not R x$
$\therefore$ R is not symmetric.

iii If xRy and yRz, then z is older than y who is older than x.
$\therefore$ z is older than x
$\therefore$ xRz
$\therefore$ R is transitive.

c **i** xRx is true for all x since a person must live in the same country as oneself.
$\therefore$ R is reflexive.

ii If xRy, x and y live in the same country and so y and x live in the same country.
$\therefore$ yRx
$\therefore$ R is symmetric.

iii If xRy and yRz then z lives in the same country as y who in turn lives in the same country as x.
$\therefore$ xRz
$\therefore$ R is transitive.

d **i** xRx is true, as one has the same mother as oneself.
$\therefore$ xRx
$\therefore$ R is reflexive.

ii If x and y have the same mother then y and x have the same mother.
$\therefore$ $xRy \Rightarrow yRx$
$\therefore$ R is symmetric.

iii If x and y have the same mother and y and z have the same mother then x and z have the same mother.
$\therefore$ xRy and $yRz \Rightarrow xRz$
$\therefore$ R is transitive.

4 **a** Consider $3 \in \mathbb{N}$. $3 \not R 3$ as 3 and 3 have a common factor of 3.
$\therefore$ R is not reflexive.

b If xRy then x and y are both in $\mathbb{N}$ and x and y have 1 only as a common factor.
$\therefore$ y and x have 1 only as a common factor
$\therefore$ $xRy \Rightarrow yRx$
$\therefore$ R is symmetric.

c As $2R3$ and $3R4$ but $2 \not R 4$
{as 2 and 4 share a common factor of 2}
then $xRy, yRz \not\Rightarrow xRz$
$\therefore$ R is not transitive.

5 **a** $ARB \Leftrightarrow A$ and B are disjoint
$\Leftrightarrow A \cap B = \varnothing$

i In general $A \cap A = A$ which is not $\varnothing$
$\therefore$ $A \not R A$
$\therefore$ R is not reflexive.

ii If ARB then $A \cap B = \varnothing$
$\Rightarrow B \cap A = \varnothing$
$\Rightarrow BRA$
$\therefore$ R is symmetric.

iii If $A = \{1, 2\}$, $B = \{3, 4\}$, $C = \{2, 5\}$ then
$A \cap B = \varnothing$ and $B \cap C = \varnothing$
$\Rightarrow ARB$ and BRC
But $A \cap C = \{2\} \neq \varnothing$
$\therefore$ ARB and $BRC \not\Rightarrow ARC$
$\therefore$ R is not transitive.

b $ARB \Leftrightarrow A \subseteq B$

i Since $A \subseteq A$ for all sets A, then ARA for all A
$\therefore$ R is reflexive.

ii If $A = \{1, 2\}$ and $B = \{1, 2, 3\}$ then
$A \subseteq B$ but $B \not\subseteq A$
$\therefore$ $ARB \not\Rightarrow BRA$
$\therefore$ R is not symmetric.

iii If ARB and BRC then
$A \subseteq B$ and $B \subseteq C \Rightarrow A \subseteq C$
$\therefore$ R is transitive.

c $ARB \Leftrightarrow n(A) = n(B)$

i Since for all sets A, $n(A) = n(A)$
$\therefore$ ARA
$\therefore$ R is reflexive.

ii If ARB then $n(A) = n(B)$
$\therefore$ $n(B) = n(A)$
$\therefore$ BRA
$\therefore$ R is symmetric.

iii If ARB and BRC then
$n(A) = n(B)$ and $n(B) = n(C)$
$\Rightarrow n(A) = n(C)$
$\Rightarrow ARC$
Thus R is transitive.

EXERCISE B.3

1 **a** Since (1, 1), (2, 2), (3, 3), and (4, 4) are all in R then R is reflexive.

b Since, for example, $(1, 2) \in R$ but $(2, 1) \notin R$ then R is not symmetric.

c Since, for example, (1, 2) and $(2, 3) \in R$, but $(1, 3) \notin R$ then R is not transitive.

2 It must be reflexive,
$\therefore$ (1, 1), (2, 2), (3, 3) must be included.
It must be symmetric,
$\therefore$ (2, 1) and (3, 2) must be included.
It must be transitive,
$\therefore$ (1, 3) must be included. {(1, 2) and (2, 3) are included}
Thus (3, 1) must also be there.
$\therefore$ (1, 1), (2, 2), (3, 3), (2, 1), (3, 2), (1, 3), and (3, 1) must be included.

3 Other solutions are possible.
a $\{(a, a), (b, b), (c, c), (a, b), (b, c)\}$
b $\{(a, b), (b, a), (a, c), (c, a)\}$ **c** $\{(a, b), (b, c), (a, c)\}$
d $\{(a, a), (b, b), (c, c), (b, c), (c, b), (a, c), (c, a)\}$
e $\{(a, a), (b, b), (c, c), (b, c), (c, a), (b, a)\}$
f $\{(a, a), (b, b)\}$

4 $xRy \Leftrightarrow x$ and y have the same gradient.

a (1) xRx is true as x has the same gradient as itself.
$\therefore$ R is reflexive.
(2) $xRy \Rightarrow yRx$ {If line x has the same gradient as line y, then line y has the same gradient as line x.}
$\therefore$ R is symmetric.
(3) If line x has the same gradient as line y which has the same gradient as line z, then line x has the same gradient as line z.
$\therefore$ xRy and $yRz \Rightarrow xRz$
$\therefore$ R is transitive.
$\therefore$ R is an equivalence relation.

b Every line in the plane will be in an equivalence class.
For any given line, the equivalence class will consist of all lines parallel to the given line.

5 $xRy \Leftrightarrow x - y$ is divisible by 7.

(1) $x - x = 0$ for all $x \in \mathbb{N}$ and 0 is a multiple of 7

$\therefore$ xRx for all $x \in \mathbb{N}$

$\therefore$ R is reflexive.

(2) $xRy \Rightarrow x - y = 7k$ for $k \in \mathbb{Z}$

$\Rightarrow y - x = 7(-k)$ for $k \in \mathbb{Z}$

$\Rightarrow y - x$ is divisible by 7

$\Rightarrow yRx$

$\therefore$ R is symmetric.

(3) If xRy and yRz then

$x - y = 7k$ and $y - z = 7l$ for $k, l \in \mathbb{Z}$

$$\therefore\ x - z = (7k + y) + (7l - y) = 7k + 7l = 7(k + l) \text{ where } k + l \in \mathbb{Z}$$

$\Rightarrow xRz$

$\therefore$ R is transitive.

Thus from (1), (2), (3), R is an equivalence relation.

6 $xRy \Leftrightarrow x$ is similar to y.

a (1) Every regular polygon is similar to itself.

$\therefore$ xRx for all $x \in S$.

$\therefore$ R is reflexive.

(2) Two regular polygons are similar if they have the same number of sides.

Thus, if xRy then yRx for all $x, y \in S$.

$\therefore$ R is symmetric.

(3) If xRy and yRz then the number of sides of x and y are equal and the number of sides of y and z are equal.

Thus, the number of sides of x and z are equal.

$\therefore$ xRz

$\therefore$ xRy and $yRz \Rightarrow xRz$

$\therefore$ R is transitive.

Hence, as R is reflexive, symmetric, and transitive, it is an equivalence relation.

b The equivalence classes would be $S_3, S_4, S_5, S_6, ...$, where S_n is the set of regular n-sided polygons.

c S_3 is the set of all equilateral triangles,

S_4 is the set of all squares, etc.

These sets are pairwise disjoint and every regular polygon will be in one of these sets.

Thus, $S_i \cap S_j = \varnothing$ for all $i \neq j$, $i, j \in \mathbb{Z}^+$, $i \geqslant 3$, $j \geqslant 3$, and $S = S_3 \cup S_4 \cup S_5 \cup S_6 \cup ....$

$\therefore$ $\{S_n\}$ partitions S.

7 xRy if $x \leqslant y$ for all $x, y \in \mathbb{Z}$.

To show that R is not an equivalence relation, we need to show that one of reflexive, symmetric, or transitive properties does not apply. In this example, R is *not* symmetric.

For example, $2 \leqslant 3$ but $3 \nleqslant 2$.

8 **a** $(a, b)R(x, y) \Leftrightarrow x = a$ for $(a, b), (x, y) \in \mathbb{Z} \times \mathbb{Z}$.

(1) $(a, b)R(a, b)$ since $a = a$ is clearly true

$\therefore$ R is reflexive.

(2) If $(a, b)R(x, y)$ then $x = a$

$\Rightarrow a = x$

$\Rightarrow (x, y)R(a, b)$

$\therefore$ R is symmetric.

(3) If $(a, b)R(x, y)$ and $(x, y)R(p, q)$ say, then $x = a$ and $p = x$

$\Rightarrow p = a$

$\Rightarrow (a, b)R(p, q)$

$\therefore$ R is transitive.

Thus from (1), (2), and (3), R is an equivalence relation.

b Each point in $\mathbb{Z} \times \mathbb{Z}$ is related to all points above or below it. The equivalence classes are sets of integer grid points lying on vertical lines defined by $x = a$, $a \in \mathbb{Z}$.

9 $(a, b)R(x, y) \Leftrightarrow ay = bx$ in $\mathbb{R} \times \mathbb{R} \setminus \{(0, 0)\}$.

a (1) $(a, b)R(a, b)$ since $ab = ba$ is clearly true.

$\therefore$ R is reflexive.

(2) If $(a, b)R(x, y)$, then $ay = bx$

$\Rightarrow ya = xb$

$\Rightarrow (x, y)R(a, b)$

$\therefore$ R is symmetric.

(3) If $(a, b)R(x, y)$ and $(x, y)R(c, d)$ say, then $ay = bx$ and $dx = cy$

$\Rightarrow \dfrac{y}{x} = \dfrac{b}{a}$ and $\dfrac{y}{x} = \dfrac{d}{c}$

$\Rightarrow \dfrac{b}{a} = \dfrac{d}{c}$

$\Rightarrow ad = bc$

$\Rightarrow (a, b)R(c, d)$

$\therefore$ R is transitive.

Thus from (1), (2), and (3), R is an equivalence relation.

b Any point of $\mathbb{R} \times \mathbb{R} \setminus \{(0, 0)\}$ is related to all points on the line passing through the point and the origin. Each point is an element of exactly one equivalence class which consists of all points (excluding $(0, 0)$) lying on the line passing through O and the point.

10 $(a, b)R(x, y) \Leftrightarrow y - b = 3x - 3a$ for $(a, b), (x, y) \in \mathbb{R} \times \mathbb{R}$

a (1) $(a, b)R(a, b)$ since $b - b = 3a - 3a$

$\therefore$ R is reflexive.

(2) If $(a, b)R(x, y)$ then $y - b = 3x - 3a$

$\Rightarrow b - y = -(3x - 3a)$

$\Rightarrow b - y = 3a - 3x$

$\Rightarrow (x, y)R(a, b)$

$\therefore$ R is symmetric.

(3) If $(a, b)R(x, y)$ and $(x, y)R(m, n)$ say, then $y - b = 3x - 3a$ and $n - y = 3m - 3x$

$\Rightarrow (n - y) + (y - b) = (3m - 3x) + (3x - 3a)$

$\Rightarrow n - b = 3m - 3a$

$\Rightarrow (a, b)R(m, n)$

$\therefore$ R is transitive.

From (1), (2), and (3), R is an equivalence relation.

b Any point $\mathbb{R} \times \mathbb{R}$ is related to all points on the line through the point with gradient 3. Each point is an element of exactly one equivalence class containing all points which lie on the line through that point, with gradient 3.

11 $aRb \Leftrightarrow 3ab \geqslant 0$ for $a, b \in \mathbb{Z}$

a **i** For any $a \in \mathbb{Z}$, $a^2 \geqslant 0$

$\therefore$ $3a^2 \geqslant 0$

$\therefore$ aRa

$\therefore$ R is reflexive.

ii If aRb then $3ab \geqslant 0$

$\Rightarrow 3ba \geqslant 0$

$\Rightarrow bRa$

$\therefore$ R is symmetric.

iii If aRb and bRc then $3ab \geqslant 0$ and $3bc \geqslant 0$.

But this *does not* $\Rightarrow 3ac \geqslant 0$

For example, if $a = 2$, $b = 0$, $c = -2$,

then $3ab = 0 \geqslant 0$

$3bc = 0 \geqslant 0$

$3ac = -12 \ngeqslant 0$

$\therefore$ R is **not** transitive.

b As R is **not** transitive,

R is not an equivalence relation.

12 $(a,\ b)R(c,\ d) \Leftrightarrow a-c$ is a multiple of 2 **and** $b-d$ is a multiple of 3

for all $(a,\ b),\ (c,\ d) \in \mathbb{Z}\times\mathbb{Z}$.

a (1) For any $(a,\ b)\in\mathbb{Z}\times\mathbb{Z}$,

$a-a=0$ which is a multiple of 2

$b-b=0$ which is a multiple of 3

$\therefore\ (a,\ b)R(a,\ b)$

$\therefore\ R$ is reflexive.

(2) Suppose $(a,\ b)R(c,\ d)$

$\therefore\ a-c=2s$ and $b-d=3t$ for some $s,\ t\in\mathbb{Z}$

$\Rightarrow\ c-a=2(-s)$ and $d-b=3(-t)$

$\Rightarrow\ c-a$ is a multiple of 2 and $d-b$ is a multiple of 3

$\Rightarrow\ (c,\ d)R(a,\ b)$

$\therefore\ R$ is symmetric.

(3) Suppose $(a,\ b)R(c,\ d)$ and $(c,\ d)R(e,\ f)$ for some $(a,\ b),\ (c,\ d),\ (e,\ f)\in\mathbb{Z}\times\mathbb{Z}$

$\therefore\ a-c=2s_1$ and $c-e=2s_2$

$b-d=3t_1$ $\quad d-f=3t_2$

for some $s_1,\ t_1,\ s_2,\ t_2\in\mathbb{Z}$.

Now $a-e=(a-c)+(c-e)=2(s_1+s_2)$ and

$b-f=(b-d)+(d-f)=3(t_1+t_2)$

$\{s_1+s_2\in\mathbb{Z}$ and $t_1+t_2\in\mathbb{Z}\}$

$\Rightarrow\ (a,\ b)R(e,\ f)$

Thus R is transitive.

From (1), (2), and (3), R is an equivalence relation.

b **i** $(a,\ b)R(0,\ 0)$

$\Leftrightarrow a-0=2s$ and $b-0=3t,\ s,\ t\in\mathbb{Z}$

$\Leftrightarrow a=2s$ and $b=3t$

The equivalence class containing $(0,\ 0)$

$=\{(a,\ b)\mid a=2s,\ b=3t,\ s,\ t\in\mathbb{Z}\}$

ii $(a,\ b)R(1,\ 3)$

$\Leftrightarrow a-1=2s$ and $b-3=3t$

$\Leftrightarrow a=1+2s$ and $b=3+3t=3(1+t)$

The equivalence class containing $(1,\ 3)$

$=\{(a,\ b)\mid a=2s+1,\ b=3t,\ s,\ t\in\mathbb{Z}\}$

c For $(a,\ b)\in\mathbb{Z}\times\mathbb{Z}$,

$a=2s$ or $2s+1,\ s\in\mathbb{Z}$

$b=3t,\ 3t+1,$ or $3t+2,\ t\in\mathbb{Z}$

There are $2\times3=6$ distinct equivalence classes.

Two of them are listed in **b**, and the other 4 are:

$\{(a,\ b)\mid a=2s,\ b=3t+1,\ s,\ t\in\mathbb{Z}\}$

$\{(a,\ b)\mid a=2s,\ b=3t+2,\ s,\ t\in\mathbb{Z}\}$

$\{(a,\ b)\mid a=2s+1,\ b=3t+1,\ s,\ t\in\mathbb{Z}\}$

$\{(a,\ b)\mid a=2s+1,\ b=3t+2,\ s,\ t\in\mathbb{Z}\}$

EXERCISE B.4

1 **a** If $a\equiv b \pmod n$, then $a-b=kn,\ k\in\mathbb{Z}$ .... (1)

If $c\equiv d \pmod n$, then $c-d=ln,\ l\in\mathbb{Z}$ .... (2)

$\therefore\ a-b+c-d=(k+l)n$

$\therefore\ (a+c)-(b+d)=(k+l)n,\ \ k+l\in\mathbb{Z}$

$\therefore\ a+c\equiv b+d \pmod n$

b From (1) and (2) in **a**, $ac=(b+kn)(d+ln)$

$\therefore\ ac=bd+lbn+kdn+kln^2$

$\therefore\ ac-bd=n(lb+kd+kln)$

where $lb+kd+kln\in\mathbb{Z}$

$\{$since $b,\ d,\ l,\ k,\ n\in\mathbb{Z}\}$

Hence $ac\equiv bd \pmod n$

2 If $a=1$, $x\equiv1 \pmod{11}$

$\therefore\ x=1+11k,\ k\in\mathbb{Z}$

$\therefore\ x=1$ is smallest in $\mathbb{Z}^+$

If $a=2$, $2x\equiv1 \pmod{11}$

$\therefore\ 2x=1+11k,\ k\in\mathbb{Z}$

$\therefore\ 2x=12$ {smallest}

$\therefore\ x=6$

If $a=3$, $3x=1+11k,\ k\in\mathbb{Z}$

$\therefore\ 3x=12$ {smallest}

$\therefore\ x=4$

If $a=4$, $4x=1+11k,\ k\in\mathbb{Z}$

$\therefore\ 4x=12$ {smallest}

$\therefore\ x=3$

If $a=5$, $5x=1+11k,\ k\in\mathbb{Z}$

$\therefore\ 5x=45$ {smallest}

$\therefore\ x=9$

If $a=6$, $6x=1+11k,\ k\in\mathbb{Z}$

$\therefore\ 6x=12$ {smallest}

$\therefore\ x=2$

If $a=7$, $7x=1+11k,\ k\in\mathbb{Z}$

$\therefore\ 7x=56$ {smallest}

$\therefore\ x=8$

If $a=8$, $8x=1+11k,\ k\in\mathbb{Z}$

$\therefore\ 8x=56$ {smallest}

$\therefore\ x=7$

If $a=9$, $9x=1+11k,\ k\in\mathbb{Z}$

$\therefore\ 9x=45$ {smallest}

$\therefore\ x=5$

If $a=10$, $10x=1+11k,\ k\in\mathbb{Z}$

$\therefore\ 10x=100$ {smallest}

$\therefore\ x=10$

3 $xRy \Leftrightarrow x\equiv y \pmod n,\ x,\ y\in\mathbb{Z}$

(1) As $x-x=0$ and 0 is a multiple of n then $x\equiv x \pmod n$

$\therefore\ xRx$ for all $x\in\mathbb{Z}$

$\therefore\ R$ is reflexive.

(2) If xRy then $x\equiv y \pmod n$

$\Rightarrow x=y+kn,\ k\in\mathbb{Z}$

$\Rightarrow y=x+(-k)n,\ -k\in\mathbb{Z}$

$\Rightarrow y\equiv x \pmod n$

$\Rightarrow yRx$

Thus $xRy\Rightarrow yRx$ for all $x,\ y\in\mathbb{Z}$

$\therefore\ R$ is symmetric.

(3) If xRy and yRz for $x,\ y,\ z\in\mathbb{Z}$ then

$x\equiv y \pmod n$ and $y\equiv z \pmod n$

$\therefore\ x=y+sn$ and $y=z+tn,\ s,\ t\in\mathbb{Z}$

$\Rightarrow\ x=z+tn+sn$

$\Rightarrow\ x=z+(t+s)n$ with $t+s\in\mathbb{Z}$

$\Rightarrow\ x\equiv z \pmod n$

$\Rightarrow\ xRz$

Thus xRy and $yRz\Rightarrow xRz$ for all $x,\ y,\ z\in\mathbb{Z}$

$\therefore\ R$ is transitive.

From (1), (2), and (3), R is an equivalence relation.

There are n residue classes possible. These leave a remainder of $0,\ 1,\ 2,\ 3,\ 4,\ ....,\ n-1$ when divided by n.

These are [0], [1], [2], [3], ..., $[n-1]$ {See **Example 19**}

Note: $\mathbb{Z}_n = \{0, 1, 2, 3, ..., n-1\}$ is the set of residues modulo n, $n \in \mathbb{Z}^+$.

4 $xRy \Leftrightarrow x^2 \equiv y^2 \pmod 3$, $x, y \in \mathbb{N}$

(1) As $x^2 = x^2$, then $x^2 \equiv x^2 \pmod 3$

$\therefore\ xRx$

$\therefore$ R is reflexive.

(2) If xRy, then $x^2 \equiv y^2 \pmod 3$

$\Rightarrow x^2 = y^2 + 3n,\ n \in \mathbb{Z}$

$\Rightarrow y^2 = x^2 + 3(-n),\ -n \in \mathbb{Z}$

$\Rightarrow y^2 \equiv x^2 \pmod 3$

$\Rightarrow yRx$

$\therefore$ R is symmetric.

(3) If xRy and yRz then

$x^2 = y^2 + 3n$ and $y^2 = z^2 + 3m$, $n, m \in \mathbb{Z}$

$\therefore\ x^2 = z^2 + 3m + 3n$

$\Rightarrow x^2 = z^2 + 3(m+n)$ where $m + n \in \mathbb{Z}$

$\Rightarrow x^2 \equiv z^2 \pmod 3$

$\Rightarrow xRz$

$\therefore$ R is transitive.

From (1), (2), and (3), R is an equivalence relation.

5 **a** $a \equiv b \pmod n$

$\Rightarrow a = b + kn,\ k \in \mathbb{Z}$

$\Rightarrow a^2 = b^2 + 2bkn + k^2n^2$

$\Rightarrow a^2 = b^2 + (2bk + k^2n)n$ where $2bk + k^2n \in \mathbb{Z}$

$\Rightarrow a^2 \equiv b^2 \pmod n$

b Consider $a = 2$, $b = 0$, $n = 4$.

Then $a^2 = 4 \equiv 0 \pmod 4$

and $b^2 = 0 \equiv 0 \pmod 4$

But $2 \not\equiv 0 \pmod 4$

Thus, $a^2 \equiv b^2 \pmod n \not\Rightarrow a \equiv b \pmod n$

EXERCISE C.1

1 **a** Is a function {for each $x \in A$, there is at most one $y \in B$}.
domain $= \{1, 2, 3\} = A$, codomain $= \{1, 2, 3\} = B$,
range $= \{1, 2\}$

b Is **not** a function {as $f(2) = 1$ or 3}.

c Is a function {for each $x \in A$, there is at most one $y \in B$}.
domain $= \{1, 2, 3\} = A$, codomain $= \{1, 2, 3\} = B$,
range $= \{1, 2, 3\} = B$

2 **a** Is a function {for each x, there is at most one y}.

b Is not a function as for each x there are infinitely many values of y.
{vertical line test}

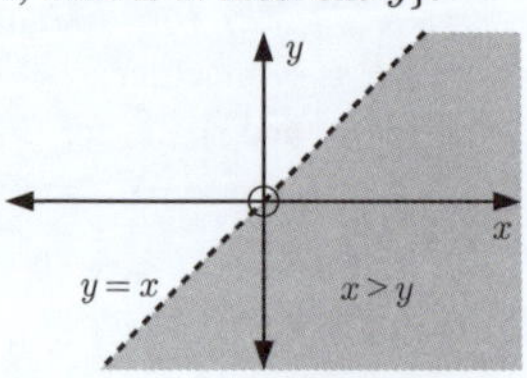

c Domain $= \{x \mid x \in \mathbb{R} \setminus \{0\}\}$
Is a function
{vertical line test}.

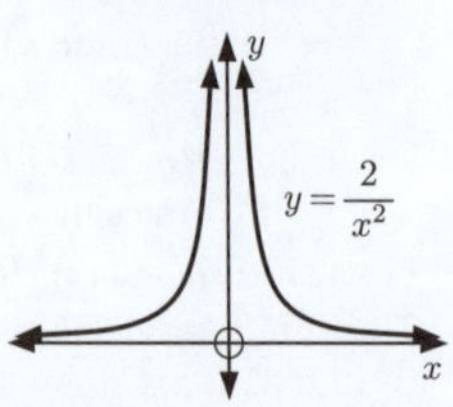

d For $x = 1$, $y = 2$ or -1
So, for $x = 1$, there is more than one y.
$\therefore$ is not a function.

3 **a** **i** Not an injection as $f(1) = f(3)$.

ii Not a surjection as range $= \{1, 2, 4\} \neq B$.

iii Not a bijection {as not both an injection and a surjection}.

b **i** Is an injection as no two elements of A are mapped onto the same element of B.

ii Is a surjection {as range $= B$}.

iii Is a bijection {as it is both an injection and a surjection}.

4 **a**

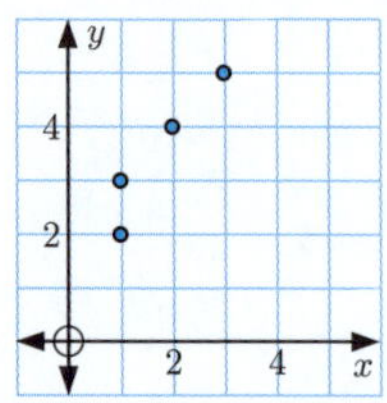

Is **not** a function as the vertical line $x = 1$ meets the graph at more than one point.

b

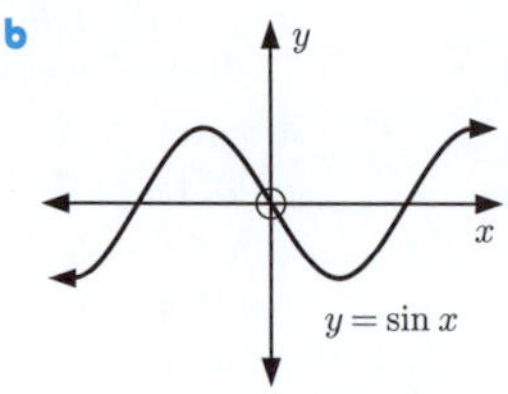

Is a function as every vertical line meets the cuve exactly one.

c

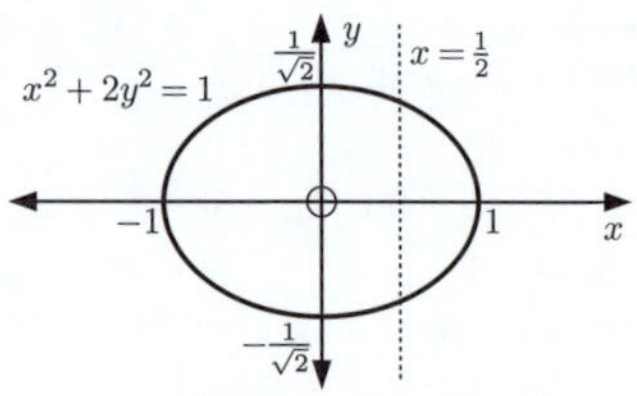

Is not a function as, for example, the vertical line $x = \frac{1}{2}$ cuts the graph twice.

5

Relation	*Function*	*Injection*	*Bijection*
a	false	false	false
b	true	false	false
c	true	true	true

6 **a**

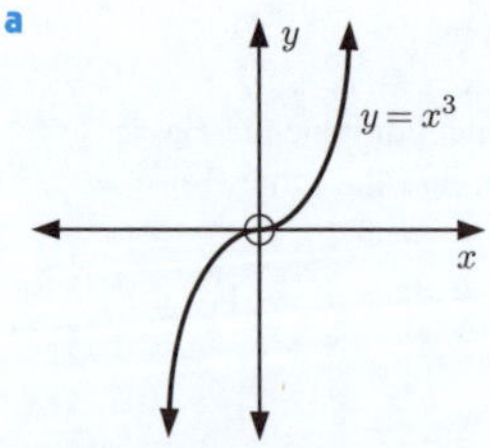

Is an injection as no horizontal line cuts it more than once.

b

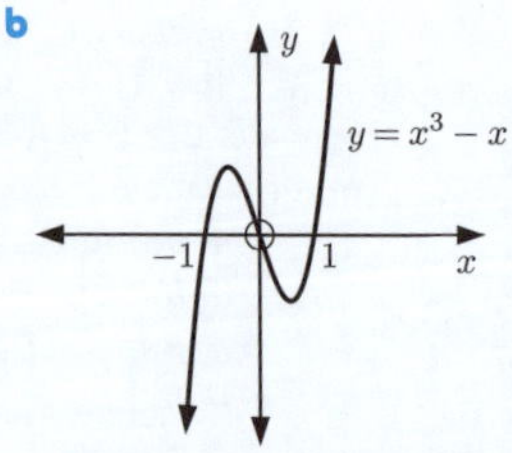

Is *not* an injection as the horizontal line $y = 0$ cuts the graph more than once.

c

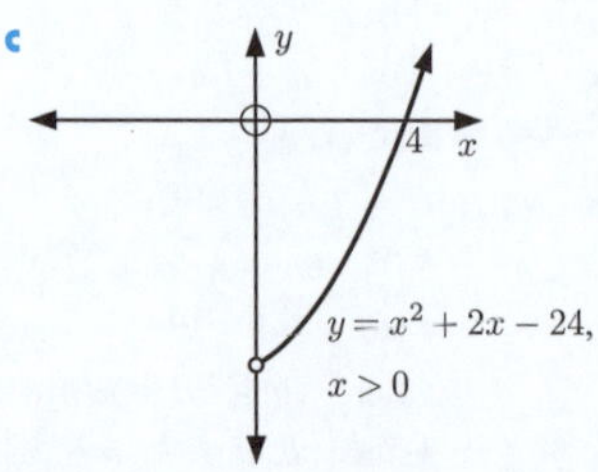

Is an injection as no horizontal line cuts the graph more than once.

7 **a** Is a function.

i Is not an injection. **ii** Is a surjection.

iii Is not a bijection.

b Is a function.

i Is not an injection. **ii** Is not a surjection.

iii Is not a bijection.

c Is not a function.

d

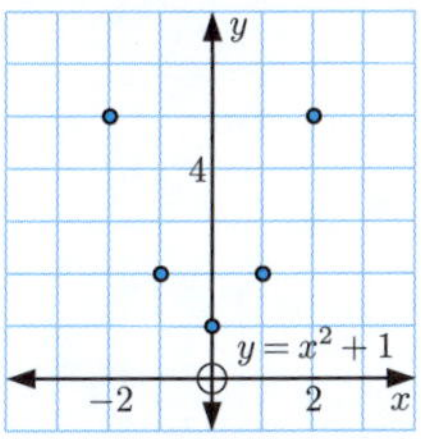

Is a function.

i Is not an injection.

ii Is not a surjection.

iii Is not a bijection.

e For every point $(x, y) \in \mathbb{R}^2$ there is exactly one value of $z \in \mathbb{R}$.

$\therefore$ R is a function.

i It is not an injection as, for example, $(1, 3) \to z = 1^2 + 3^2 = 10$ and $(-1, 3) \to z = 10$ also.

ii It is not a surjection as $x^2+y^2 \geqslant 0$ for all $(x, y) \in \mathbb{R}^2$, $\therefore$ range is $\mathbb{R}^+ \cup \{0\}$ which is *not* $\mathbb{R}$.

iii Is not both an injection and surjection. $\therefore$ cannot be a bijection.

f $(a, b)R(x, y) \Leftrightarrow y = a$ and $x = b$.

Is a function.

i Is an injection. **ii** Is a surjection. **iii** Is a bijection.

8 **a**

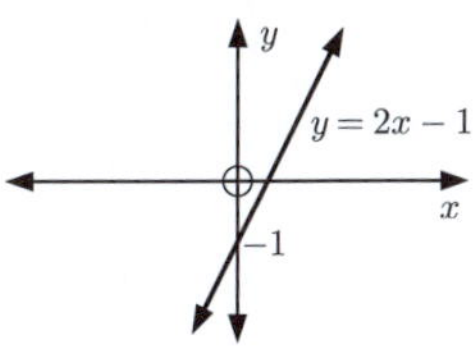

The function is an injection. {horizontal line test}
It is also a surjection as its range is $\mathbb{R}$, the codomain.
$\therefore$ the function *is* a bijection.

b

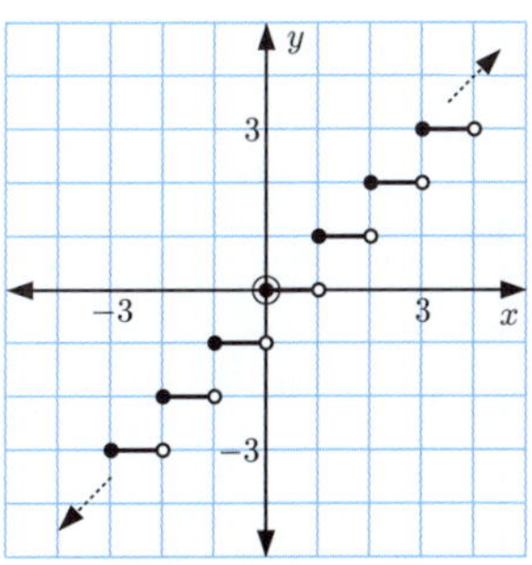

The function is *not* an injection.
$\therefore$ cannot be a bijection.

c

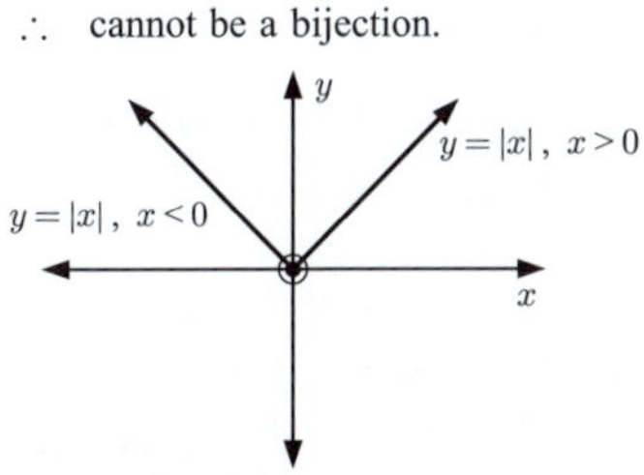

The function is not an injection.
$\therefore$ cannot be a bijection.

d The function is *not* a surjection. For example, there is no $x \in \mathbb{Q}^+$ such that $f(x) = x^2 = 3$. $(3 \in \mathbb{Q}^+)$

e

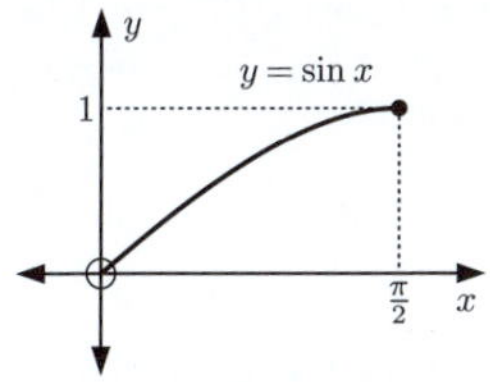

The function is an injection as no horizontal line cuts it more than once. It is also a surjection as its range is $[0, 1]$, the codomain.
$\therefore$ the function is a bijection.

f

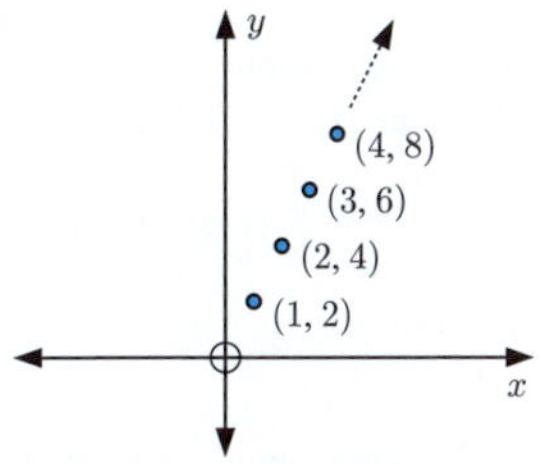

It is an injection, but not a surjection as its range is {even integers} which is not $\mathbb{Z}^+$.

9 We are given $A \subseteq B \subseteq S$

To prove: $f(A) \subseteq f(B)$

Proof: $B \subseteq S$ and $f : S \to S \Rightarrow f(B) \subseteq$ range of B

Now if $f(x) \in f(A)$
$\Rightarrow x \in A$
$\Rightarrow x \in B$ {as $A \subseteq B$}
$\Rightarrow f(x) \in f(B)$

Thus $f(x) \in f(B)$ for all $x \in A$.
Hence, $f(x) \in f(A) \Rightarrow f(x) \in f(B)$.
Thus $f(A) \subseteq f(B)$.

10 **a** $\mathbb{R} \to \mathbb{R}$, $f(x) = 2^x$
By the horizontal line test, f is an injection
$\therefore$ one-to-one
but its range is $\mathbb{R}^+$ not $\mathbb{R}$
$\therefore$ is not onto.

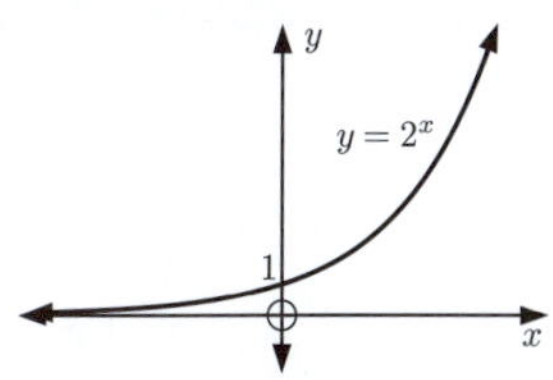

b $\mathbb{R} \to \mathbb{R}$,
$f(x) = x(x-1)(x-2)$
The horizontal line $y = 0$ cuts the graph more than once.
$\therefore$ f is not an injection
$\therefore$ not one-to-one
and its range is $\mathbb{R}$, the codomain $\therefore$ is onto.

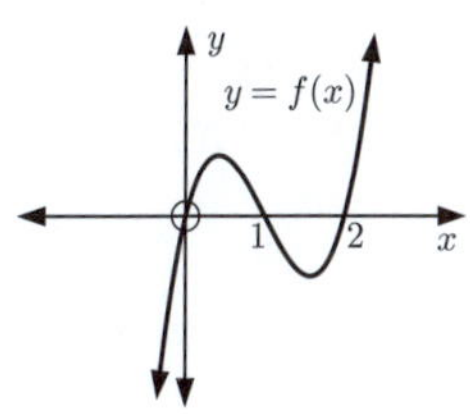

c $\mathbb{R} \to \mathbb{R}$, $f(x) = x + 1$
No horizontal line cuts the graph more than once.
$\therefore$ f is an injection or is one-to-one
and its range is $\mathbb{R}$, the codomain $\therefore$ is onto.

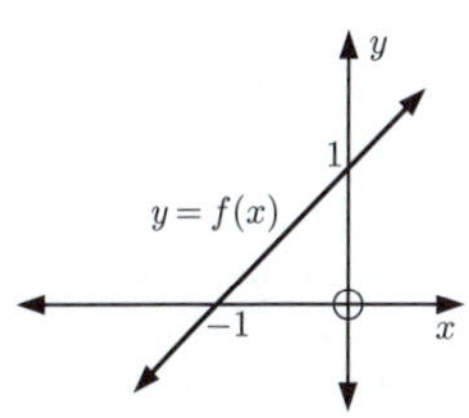

d $\mathbb{R} \to \mathbb{R}$, $f(x) = x^2$
A horizontal line can cut the graph more than once.
$\therefore$ f is not one-to-one
and its range is $\mathbb{R}^+ \cup \{0\}$, not $\mathbb{R}$
$\therefore$ is not onto.

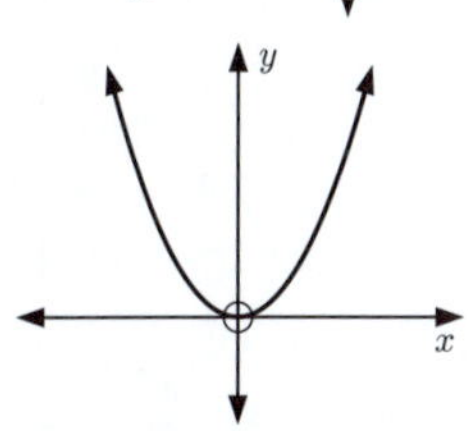

11 **a** **i** Suppose f is onto.

$\Rightarrow$ each element of $S = \{1, 2, 3\}$ lies in the range of f

$\Rightarrow f(1) = a,\ f(2) = b,$ and $f(3) = c$ where $\{a, b, c\} \subseteq \{1, 2, 3\}$ {as f is a function}

$\Rightarrow a, b,$ and c are distinct

$\Rightarrow f$ is one-to-one

So, no example exists where f is onto, but not one-to-one.

ii Suppose f is one-to-one.

$\Rightarrow$ the elements 1, 2, 3 are mapped to distinct elements in the range

$\Rightarrow$ the range has 3 distinct elements

$\Rightarrow R(f) = \{1, 2, 3\}$ {as $R(f) \subseteq \{1, 2, 3\}$}

$\Rightarrow f$ is onto

So, no example exists where f is one-to-one, but not onto.

b .... f is one-to-one $\Leftrightarrow$ f is onto.

12 $f(p(x)) = p'(x)$

a As every polynomial $p(x)$, has a unique derivative $p'(x)$, then f is a function.

b It is *not* an injection. For example,

$f(x^2) = 2x$ and $f(x^2+1) = 2x$

$\{f(x_1) = f(x_2) = 2x,$ but $x_1 \neq x_2\}$

c Each polynomial $p(x)$ has an antiderivative $q(x)$ where

$$q(x) = \int p(x)\,dx + c$$

$\therefore\ f(q(x)) = p(x)$

$\therefore$ the range of f is P, and so f is a surjection.

d As f is not an injection, f cannot be a bijection.

EXERCISE C.2

1 **a**

$(g\circ f)(1)$	$(g\circ f)(2)$	$(g\circ f)(3)$
$= g(f(1))$	$= g(f(2))$	$= g(f(3))$
$= g(5) = 0$	$= g(6) = 1$	$= g(4) = 1$

$\therefore\ g\circ f = \{(1, 0), (2, 1), (3, 1)\}$

b $f\circ g$ is not defined, as the domain of f does not contain the range of g.

2 **a** **i** $(f\circ g)(1) = f(g(1)) = f(3) = 3$

ii $(g\circ f)(1) = g(f(1)) = g(2) = 0$

iii $(f\circ g)(3) = f(g(3)) = f(1) = 2$

iv $(g\circ f)(3) = g(f(3)) = g(3) = 1$

b **i** As f is a bijection, f^{-1} exists and $f^{-1} = \{(0, 2), (1, 0), (2, 1), (3, 3)\}$.

ii As g is a bijection g^{-1} exists and $g^{-1} = \{(0, 2), (1, 3), (2, 0), (3, 1)\}$.

iii $(g\circ f)(0) = g(f(0)) = g(1) = 3$ $\quad$ $(g\circ f)(1) = g(f(1)) = g(2) = 0$

$(g\circ f)(2) = g(f(2)) = g(0) = 2$ $\quad$ $(g\circ f)(3) = g(f(3)) = g(3) = 1$

$\therefore\ g\circ f = \{(0, 3), (1, 0), (2, 2), (3, 1)\}$

$\therefore\ (g\circ f)^{-1} = \{(0, 1), (1, 3), (2, 2), (3, 0)\}$

iv $(f^{-1}\circ g^{-1})(0) = f^{-1}(g^{-1}(0)) = f^{-1}(2) = 1$ $\quad$ $(f^{-1}\circ g^{-1})(1) = f^{-1}(g^{-1}(1)) = f^{-1}(3) = 3$

$(f^{-1}\circ g^{-1})(2) = f^{-1}(g^{-1}(2)) = f^{-1}(0) = 2$ $\quad$ $(f^{-1}\circ g^{-1})(3) = f^{-1}(g^{-1}(3)) = f^{-1}(1) = 0$

$\therefore\ f^{-1}\circ g^{-1} = \{(0, 1), (1, 3), (2, 2), (3, 0)\}$

3 **a** $f:\mathbb{Z}\to\mathbb{Z},\ f(x) = 1 - x$

f is both an injection and surjection

$\therefore$ it is a bijection and so f^{-1} exists.

Let $y = 1 - x$ which has inverse $x = 1 - y$ or $y = 1 - x$

$\therefore\ f^{-1}(x) = 1 - x,\ x\in\mathbb{Z}$

{This function is its own inverse.}

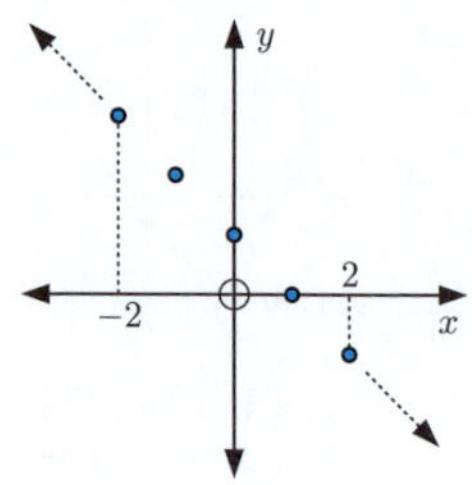

b $f:\mathbb{R}\to\mathbb{R},\ f(x) = x^3 - 2$

f is both an injection and a surjection

$\therefore$ it is a bijection and f^{-1} exists.

Let $y = x^3 - 2$ which has inverse

$x = y^3 - 2$

$\therefore\ y^3 = x + 2$

$\therefore\ y = \sqrt[3]{x+2}$

$\therefore\ f^{-1}(x) = \sqrt[3]{x+2},\ x\in\mathbb{R}$

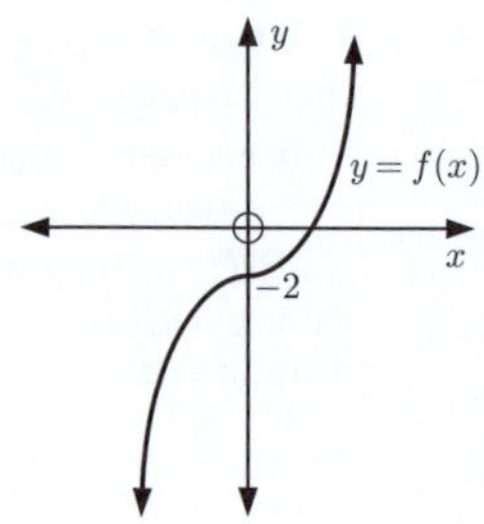

4 $f(x) = \sqrt{1-x}$

a For $f(x)$ to exist, $1 - x \geqslant 0$

$\therefore\ x \leqslant 1$

$\therefore\ f$ has domain $]-\infty, 1]$.

If $y = \sqrt{1-x}$

then $y \geqslant 0$ $\{\sqrt{a} \geqslant 0$ for all $a\in\mathbb{R}\}$

$\therefore\ f$ has range $[0, \infty[$.

b, e

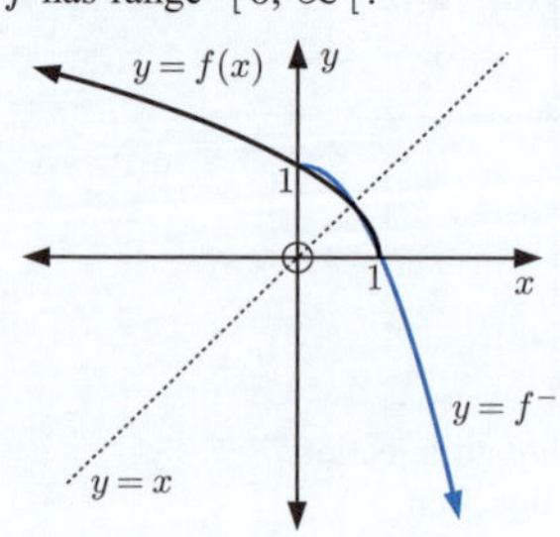

c As $y = \sqrt{1-x}$ then the inverse function is

$x = \sqrt{1-y}$

$\therefore\ 1 - y = x^2$

$\therefore\ y = 1 - x^2$

But, as $y \geqslant 0$ for f, $x \geqslant 0$ for f^{-1}.

$\therefore\ f^{-1}(x) = 1 - x^2,\ x \geqslant 0$

d Domain of $f^{-1}(x)$ is $[0, \infty[$.
Range of $f^{-1}(x)$ is $]-\infty, 1]$.

e On graph above.

5 $f(x) = \ln(x+1)$, $g(x) = x^2$. Domain $\mathbb{R}^+$.

a $(g \circ f)(x) = g(f(x))$
$= g(\ln(x+1))$
$= [\ln(x+1)]^2$

b $(f \circ g)(x) = f(g(x))$
$= f(x^2)$
$= \ln(x^2+1)$

c

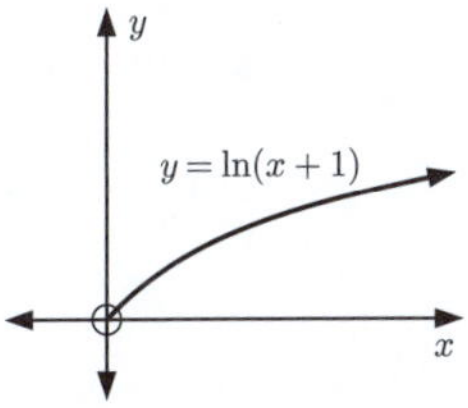

$y = \ln(x+1)$ has inverse $x = \ln(y+1)$

$\therefore\ y + 1 = e^x$

$\therefore\ y = e^x - 1$

$\therefore\ f^{-1}(x) = e^x - 1$

d

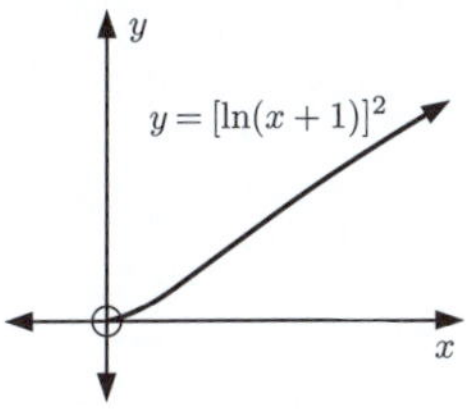

$(g \circ f)(x) = [\ln(x+1)]^2$

$\therefore\ y = [\ln(x+1)]^2$ has inverse

$x = [\ln(y+1)]^2,\ x \geqslant 0$

$\therefore\ \ln(y+1) = \sqrt{x}$

$y + 1 = e^{\sqrt{x}}$

$y = e^{\sqrt{x}} - 1$

$\therefore\ (g \circ f)^{-1}(x) = e^{\sqrt{x}} - 1$

e $g(x) = x^2,\ x \in \mathbb{R}^+$

$\therefore\ y = x^2,\ x > 0$ and has inverse $x = y^2,\ y > 0$

$\therefore\ y = \sqrt{x}$ $\quad \{y = -\sqrt{x}$ has $y < 0\}$

$\therefore\ g^{-1}(x) = \sqrt{x}$

Hence $(f^{-1} \circ g^{-1})(x) = f^{-1}(g^{-1}(x))$
$= f^{-1}(\sqrt{x})$
$= e^{\sqrt{x}} - 1$ $\quad$ {from **c**}

6 For a function $h(x)$ with inverse $h^{-1}(x)$,
$h(h^{-1}(x)) = h^{-1}(h(x)) = x$, the identity function. .... (1)

Now $((g \circ f) \circ (f^{-1} \circ g^{-1}))(x)$
$= g(f(f^{-1}(g^{-1}(x))))$
$= g(g^{-1}(x))$ $\quad$ {as $f(f^{-1}(x)) = x$, by (1)}
$= x$ $\quad$ {same reason}

Also $((f^{-1} \circ g^{-1}) \circ (g \circ f))(x)$
$= f^{-1}(g^{-1}(g(f(x))))$
$= f^{-1}(f(x))$ $\quad$ {as $g^{-1}(g(x)) = x$}
$= x$ $\quad$ {as $f^{-1}(f(x)) = x$}

Thus $g \circ f$ and $f^{-1} \circ g^{-1}$ are inverses.

$\therefore\ (g \circ f)^{-1} = f^{-1} \circ g^{-1}$

7 **a** $f(x) = e^x + 3e^{-x}$

From the graph of $y = f(x)$, a horizontal line can cut the graph more than once.

$\therefore$ f is not one-to-one

$\therefore$ f is not invertible.

Thus $f^{-1}(x)$ does not exist.

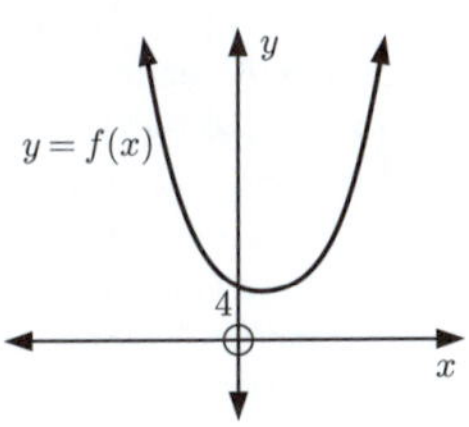

b As $x \to \infty$, $e^x \to \infty$, $e^{-x} \to 0$

$\therefore\ f(x) \to \infty$

As $x \to -\infty$, $e^x \to 0$, $-e^{-x} \to -\infty$

$\therefore\ f(x) \to -\infty$

$f(x)$ appears to be one-to-one and onto.

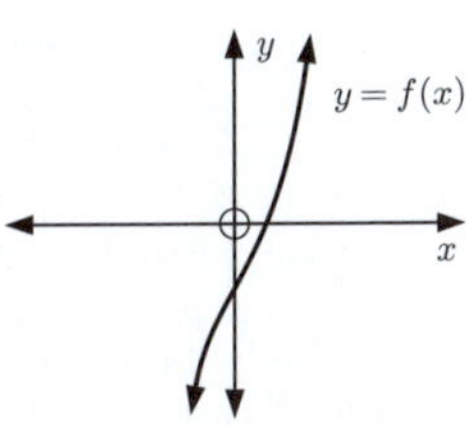

Check: (algebraic)

Suppose $f(x_1) = f(x_2)$ for some $x_1, x_2 \in \mathbb{R}$ and $x_1 \neq x_2$.

$\Rightarrow\ e^{x_1} - 3e^{-x_1} = e^{x_2} - 3e^{-x_2}$

$\Rightarrow\ e^{x_1} - e^{x_2} = 3\left(e^{-x_1} - e^{-x_2}\right)$

$\Rightarrow\ e^{x_1} - e^{x_2} = 3\left(\dfrac{1}{e^{x_1}} - \dfrac{1}{e^{x_2}}\right)$

$\Rightarrow\ e^{x_1} - e^{x_2} = 3\left(\dfrac{e^{x_2} - e^{x_1}}{e^{x_1}e^{x_2}}\right)$

$\Rightarrow\ e^{x_1} - e^{x_2} = \dfrac{-3\left(e^{x_1} - e^{x_2}\right)}{e^{x_1+x_2}}$

$\Rightarrow\ e^{x_1+x_2} = -3$ $\quad$ $\{x_1 \neq x_2 \Rightarrow e^{x_1} \neq e^{x_2} \Rightarrow e^{x_1} - e^{x_2} \neq 0\}$

which is a contradiction $\{$as $e^a > 0$ for all $a \in \mathbb{R}\}$

$\therefore$ the supposition is false.

$\therefore\ f(x_1) = f(x_2) \Rightarrow x_1 = x_2$

Thus f is a one-to-one function and as the range of f is $\mathbb{R}$, f is onto.

$\therefore\ f^{-1}$ exists.

Let $y = e^x - 3e^{-x} = e^x - \dfrac{3}{e^x}$

$\therefore\ e^x y = (e^x)^2 - 3$ or $(e^x)^2 - e^x y - 3 = 0$

$\therefore\ e^x = \dfrac{-(-y) \pm \sqrt{y^2 - 4(1)(-3)}}{2}$

$\therefore\ e^x = \dfrac{y \pm \sqrt{y^2+12}}{2}$

Hence $e^x = \dfrac{y + \sqrt{y^2+12}}{2}$ $\quad$ $\{$as $e^x > 0$ for all $y \in \mathbb{R}\}$

$\therefore\ x = \ln\left(\dfrac{y + \sqrt{y^2+12}}{2}\right)$

which has inverse $y = \ln\left(\dfrac{x + \sqrt{x^2+12}}{2}\right)$

$\therefore\ f^{-1}(x) = \ln\left(\dfrac{x + \sqrt{x^2+12}}{2}\right)$

8 $f(x, y) = \left(\dfrac{y}{x}, xy\right)$ $\qquad f: \mathbb{R}^+ \times \mathbb{R}^+ \to \mathbb{R}^+ \times \mathbb{R}^+$

a Suppose $f(x_1, y_1) = f(x_2, y_2)$ for $(x_1, y_1), (x_2, y_2) \in \mathbb{R}^+ \times \mathbb{R}^+$

$\therefore\ \left(\dfrac{y_1}{x_1}, x_1y_1\right) = \left(\dfrac{y_2}{x_2}, x_2y_2\right)$

$\therefore\ \dfrac{y_1}{x_1} = \dfrac{y_2}{x_2}$ and $x_1y_1 = x_2y_2$

$\therefore\ \dfrac{x_1}{x_2} = \dfrac{y_1}{y_2} = \dfrac{y_2}{y_1}$

$\therefore\ {y_1}^2 = {y_2}^2$

$\therefore\ y_1 = \pm y_2$

$\therefore\ y_1 = y_2$ $\quad$ $\{y_1, y_2 \in \mathbb{R}^+\}$

and since $x_1y_1 = x_2y_2$, $x_1 = x_2$ also.

Thus $(x_1, y_1) = (x_2, y_2)$

Hence f is one-to-one. .... (1)

Now let $\dfrac{y}{x} = a$ and $xy = b$, $(a, b) \in \mathbb{R}^+ \times \mathbb{R}^+$

$\therefore\ y = ax$ and $y = \dfrac{b}{x}$

$\therefore\ ax = \dfrac{b}{x}$

$\therefore\ x^2 = \dfrac{b}{a}$

$\therefore\ x = \sqrt{\dfrac{b}{a}}$ $\quad \{x \in \mathbb{R}^+, \ \therefore\ x \neq -\sqrt{\dfrac{b}{a}}\}$

and so $y = ax = a\sqrt{\dfrac{b}{a}} = \sqrt{ab}$

Consequently, $(a, b) = f\left(\sqrt{\dfrac{b}{a}}, \sqrt{ab}\right)$.

Hence, each point in $\mathbb{R}^+ \times \mathbb{R}^+$ is in the range of f

$\therefore$ f is onto .... (2)

Since f is one-to-one, (1), and f is onto, (2), f is a bijection.

b $f^{-1}(x, y) = \left(\sqrt{\dfrac{y}{x}}, \sqrt{xy}\right)$

Check: (must be done)

$$\begin{aligned}(f \circ f^{-1})(x, y) &= f(f^{-1}(x, y))\\ &= f\left(\sqrt{\frac{y}{x}}, \sqrt{xy}\right)\\ &= \left(\frac{\sqrt{xy}}{\sqrt{\frac{y}{x}}}, \sqrt{\frac{y}{x}} \times \sqrt{xy}\right)\\ &= \left(\sqrt{xy}\,\frac{\sqrt{x}}{\sqrt{y}}, \frac{\sqrt{y}}{\sqrt{x}}\sqrt{x}\sqrt{y}\right)\\ &= (x, y)\end{aligned}$$

and
$$\begin{aligned}(f^{-1} \circ f)(x, y) &= f^{-1}(f(x, y))\\ &= f^{-1}\left(\frac{y}{x}, xy\right)\\ &= \left(\sqrt{\frac{xy}{\frac{y}{x}}}, \sqrt{\frac{y}{x}xy}\right)\\ &= \left(\sqrt{xy\left(\frac{x}{y}\right)}, \sqrt{y^2}\right)\\ &= \left(\sqrt{x^2}, \sqrt{y^2}\right)\\ &= (x, y) \text{ as } (x, y) \in \mathbb{R}^+ \times \mathbb{R}^+\end{aligned}$$

Thus $(f \circ f^{-1})(x, y) = (f^{-1} \circ f)(x, y) = (x, y)$

Hence, $f^{-1}(x, y) = \left(\sqrt{\dfrac{y}{x}}, \sqrt{xy}\right)$

EXERCISE D.1

1 $a * b = a - b + 1$ and $a \diamondsuit b = ab - a$ are defined on $\mathbb{Q}$.

a **i** $3 * 4 = 3 - 4 + 1 = 0$

ii $4 * 3 = 4 - 3 + 1 = 2$

iii $(-2) \diamondsuit 3 = -2 \times 3 - (-2) = -6 + 2 = -4$

iv $6 \diamondsuit 0 = 6 \times 0 - 6 = -6$

v $0 \diamondsuit 7 = 0 \times 7 - 0 = 0$

vi $4 * ((-5) \diamondsuit 2) = 4 * (-10 - (-5)) = 4 * (-5) = 4 - (-5) + 1 = 10$

vii $(4 * (-5)) \diamondsuit 2 = (4 - (-5) + 1) \diamondsuit 2 = 10 \diamondsuit 2 = 20 - 10 = 10$

b **i** $4 * x = 7$
$\therefore\ 4 - x + 1 = 7$
$\therefore\ 5 - x = 7$
$\therefore\ x = -2$

ii $x \diamondsuit 3 = -2$
$\therefore\ 3x - x = -2$
$\therefore\ 2x = -2$
$\therefore\ x = -1$

2

	+	−	×	÷
$\mathbb{Z}^+$	T	F (1)	T	F (2)
$\mathbb{Z}$	T	T	T	F (3)
$\mathbb{Q}^+$	T	F (4)	T	T
$\mathbb{Q}$	T	T	T	F (5)
$\mathbb{R}$	T	T	T	F (6)

(1) is false, as for example, $3 - 5 = -2 \notin \mathbb{Z}^+$

(2) is false, as for example, $4 \div 5 = \frac{4}{5} \notin \mathbb{Z}^+$

(3) is false, as for example, $5 \div 6 = \frac{5}{6} \notin \mathbb{Z}$

(4) is false, as for example, $\frac{3}{4} - \frac{7}{8} \notin \mathbb{Q}^+$

(5) is false, as for example, $\frac{2}{3} \div \frac{0}{1} \notin \mathbb{Q}$

(6) is false, as for example, $6 \div 0 \notin \mathbb{R}$

3 **a** This set is *not closed* as, for example, $1+i$ and $1-i$ are in the set but their product is 2 or $2 + 0i \notin$ the set.

b This set is *not closed* as, for example, $2+i$ and $1+2i$ are in the set but $(2+i)(1+2i) = 2 + 4i + i - 2 = 0 + 5i$ which is $\notin$ the set.

c Let $S = \{a + bi \mid a, b \in \mathbb{Q},\ a \text{ and } b \text{ not both } 0\}$
$= \{a + bi \mid a, b \in \mathbb{Q},\ |a + bi| \neq 0\}$

If $a + bi$ and $c + di \in S$,
$(a + bi)(c + di) = (ac - bd) + (bc + ad)i$.
Now $ac - bd$ and $bc + ad \in \mathbb{Q}$ since $a, b, c, d \in \mathbb{Q}$,
and $|(a + bi)(c + di)| = |a + bi|\,|c + di| \neq 0$.
$\therefore$ $(a + bi)(c + di) \in S$
$\therefore$ the set is closed under $\times$.

4 **a** $S = \{2, 4, 6, 8, ...\}$ under $+$
Let $2m$ and $2n$ be in S; $m, n \in \mathbb{Z}^+$
$\therefore$ $2m + 2n = 2(m + n)$ where $m + n \in \mathbb{Z}^+$
$\therefore$ $2m + 2n$ is even.
Thus S is closed under $+$.

b $S = \{2, 4, 6, 8, ...\}$ under $\times$
Let $2m$ and $2n$ be in S; $m, n \in \mathbb{Z}^+$
$\therefore$ $2m \times 2n = 2(2mn)$ where $2mn \in \mathbb{Z}^+$
$\therefore$ $2m \times 2n$ is even.
Thus S is closed under $\times$.

c S is *not* closed under $+$ as, for example, $1,\ 3 \in S$ but $1+3=4 \notin S$.

d Any members of S can be written as $2a-1$ and $2b-1$ where $a, b \in \mathbb{Z}^+$.

$(2a-1)(2b-1) = 4ab - 2a - 2b + 1$
$= 2(2ab - a - b) + 1$
where $2ab - a - b \in \mathbb{Z}^+$

$\therefore$ $(2a-1)(2b-1) \in S$ also

$\therefore$ S is closed under $\times$.

5 **a** $a * b = a^2 - b$

i If $a, b \in \mathbb{Z}$, $a^2 \in \mathbb{Z}$ and $b \in \mathbb{Z}$
$\therefore$ $a^2 - b \in \mathbb{Z}$
$\therefore$ $a * b$ is closed on $\mathbb{Z}$.

ii If $a, b \in \mathbb{Q}$, $a^2 \in \mathbb{Q}$ and $b \in \mathbb{Q}$
$\therefore$ $a^2 - b \in \mathbb{Q}$
$\therefore$ $a * b$ is closed on $\mathbb{Q}$.

b $a * b = \dfrac{a+b}{a}$

i As $3 * 2 = \frac{5}{3} \notin \mathbb{Z}$, $a * b$ is not closed on $\mathbb{Z}$.

ii As $0 * 2 = \frac{2}{0} \notin \mathbb{Q}$, $a * b$ is not closed on $\mathbb{Q}$.

c $a * b = \sqrt{a^2 b^2} = |ab|$

i If $a, b \in \mathbb{Z}$, $|ab| \in \mathbb{Z} \Rightarrow a * b \in \mathbb{Z}$
$\therefore$ $a * b$ is closed on $\mathbb{Z}$.

ii If $a, b \in \mathbb{Q}$, $|ab| \in \mathbb{Q} \Rightarrow a * b \in \mathbb{Q}$
$\therefore$ $a * b$ is closed on $\mathbb{Q}$.

d $a * b = \sqrt{|ab|}$ where $|ab| \geqslant 0$

If $a = 1$, $b = 2$, $a * b = \sqrt{2}$ which is not in $\mathbb{Z}$ and not in $\mathbb{Q}$.

$\therefore$ $a * b$ is not closed on **i** $\mathbb{Z}$ **ii** $\mathbb{Q}$

EXERCISE D.2

1 **a** $a * b = a + 2b$

i If $a = 1$, $b = 2$
$1 * 2 = 1 + 2(2)$
$= 5$
and
$2 * 1 = 2 + 2(1)$
$= 4$
$\neq 1 * 2$
$\therefore$ $*$ is not commutative on $\mathbb{R}$.

ii If $a = 1$, $b = 2$, $c = 1$
$(a * b) * c$
$= 5 * 1$
$= 7$
and
$a * (b * c)$
$= 1 * (2 * 1)$
$= 1 * 4$
$= 9$
$\neq (a * b) * c$
$\therefore$ $*$ is not associative on $\mathbb{R}$.

b $a * b = a^2 + b^2$

i $b * a = b^2 + a^2$
$= a^2 + b^2$
$= a * b$ for all $a, b \in \mathbb{R}$
$\therefore$ $*$ is commutative on $\mathbb{R}$.

ii If $a = 2$, $b = 1$, $c = 1$

$(a * b) * c$	and	$a * (b * c)$
$= (2^2 + 1^2) * 1$		$= 2 * (1^2 + 1^2)$
$= 5 * 1$		$= 2 * 2$
$= 5^2 + 1^2$		$= 2^2 + 2^2$
$= 26$		$= 8$
		$\neq (a * b) * c$

$\therefore$ $*$ is not associative on $\mathbb{R}$.

c $a * b = ab - a - b$

i $b * a = ba - b - a$
$= ab - a - b$
$= a * b$ for all $a, b \in \mathbb{R}$
$\therefore$ $*$ is commutative on $\mathbb{R}$.

ii If $a = 1$, $b = 2$, $c = 3$

$(a * b) * c$	and	$a * (b * c)$
$= (1 * 2) * 3$		$= 1 * (2 * 3)$
$= (2 - 1 - 2) * 3$		$= 1 * (6 - 2 - 3)$
$= -1 * 3$		$= 1 * 1$
$= -3 - (-1) - 3$		$= 1 - 1 - 1$
$= -5$		$= -1$
		$\neq (a * b) * c$

$\therefore$ $*$ is not associative on $\mathbb{R}$.

d $a * b = \dfrac{1}{a+b}$

i $b * a = \dfrac{1}{b+a}$
$= \dfrac{1}{a+b}$
$= a * b$ for all $a, b \in \mathbb{R}$
$\therefore$ $*$ is commutative on $\mathbb{R}$.

ii If $a = 1$, $b = 2$, $c = 3$

$(a * b) * c$	and	$a * (b * c)$
$= \dfrac{1}{1+2} * 3$		$= 1 * (2 * 3)$
$= \frac{1}{3} * 3$		$= 1 * \dfrac{1}{2+3}$
$= \dfrac{1}{\frac{1}{3} + 3}$		$= 1 * \frac{1}{5}$
$= \dfrac{1}{\frac{10}{3}}$		$= \dfrac{1}{1 + \frac{1}{5}}$
$= \frac{3}{10}$		$= \dfrac{1}{\frac{6}{5}}$
		$= \frac{5}{6} \neq (a * b) * c$

$\therefore$ $*$ is not associative on $\mathbb{R}$.

2 P_n is: "If $*$ is associative and commutative on a set S, then $(a * b)^n = a^n * b^n$" for all $n \in \mathbb{Z}^+$.

Proof by induction on n

(1) If $n = 1$, $(a * b)^1 = a * b = a^1 * b^1$
$\therefore$ P_1 is true.

(2) If P_k is true $(k \in \mathbb{Z}^+)$ then
$(a * b)^k = a^k * b^k$
$\therefore$ $(a * b)^{k+1} = (a * b)^k * (a * b)^1$
$= a^k * b^k * a * b$
$= a^k * a * b^k * b$ {commutative law}
$= a^{k+1} * b^{k+1}$

Thus P_1 is true and P_{k+1} is true whenever P_k is true
$\therefore$ P_n is true. {Principle of mathematical induction}

3 $a * b = a - b$, $a \diamondsuit b = ab$

a If $a = 1$, $b = 2$, $c = 3$

$a * (b \diamondsuit c)$	and	$(a * b) \diamondsuit (a * c)$
$= 1 * (2 \diamondsuit 3)$		$= (1 * 2) \diamondsuit (1 * 3)$
$= 1 * 6$		$= -1 \diamondsuit -2$
$= -5$		$= 2$

$\therefore$ $*$ is *not* distributive over $\diamondsuit$.

b $a \diamondsuit (b * c) = a \diamondsuit (b - c)$ and $(a \diamondsuit b) * (a \diamondsuit c)$

$= a(b - c)$ $\quad = ab * ac$

$= ab - ac$ $\quad = ab - ac$

$\therefore \; a \diamondsuit (b * c) = (a \diamondsuit b) * (a \diamondsuit c)$ for all $a, b, c \in \mathbb{R}$

$\therefore \; \diamondsuit$ is distributive over $*$.

4 $a * b = 3a + b, \; a \diamondsuit b = 2ab$

a If $a = 1, \; b = 0, \; c = 2$

$a * (b \diamondsuit c)$ and $(a * b) \diamondsuit (a * c)$

$= 1 * (0 \diamondsuit 2)$ $\quad = (1 * 0) \diamondsuit (1 * 2)$

$= 1 * 0$ $\quad = 3 \diamondsuit 5$

$= 3$ $\quad = 30$

$\therefore \; *$ is *not* distributive over $\diamondsuit$.

b $a \diamondsuit (b * c)$ and $(a \diamondsuit b) * (a \diamondsuit c)$

$= a \diamondsuit (3b + c)$ $\quad = 2ab * 2ac$

$= 2a(3b + c)$ $\quad = 3(2ab) + 2ac$

$= 6ab + 2ac$ $\quad = 6ab + 2ab$

$\therefore \; a \diamondsuit (b * c) = (a \diamondsuit b) * (a \diamondsuit c)$ for all $a, b, c \in \mathbb{R}^+$

$\therefore \; \diamondsuit$ is distributive over $*$.

5 **a** $A \backslash B = A \cap B'$

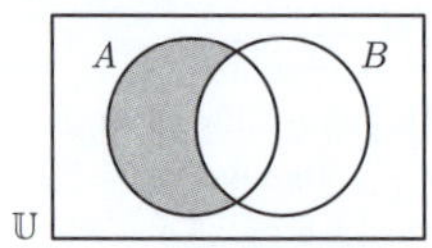

Consider $(A \backslash B) \backslash C$

It is shaded.

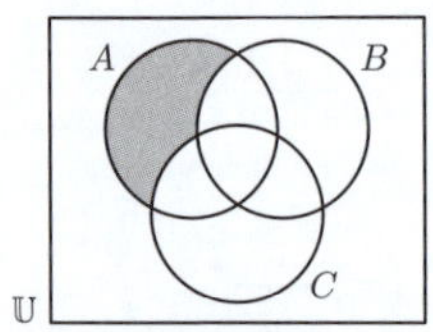

Consider $A \backslash (B \backslash C)$

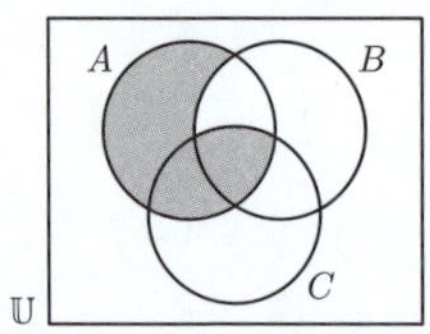

Thus $(A \backslash B) \backslash C \neq A \backslash (B \backslash C)$

$\therefore$ set difference is not associative.

$\therefore$ statement is true.

b $A \Delta B = (A \backslash B) \cup (B \backslash A)$

$= (A \cap B') \cup (B \cap A')$

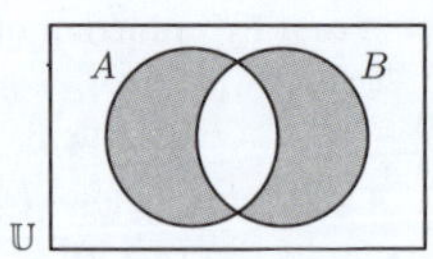

Consider $(A \Delta B) \Delta C = (A \Delta B) \backslash C \cup C \backslash (A \Delta B)$

$(A \Delta B) \backslash C$ $\qquad C \backslash A \Delta B$

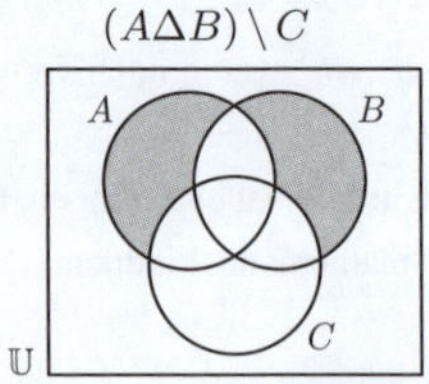

$\cup$

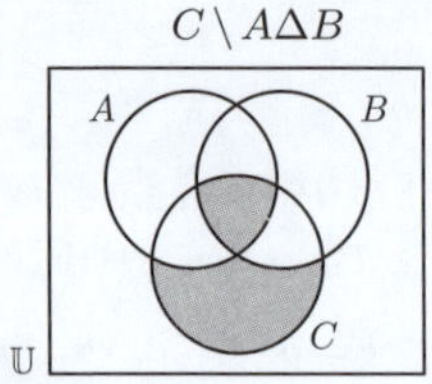

which is

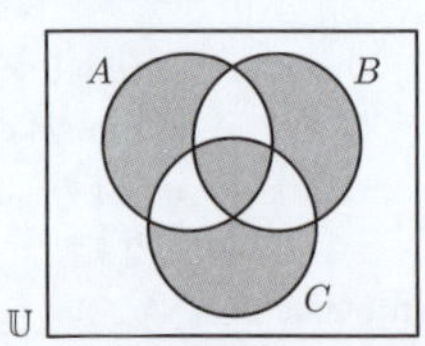

$(A \Delta B) \Delta C$

Now consider

$A \Delta (B \Delta C) = A \backslash (B \Delta C) \cup (B \Delta C) \backslash A$

$A \backslash (B \Delta C)$ $\qquad (B \Delta C) \backslash A$

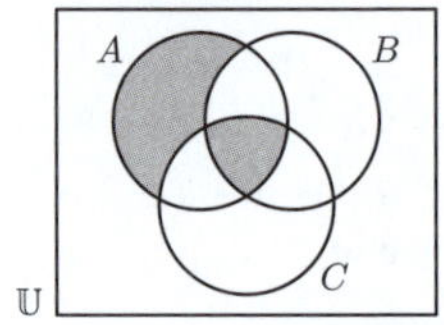

$\cup$

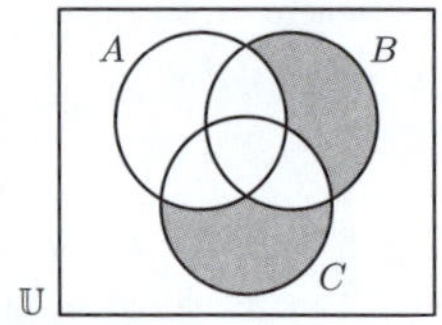

which is

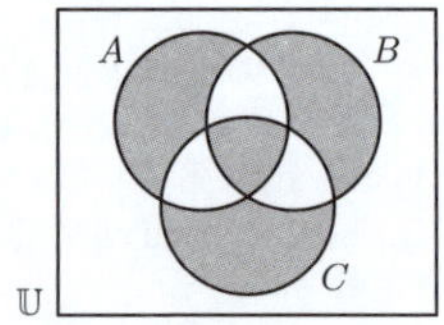

$A \Delta (B \Delta C)$

Thus $(A \Delta B) \Delta C = A \Delta (B \Delta C)$

$\therefore$ symmetric difference is associative

$\therefore$ statement is true.

EXERCISE D.3

1 **a** 0 as $a + 0 = 0 + a = a$ for all $a \in \mathbb{R}$

b 1 as $a \times 1 = 1 \times a = a$ for all $a \in \mathbb{Z}$

c If x is the identity, then $a * x = x * a = a$ for all $a \in \mathbb{R}$

$\therefore \; a = x = a$

$\therefore \; x = a$ for all $a \in \mathbb{R}$

$\therefore \; x$ is not unique

$\therefore$ an identity element does not exist.

d If x is the identity, then $a * x = x * a = a$ for all $a \in \mathbb{R}$

$\therefore \; 5ax = 5xa = a$

$\therefore \; 5x = 1$

$\therefore \; x = \frac{1}{5}$

$\therefore$ the identity element is $\frac{1}{5}$.

e If x is the identity, then $a * x = a$ for all $a \in \mathbb{R}$

$\therefore \; 2a + ax + 2x = a$

$\therefore \; x(a + 2) = -a$

$\therefore \; x = \dfrac{-a}{a + 2}$ for all $a \in \mathbb{R}$

$\therefore \; x$ is not unique

$\therefore$ an identity element does not exist.

f If x is the identity, then $x \div a = a \div x = a$ for all $a \in \mathbb{R}$

$\therefore \; \dfrac{x}{a} = \dfrac{a}{x} = a$

$\therefore \; x^2 = a^2, \; x = 1, \; x = a^2$

$\therefore \; x$ is not unique

$\therefore$ an identity element does not exist.

2 **a** For $\mathbb{Q}$ under $+$, the identity is 0 as $a + 0 = 0 + a = a$ for every $a \in \mathbb{Q}$.

The inverse of $a \in \mathbb{Q}$ is $-a \in \mathbb{Q}$ as $a + (-a) = (-a) + a = 0$.

b For $\mathbb{Q}$ under $\times$, the identity is 1 as $a \times 1 = 1 \times a = a$ for all $a \in \mathbb{Q}$. However $0 \in \mathbb{Q}$ but 0 does not have an inverse under $\times$. All other members of $\mathbb{Q}$ have an inverse $\dfrac{1}{a}$.

c For $\mathbb{Z}^+$ under $\times$, the identity is 1 as $a \times 1 = 1 \times a = a$ for all $a \in \mathbb{Z}^+$. However, only the element 1 has an inverse in $\mathbb{Z}^+$, and its inverse is 1. All other elements $a \in \mathbb{Z}^+$, $a \neq 1$, do not have an inverse in $\mathbb{Z}^+$.

d For $\mathbb{R}$ under $*$ where $a*b=2ab$

Suppose $a*e=e*a=a$ for all $a\in\mathbb{R}$

$\therefore\ 2ae=2ea=a$

$\therefore\ a(2e-1)=0$ for all $a\in\mathbb{R}$

$\therefore\ e=\frac{1}{2}$

Now, if $a*x=x*a=\frac{1}{2}$

$\therefore\ 2ax=2xa=\frac{1}{2}$

$\therefore\ x=\dfrac{1}{4a}$

So, if $a\neq 0$, $\dfrac{1}{4a}$ is the inverse of $a\in\mathbb{R}$.

$a=0$ does not have an inverse.

3 **a**

$2\times_{10}2=4$	$4\times_{10}2=8$	$6\times_{10}2=2$
$2\times_{10}4=8$	$4\times_{10}4=6$	$6\times_{10}4=4$
$2\times_{10}6=2$	$4\times_{10}6=4$	$6\times_{10}6=6$
$2\times_{10}8=6$	$4\times_{10}8=2$	$6\times_{10}8=8$

$8\times_{10}2=6$
$8\times_{10}4=2$
$8\times_{10}6=8$
$8\times_{10}8=4$

All $4\times 4=16$ possibilities show that $a\times_{10}b\in S$

$\therefore\ S$ is closed under $\times_{10}$.

b $x\times_{10}6=6\times_{10}x=x$ for all $x\in S$

$\therefore$ 6 is the identity.

4 **a** **i** If $a, b\in\mathbb{Q}\setminus\{1\}$, then a and b are rationals. Since $a+b$ and ab are rationals $\{\mathbb{Q}$ is closed under $+, \times\}$, $a+b-ab\ \in\mathbb{Q}$ as $\mathbb{Q}$ is closed under $-$.
So it remains to show that

$a-ab+b\neq 1$ for $a\neq 1$, $b\neq 1$

Now $a-1\neq 0$ and $b-1\neq 0$

$\therefore\ (a-1)(b-1)\neq 0$

$\therefore\ ab-a-b+1\neq 0$

$\therefore\ a-ab+b\neq 1$

ii Suppose $a, b, c\in\mathbb{Q}\setminus\{1\}$.

$\therefore\ (a\diamondsuit b)\diamondsuit c$
$=(a-ab+b)\diamondsuit c$
$=a-ab+b-(a-ab+b)c+c$
$=a-ab+b-ac+abc-bc+c$
$=a+b+c-ab-ac-bc+abc$

and $a\diamondsuit(b\diamondsuit c)$
$=a\diamondsuit(b-bc+c)$
$=a-a(b-bc+c)+b-bc+c$
$=a-ab+abc-ac+b-bc+c$
$=a+b+c-ab-ac-bc+abc$

$\therefore\ (a\diamondsuit b)\diamondsuit c=a\diamondsuit(b\diamondsuit c)$ for all $a, b, c\in\mathbb{Q}\setminus\{1\}$

$\therefore\ \diamondsuit$ is associative on $\mathbb{Q}\setminus\{1\}$.

iii Suppose $a\diamondsuit e=a$ for all $a\in\mathbb{Q}\setminus\{1\}$

$\therefore\ a-ae+e=a$ for all $a\in\mathbb{Q}\setminus\{1\}$

$\therefore\ e(1-a)=0$ for all $a\in\mathbb{Q}\setminus\{1\}$

$\therefore\ e=0$ as $a\neq 1$

Thus $a\diamondsuit 0=a$

Also $0\diamondsuit a=0-0+a=a$

Hence the identity is $e=0$.

iv Consider $a\diamondsuit x=e$

$\therefore\ a-ax+x=0$

$\therefore\ x(1-a)=-a$ and so $x=\dfrac{a}{a-1}$

Also, $x\diamondsuit a=\dfrac{a}{a-1}\diamondsuit a$

$=\dfrac{a}{a-1}-\dfrac{a^2}{a-1}+a$

$=\dfrac{a-a^2}{a-1}+a$

$=-a\left(\dfrac{a-1}{a-1}\right)+a$

$=-a+a$ provided that $a\neq 1$

$=0$ as $a\neq 1$

Thus the inverse of $a\in\mathbb{Q}\setminus\{1\}$ is $\dfrac{a}{a-1}\in\mathbb{Q}\setminus\{1\}$.

b Yes.

5 $(a, b)*(c, d)=(ac-bd, ad+bc)$ on $\mathbb{R}^2$

a $[(a, b)*(c, d)]*(g, h)$
$=(ac-bd, ad+bc)*(g, h)$
$=((ac-bd)g-(ad+bc)h, (ac-bd)h+(ad+bc)g)$
$=(acg-bdg-adh-bch, ach-bdh+adg+bcg)$

and

$(a, b)*[(c, d)*(g, h)]$
$=(a, b)*(cg-dh, ch+dg)$
$=(a(cg-dh)-b(ch+dg), a(ch+dg)+b(cg-dh))$
$=(acg-adh-bch-bdg, ach+adg+bcg-bdh)$
$=[(a, b)*(c, d)]*(g, h)$ for all elements in $\mathbb{R}^2$

$\therefore\ *$ is associative on $\mathbb{R}^2$.

b $(a, b)*(c, d)=(ac-bd, ad+bc)$
$=(ca-db, cb+da)$
$=(c, d)*(a, b)$

$\{\times$ and $+$ all commutative for reals$\}$

$\therefore\ *$ is commutative.

c Suppose $(a, b)*(e, f)=(a, b)$

$\therefore\ (ae-bf, af+be)=(a, b)$

$\therefore\ \left.\begin{array}{l} ae-bf=a \\ be+af=b \end{array}\right\}$ for all $a, b\in\mathbb{R}$

$\therefore\ e=1$ and $f=0$ {equating coefficients}

Thus $(a, b)*(1, 0)=(a, b)$ and also
$(1, 0)*(a, b)=(a-0, b+0)$
$=(a, b)$

$\therefore$ the identity element is $(1, 0)$.

d $(0, 0)$ has no inverse as $(a, b)*(0, 0)=(0, 0)\neq(1, 0)$

e For $(a, b)\neq(0, 0)$

Consider $(a, b)*(p, q)=(1, 0)$ for all $a, b\in\mathbb{R}$

$\therefore\ (ap-bq, aq+bp)=(1, 0)$

$\therefore\ \left\{\begin{array}{l} ap-bq=1 \\ bp+aq=0 \end{array}\right.$ for all $a, b\in\mathbb{R}$

Hence $\left\{\begin{array}{l} a^2p-abq=a \\ b^2p+abq=0 \end{array}\right.$

$\therefore\ (a^2+b^2)p=a$

$p = \dfrac{a}{a^2+b^2}$ and $aq = -bp$

$\therefore\ aq = \dfrac{-ab}{a^2+b^2}$

$\therefore\ p = \dfrac{a}{a^2+b^2}$ and $q = \dfrac{-b}{a^2+b^2}$

Since $(a, b) \neq (0, 0)$, then $a^2+b^2 \neq 0$

$\therefore$ the inverse of (a, b) is $\left(\dfrac{a}{a^2+b^2}, \dfrac{-b}{a^2+b^2}\right)$

as $\left(\dfrac{a}{a^2+b^2}, \dfrac{-b}{a^2+b^2}\right) * (a, b)$

$= \left(\dfrac{a^2}{a^2+b^2} - \dfrac{-b^2}{a^2+b^2}, \dfrac{ab}{a^2+b^2} + \dfrac{-ab}{a^2+b^2}\right)$

$= \left(\dfrac{a^2+b^2}{a^2+b^2}, 0\right)$

$= (1, 0)$ also

6 **a** For $p(x)$ and $q(x) \in P$, $p(x)+q(x)$ is a polynomial with coefficients in $\mathbb{R}$ and with degree equal to the maximum of {degree of $p(x)$, degree of $q(x)$}

$\therefore\ p(x)+q(x) \in P$

$\therefore\ P$ is closed under $+$.

b For all $p(x), q(x) \in P$, $p(x)+q(x) = q(x)+p(x)$ {as $+$ in $\mathbb{R}$ is commutative for each coefficient of x}

c $[p(x)+q(x)]+r(x) = p(x)+[q(x)+r(x)]$ for all p, q, and $r \in P$.
{as addition in $\mathbb{R}$ is associative for each coefficient of x}

d The zero polynomial is the identity as $p(x)+0 = 0+p(x) = p(x)$ for every $p(x) \in P$.

e The inverse of $p(x)$ is $-p(x)$ under $+$
{$-p(x)$ is the polynomial of $p(x)$ multiplied by -1}
as $p(x)+(-p(x)) = -p(x)+p(x) = 0$, the identity.

EXERCISE D.4

1 **a**

$\times_5$	1	2	3	4
1	1	2	3	4
2	2	4	1	3
3	3	1	4	2
4	4	3	2	1

i $x = 3$
ii $x = 2$
iii $x = 3$
iv $x = 4$

b The identity is 1 as $a \times_5 1 = 1 \times_5 a = a$ for all $a \in \{1, 2, 3, 4\}$.

c As $3 \times_5 2 = 2 \times_5 3 = 1$

$\therefore$ 3 and 2 are inverses, and so $3^{-1} = 2$

d As $1 \times_5 1 = 1$ and $4 \times_5 4 = 1$,
$1^{-1} = 1$, $2^{-1} = 3$, $3^{-1} = 2$, and $4^{-1} = 4$

2 **a**

$*$	a	b	c
a	a	b	c
b	b	c	a
c	c	a	b

i a is clearly the identity.

ii As $a*a = a$ and $b*c = c*b = a$, a is its own inverse, and b and c are inverses.

iii $*$ is commutative as there is symmetry about the leading diagonal.

iv We need to check all 27 $(= 3\times 3\times 3)$ possibilities.

(1) $(a*a)*a = a*a = a$ and $a*(a*a) = a*a = a$ ✓
(2) $(a*a)*b = a*b = b$ and $a*(a*b) = a*b = b$ ✓
(3) $(a*a)*c = a*c = c$ and $a*(a*c) = a*c = c$ ✓

(4) $(a*b)*a = b*a = b$ and $a*(b*a) = a*b = b$ ✓
(5) $(a*b)*b = b*b = c$ and $a*(b*b) = a*c = c$ ✓
(6) $(a*b)*c = b*c = a$ and $a*(b*c) = a*a = a$ ✓

(7) $(a*c)*a = c*a = c$ and $a*(c*a) = a*c = c$ ✓
(8) $(a*c)*b = c*b = a$ and $a*(c*b) = a*a = a$ ✓
(9) $(a*c)*c = c*c = b$ and $a*(c*c) = a*b = b$ ✓

(10) $(b*a)*a = b*a = b$ and $b*(a*a) = b*a = b$ ✓
(11) $(b*a)*b = b*b = c$ and $b*(a*b) = b*b = c$ ✓
(12) $(b*a)*c = b*c = a$ and $b*(a*c) = b*c = a$ ✓

(13) $(b*b)*a = c*a = c$ and $b*(b*a) = b*b = c$ ✓
(14) $(b*b)*b = c*b = a$ and $b*(b*b) = b*c = a$ ✓
(15) $(b*b)*c = c*c = b$ and $b*(b*c) = b*a = b$ ✓

(16) $(b*c)*a = a*a = a$ and $b*(c*a) = b*c = a$ ✓
(17) $(b*c)*b = a*b = b$ and $b*(c*b) = b*a = b$ ✓
(18) $(b*c)*c = a*c = c$ and $b*(c*c) = b*b = c$ ✓

(19) $(c*a)*a = c*a = c$ and $c*(a*a) = c*a = c$ ✓
(20) $(c*a)*b = c*b = a$ and $c*(a*b) = c*b = a$ ✓
(21) $(c*a)*c = c*c = b$ and $c*(a*c) = c*c = b$ ✓

(22) $(c*b)*a = a*a = a$ and $c*(b*a) = c*b = a$ ✓

(23) $(c*b)*b = a*b = b$ and $c*(b*b) = c*c = b$ ✓

(24) $(c*b)*c = a*c = c$ and $c*(b*c) = c*a = c$ ✓

(25) $(c*c)*a = b*a = b$ and $c*(c*a) = c*c = b$ ✓

(26) $(c*c)*b = b*b = c$ and $c*(c*b) = c*a = c$ ✓

(27) $(c*c)*c = b*c = a$ and $c*(c*c) = c*b = a$ ✓

Thus $(x*y)*z = x*(y*z)$ for all $x, y, z \in S$

$\therefore$ $*$ is associative on S.

v It **is** a Latin square.

b

$*$	a	b	c
a	a	a	a
b	a	b	c
c	a	c	b

i b is the identity.

ii a has no inverse
b is its own inverse
c is its own inverse.

iii $*$ is commutative as there is symmetry about the leading diagonal.

iv We need to check all 27 possibilities as we did in part **a**. When this is done we find that $*$ is associative.

v It **is not** a Latin square.

c

$*$	a	b	c
a	a	c	b
b	c	b	a
c	b	a	c

i No identity exists.

ii Without an identity, no inverses are possible.

iii $*$ is commutative as there is symmetry about the leading diagonal.

iv $*$ is not associative, for example

$(a*b)*c = c*c = c$ whereas $a*(b*c) = a*a = a$

So, in general, $(x*y)*z \neq x*(y*z)$ for all $x, y, z \in S$. Thus $*$ is not associative.

v It **is** a Latin square.

d

$*$	a	b	c
a	c	a	b
b	a	b	c
c	b	c	c

i b is the identity.

ii b is its own inverse
a and c are inverses.

iii $*$ is commutative as there is symmetry about the leading diagonal.

iv $*$ is not associative as, for example,

$(a*c)*c = b*c = c$ whereas $a*(c*c) = a*c = b$

So, in general, $(x*y)*z \neq x*(y*z)$ for all $x, y, z \in S$.

v It **is not** a Latin square.

e

$*$	a	b	c
a	b	c	a
b	a	b	c
c	c	a	b

i No identity exists.

ii Without an identity, no inverses are possible.

iii $*$ is not commutative as there is no symmetry about the leading diagonal.

iv $*$ is not associative as, for example,

$(c*b)*a = a*a = b$ whereas $c*(b*a) = c*a = c$

So, in general, $(x*y)*z \neq x*(y*z)$ for all $x, y, z \in S$. Thus $*$ is not associative.

v It **is** a Latin square.

3 a i

$\times$	1	i	$-i$	-1
1	1	i	$-i$	-1
i	i	-1	1	$-i$
$-i$	$-i$	1	-1	i
-1	-1	$-i$	i	1

ii The table is symmetric about the leading diagonal.
$\therefore$ $\times$ is commutative.
As $\times$ is associative in $\mathbb{C}$, it is associative in U_4.

iii The identity is 1 as $1 \times a = a \times 1 = a$ for all $a \in U_4$.

iv As $1 \times 1 = 1$, $1^{-1} = 1$
As $-1 \times -1 = 1$, $(-1)^{-1} = -1$
As $i \times -i = 1$, $i^{-1} = -i$ and $i^{-1} = -i$
So, 1 and -1 are there own inverses and i and $-i$ are inverses.

b i U_n is the set of nth roots of unity in $\mathbb{C}$.

ii For $n = 4$, $1 = 1$

$$\alpha = \text{cis}\left(\tfrac{2\pi}{4}\right) = \text{cis}\left(\tfrac{\pi}{2}\right) = i$$
$$\alpha^2 = \text{cis}\left(\tfrac{2\pi}{2}\right) = \text{cis}\,\pi = -1$$
$$\alpha^3 = \text{cis}\left(\tfrac{3\pi}{2}\right) = -i$$

$\therefore$ $U_4 = \{1, \alpha, \alpha^2, \alpha^3\} = \{1, i, -1, -i\}$.

EXERCISE E.1

1 a i As the Cayley table contains only elements of S, then S is closed under $*$.

ii The identity element is e as $x*e = e*x = x$ for all $x \in S$.

iii e is its own inverse.
Likewise, a, b, c, and d are their own inverses.

b As $n(S) = 5$ and we need to check that $(x*y)*z = x*(y*z)$ for all $x, y, z \in S$, there are $5 \times 5 \times 5 = 125$ checks to be made.

2 To prove: In group $\{G, *\}$ where a, b, and $c \in G$,
$a*b = a*c \Rightarrow b = c$.

Proof: $a*b = a*c$

$\therefore$ $a^{-1}*(a*b) = a^{-1}*(a*c)$

{We are pre-multiplying both sides by a^{-1} which exists as $a \in G$.}

$\Rightarrow$ $(a^{-1}*a)*b = (a^{-1}*a)*c$ {Associative law}

$\Rightarrow$ $e*b = e*c$ {Inverse law}

$\Rightarrow$ $b = c$ {Identity law}

3 **a** $\{\mathbb{Q}, \times\}$ is not a group.

The identity is 1 since $1 \times \frac{a}{b} = \frac{a}{b} \times 1 = \frac{a}{b}$ for all $\frac{a}{b} \in \mathbb{Q}$.

However $0 \in \mathbb{Q}$ has no inverse in $\mathbb{Q}$ as $0 \times \frac{a}{b} = 0$ for all $\frac{a}{b} \in \mathbb{Q}$.

b $\{\mathbb{Q} \setminus \{0\}, \times\}$ is an Abelian group.

Closure: If $\frac{a}{b}$ and $\frac{c}{d} \in \mathbb{Q} \setminus \{0\}$ then $\frac{a}{b} \times \frac{c}{d} = \frac{ac}{bd}$ where $bd \neq 0$ and $ac \neq 0$.
$\therefore \frac{a}{b} \times \frac{c}{d} \in \mathbb{Q} \setminus \{0\}$
$\therefore \mathbb{Q} \setminus \{0\}$ is closed under $\times$.

Associative: As $\times$ is associative in $\mathbb{R}$, $\times$ is associative in the subset $\mathbb{Q} \setminus \{0\}$.

Identity: 1 is the identity in $\mathbb{Q} \setminus \{0\}$ as
$\frac{a}{b} \times 1 = 1 \times \frac{a}{b} = \frac{a}{b}$ for all $\frac{a}{b} \in \mathbb{Q} \setminus \{0\}$.

Inverse: Since $\frac{a}{b} \times \frac{b}{a} = \frac{b}{a} \times \frac{a}{b} = 1$ for all $\frac{a}{b}, \frac{b}{a} \in \mathbb{Q} \setminus \{0\}$
$\left(\frac{a}{b}\right)^{-1} = \frac{b}{a}$

Commutative: $\frac{a}{b} \times \frac{c}{d} = \frac{c}{d} \times \frac{a}{b}$ for all $\frac{a}{b}, \frac{c}{d} \in \mathbb{Q} \setminus \{0\}$.

c {Odd integers, $\times$} **is not** a group.
Although 1 is the identity, each element does not necessarily have an inverse.
For example 3^{-1} does not exist as there is no odd integer x such that $3 \times x = x \times 3 = 1$.

d $S = \{3^n \mid n \in \mathbb{Z}\}$ is an Abelian group under $\times$ as:

Closure: For $m, n \in \mathbb{Z}$, $3^m \times 3^n = 3^{m+n}$ where $m + n \in \mathbb{Z}$.

Associative:
$$\begin{aligned}(3^a 3^b)3^c &= 3^{a+b} \times 3^c \\ &= 3^{a+b+c} \\ &= 3^a \times 3^{b+c} \\ &= 3^a(3^b 3^c) \quad \text{for all } a, b, c \in \mathbb{Z}.\end{aligned}$$

Identity: Is $3^0 = 1$ as
$3^a 3^0 = 3^0 3^a = 3^a$ for all $a \in \mathbb{Z}$.

Inverse: $3^n \times 3^{-n} = 3^0 = 1 = 3^{-n} \times 3^n$
$\therefore$ 3^{-n} is the inverse of 3^n for all $n \in \mathbb{Z}$.

Commutative: $3^a 3^b = 3^b 3^a = 3^{a+b}$ for all $a, b \in \mathbb{Z}$
$\therefore$ S is commutative under $\times$.

e $S = \{1, -\frac{1}{2} + \frac{\sqrt{3}}{2}i, -\frac{1}{2} - \frac{\sqrt{3}}{2}i\}$
$= \{1, \alpha, \alpha^2\}$
where $\alpha = \text{cis}\left(\frac{2\pi}{3}\right)$

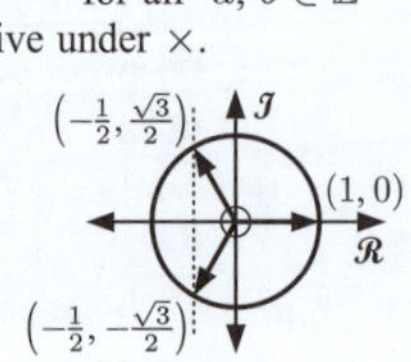

$\{S, \times\}$ is an Abelian group as:

Closure:

$\times$	1	α	α^2
1	1	α	α^2
α	α	α^2	1
α^2	α^2	1	α

Only 1, α, and α^2 are in the table
$\therefore$ S is closed under $\times$.

Associative: $S = \{1, \alpha, \alpha^2\} \subseteq \mathbb{C}$, the set of all complex numbers.
As $\times$ is associative in $\mathbb{C}$, $\times$ is associative in S (a subset of $\mathbb{C}$).

Identity: Is 1 as $a \times 1 = 1 \times a = a$ for all $a \in S$.

Inverse: $1 \times 1 = 1$ $\therefore$ 1 is its own inverse.
α and α^2 are inverses.

Commutative: As $\times$ is commutative in $\mathbb{C}$, $\times$ is commutative in S (a subset of $\mathbb{C}$).

f $S = \{3n \mid n \in \mathbb{Z}\}$ is an Abelian group under $+$ as:

Closure: As $3a \in S$ and $3b \in S$ for $a, b \in \mathbb{Z}$
$3a + 3b = 3(a + b)$ where $a + b \in \mathbb{Z}$
$\Rightarrow 3a + 3b \in S$ $\therefore$ S is closed under $+$.

Associative: $+$ is associative in $\mathbb{Z}$
$\therefore$ $+$ is associative in S as $S \subseteq \mathbb{Z}$.

Identity: $0 = 3 \times 0 \in S$ and $0 + 3a = 3a + 0 = 3a$ for all $a \in \mathbb{Z}$ $\therefore$ 0 is the identity.

Inverse: $-3n$ is the inverse of $3n$ for all $n \in \mathbb{Z}$ as $3n + (-3n) = (-3n) + 3n = 0$.

Commutative: $3m + 3n = 3n + 3m$ for all $m, n \in \mathbb{Z}$.

g $S = \{3n \mid n \in \mathbb{Z}\}$ is not a group under $\times$ as there is no identity element.
Reason: If e was the identity then $3a \times e = 3a$ for all $a \in S$ would $\Rightarrow e = 1$ but $1 \notin S$.

h $\{\mathbb{C}, +\}$ is an Abelian group.

Closure: $a + bi, c + di \in \mathbb{C}$ then a, b, c, and d are real.
Now $a + bi + c + di = (a + c) + (b + d)i$ where $a + c, b + d \in \mathbb{R}$
$\therefore$ $\mathbb{C}$ is closed under $+$.

Associative: Let $a + bi, c + di, e + fi \in \mathbb{C}$
$$\begin{aligned}&[(a + bi) + (c + di)] + (e + fi) \\ &= (a + bi) + (c + di) + (e + fi) \\ &= (a + bi) + [(c + di) + (e + fi)]\end{aligned}$$
$\therefore$ $+$ is associative on $\mathbb{C}$.

Identity: $0 = 0 + 0i$ is the identity as
$$\begin{aligned}(a + bi) + (0 + 0i) = (0 + 0i) + (a + bi) \\ = a + bi\end{aligned}$$
for all $a + bi \in \mathbb{C}$

Inverse: $-a - bi$ is the inverse of $a + bi$ for all $a + bi \in \mathbb{C}$ as
$$\begin{aligned}&(-a - bi) + (a + bi) \\ &= (a + bi) + (-a - bi) = 0\end{aligned}$$

Commutative:
$$\begin{aligned}(a + bi) + (c + di) &= (a + c) + (b + d)i \\ &= (c + di) + (a + bi)\end{aligned}$$
for all $a + bi, c + di \in \mathbb{C}$
$\therefore$ $+$ is commutative on $\mathbb{C}$.

i $\{\mathbb{C}, \times\}$ is not a group.
It has identity $1 = 1 + 0i$, but $0 = 0 + 0i$ does not have an inverse in $\mathbb{C}$ since $0 \times z = z \times 0 \neq 1$ for any $z \in \mathbb{C}$.

j Let $S = \{a + bi \mid a, b \in \mathbb{R}, \ |a + bi| = 1\}$
$= \{z \mid z \in \mathbb{C}, \ |z| = 1\}$
This is the set of all complex numbers which lie on the circle, centre $0 + 0i$ and with radius 1.

Closure: For all $z_1, z_2 \in S$
$|z_1 z_2| = |z_1||z_2| = 1 \times 1 = 1$
$\therefore$ $z_1 z_2 \in S$.

Associative: $S \subseteq \mathbb{C}$ and so $\times$ is associative on S {as $\times$ is associative on $\mathbb{C}$}.

Identity: $1 = 1 + 0i$ has $|1 + 0i| = 1$ and $z \times 1 = 1 \times z = z$ for all $z \in S$
$\therefore$ 1 is the identity under $\times$.

Inverse: $z^{-1} = \dfrac{1}{z}$ is the inverse of $z \in S$

as $z \times \dfrac{1}{z} = \dfrac{1}{z} \times z = 1$

and $\left|\dfrac{1}{z}\right| = \dfrac{1}{|z|} = \dfrac{1}{1} = 1$

$\Rightarrow \dfrac{1}{z} \in S$

Commutative: $z_1 z_2 = z_2 z_1$ for all $z_1, z_2 \in S$
{as $S \subseteq \mathbb{C}$ where commutative law holds}

$\therefore$ $\{S, \times\}$ is an Abelian group.

k $\mathbb{C} \setminus \{0\}$ under $\times$ is an Abelian group.

Closure: $\mathbb{C} \setminus \{0\} \subseteq \mathbb{C}$ where closure holds under $\times$
$\therefore$ $\mathbb{C} \setminus \{0\}$ is closed under $\times$.

Associative: Likewise to closure $\mathbb{C} \setminus \{0\}$ is associative under $\times$.

Identity: The identity is $1 = 1 + 0i$ as
$z \times 1 = 1 \times z = z$ for all $z \in \mathbb{C} \setminus \{0\}$.

Inverse: Let $z = a + bi \in \mathbb{C} \setminus \{0\}$.
The inverse of z is

$$\frac{1}{z} = \frac{1}{a+bi}\left(\frac{a-bi}{a-bi}\right) = \frac{a-bi}{a^2+b^2} = \left(\frac{a}{a^2+b^2}\right) - \left(\frac{b}{a^2+b^2}\right)i$$

Since $z \neq 0$, a and b are not both 0.

$\therefore$ $\dfrac{1}{z}$ is defined and $\in \mathbb{C} \setminus \{0\}$.

Commutative: Commutativity holds in $\mathbb{C}$ under $\times$
$\therefore$ will hold in $\mathbb{C} \setminus \{0\}$, a subset of $\mathbb{C}$.

4 $a * b = \dfrac{ab}{4}$ on $\mathbb{Q}^+$

Closure: For every $a, b \in \mathbb{Q}^+$, let $a = \dfrac{p}{q}$ and $b = \dfrac{r}{s}$; $p, q, r, s \in \mathbb{Z}^+$

$\therefore$ $a * b = \dfrac{pr}{4qs} > 0$ and $a * b \in \mathbb{Q}^+$.

Associative: $(a * b) * c$ and $a * (b * c)$

$$(a*b)*c = \frac{ab}{4} * c = \frac{\frac{ab}{4} \times c}{4} = \frac{abc}{16} \qquad a*(b*c) = a * \frac{bc}{4} = \frac{a \times \frac{bc}{4}}{4} = \frac{abc}{16}$$

$\therefore$ $(a*b)*c = a*(b*c)$ for all $a, b, c \in \mathbb{Q}^+$.

Identity: The identity is 4, as $4 \in \mathbb{Q}^+$

and $a * 4 = \dfrac{a \times 4}{4} = a$,

$4 * a = \dfrac{4 \times a}{4} = a$ for all $a \in \mathbb{Q}^+$

Inverse: Consider $a * x = x * a = 4$

$\therefore$ $\dfrac{ax}{4} = \dfrac{xa}{4} = 4$

$\therefore$ $x = \dfrac{16}{a}$ where $\dfrac{16}{a} \in \mathbb{Q}^+$

$\therefore$ $a^{-1} = \dfrac{16}{a}$ is the inverse of $a \in \mathbb{Q}^+$.

$\therefore$ $\{G, *\}$ is a group.

5 The statement is *false*.
For example, consider the Latin square under $*$ where $*$ is not associative.
For example, $(a * b) * c = c * c = c$
But $a * (b * c) = a * a = a$.

$*$	a	b	c
a	a	c	b
b	c	b	a
c	b	a	c

6 **a** $\{\mathbb{Z}_7 \setminus \{0\}, \times_7\}$

$\times_7$	1	2	3	4	5	6
1	1	2	3	4	5	6
2	2	4	6	1	3	5
3	3	6	2	5	1	4
4	4	1	5	2	6	3
5	5	3	1	6	4	2
6	6	5	4	3	2	1

Closure: Each element of the Cayley table is in $\mathbb{Z}_7 \setminus \{0\}$
$\therefore$ $\mathbb{Z}_7 \setminus \{0\}$ is closed under $\times_7$.

Associative: $\times$ is associative on $\mathbb{R}$.
Hence $\times_7$ is associative on $\mathbb{Z}_7 \setminus \{0\}$.

Identity: $1 \times_7 a = a \times_7 1 = a$ for every $a \in \mathbb{Z}_7 \setminus \{0\}$.
$\therefore$ 1 is the identity.

Inverse: 1 is its own inverse, $1^{-1} = 1$
2 and 4 are inverses, $2^{-1} = 4$, $4^{-1} = 2$
3 and 5 are inverses, $3^{-1} = 5$, $5^{-1} = 3$
6 is its own inverse, $6^{-1} = 6$

$\therefore$ $\{\mathbb{Z}_7 \setminus \{0\}, \times_7\}$ is a group.

b $\{\mathbb{Z}_9 \setminus \{0\}, \times_9\}$ is not a group as, for example, $3 \times_9 3 = 0$ and $0 \notin \mathbb{Z}_9 \setminus \{0\}$.
Thus $\mathbb{Z}_9 \setminus \{0\}$ is not closed under $\times_9$.

7 $(f + g)(x) = f(x) + g(x)$ for all $x \in \mathbb{R}$.

Closure: For f and $g \in F$, $(f+g)(x) = f(x) + g(x)$ is a real-valued function {$\mathbb{R}$ is closed under $+$}.
$\therefore$ F is closed under $+$.

Associative: For f, g, and $h \in F$,

$$\begin{aligned}(f + (g+h))(x) &= f(x) + (g+h)(x)\\ &= f(x) + g(x) + h(x)\\ &= (f+g)(x) + h(x)\\ &= ((f+g)+h)(x)\end{aligned}$$

$\therefore$ $+$ is associative in F.

Identity: $f(x) = 0$ is the identity function
{actually it is the horizontal line $y = 0$}
For any $g \in F$,

$$\begin{aligned}(f+g)(x) &= f(x) + g(x)\\ &= 0 + g(x) \quad \{\text{if } f(x) = 0\}\\ &= g(x)\end{aligned}$$

and
$$\begin{aligned}(g+f)(x) &= g(x) + f(x)\\ &= g(x) + 0\\ &= g(x)\end{aligned}$$

$\therefore$ $f(x) = 0$ is the identity function.

Inverse: For any function $f(x) \in F$, $-f(x) \in F$.
Also, $f(x) + (-f(x)) = f(x) - f(x) = 0$
and $(-f(x)) + f(x) = -f(x) + f(x) = 0$

$\therefore$ for all $f \in F$, $f^{-1} = -f$
$\therefore$ each element of F has an inverse.

$\therefore$ $\{F, +\}$ is a group.

8 **a** By the group axioms there exists an identity element in G. Suppose there are two of them, e and f, say.

$$\left.\begin{array}{ll}\therefore\ e*a=a*e=a & \text{.... (1)}\\ \text{and}\ f*a=a*f=a & \text{.... (2)}\end{array}\right\}\ \text{for all}\ a\in G$$

Since e and $f\in G$,

in (1), $e*f=f*e=f$ and

in (2), $f*e=e*f=e$

$\therefore\ e=f$

$\therefore$ the identity in $\{G,*\}$ is unique.

b By the group axioms there exists an inverse element for every a in G.

Suppose ${a_1}^{-1}$ and ${a_2}^{-2}$ are both inverses of a

$\therefore\ a*{a_1}^{-1}={a_1}^{-1}*a=e,$ the identity

and $a*{a_2}^{-1}={a_2}^{-1}*a=e$ also

Thus $a*{a_1}^{-1}=a*{a_2}^{-1}$

We now multiply on the left by a^{-1}

$\therefore\ a^{-1}*(a*{a_1}^{-1})=a^{-1}*(a*{a_2}^{-1})$

$\therefore\ (a^{-1}*a)*{a_1}^{-1}=(a^{-1}*a)*{a_2}^{-1}$ {as $*$ is associative in G}

$\therefore\ e*{a_1}^{-1}=e*{a_2}^{-1}$ {inverse axiom}

$\therefore\ {a_1}^{-1}={a_2}^{-1}$ {identity axiom}

Hence each element has a **unique inverse**.

9 Let $a,b\in G$ where $a*x=b$.

Since $a\in G$, it has a unique inverse $a^{-1}\in G$ such that $a*a^{-1}=a^{-1}*a=e$, the identity.

We now multiply $a*x=b$ on the left by a^{-1}.

$\therefore\ a^{-1}*(a*x)=a^{-1}*b$

$\therefore\ (a^{-1}*a)*x=a^{-1}*b$ {associative axiom}

$\therefore\ e*x=a^{-1}*b$ {inverse axiom}

$\therefore\ x=a^{-1}*b$ {identity axiom}

where $a^{-1}*b$ is the unique solution in G as $a^{-1},b\in G$ and $*$ is closed on G.

Similarly, for $y*a=b$

$(y*a)*a^{-1}=b*a^{-1}$

$\therefore\ y*(a*a^{-1})=b*a^{-1}$

$\therefore\ y*e=b*a^{-1}$

$\therefore\ y=b*a^{-1}$

where $b,a^{-1}\in G$ and $*$ is closed on G

$\therefore\ y=b*a^{-1}$ is the unique solution in G.

10 **a** $S=\mathbb{R}\backslash\{-2\}$ and $a*b=a+b+\dfrac{ab}{2}$

Closure:

$$\begin{aligned}& a*b=-2\\ \Leftrightarrow\ & a+b+\frac{ab}{2}=-2\\ \Leftrightarrow\ & 2a+2b+ab=-4\\ \Leftrightarrow\ & ab+2a+2b+4=0\\ \Leftrightarrow\ & (a+2)(b+2)=0\\ \Leftrightarrow\ & a=-2\ \text{or}\ b=-2\end{aligned}$$

So, if $a\neq-2,\ b\neq-2$ then $a*b\neq-2$

$\therefore$ for $a,b\in S,\ a*b\in S$

$\therefore\ *$ is closed on S.

Associative:

$$\begin{aligned}& a*(b*c)\\ &=a*\left(b+c+\frac{bc}{2}\right)\\ &=a+b+c+\frac{bc}{2}+\frac{a\left(b+c+\frac{bc}{2}\right)}{2}\\ &=a+b+c+\frac{bc}{2}+\frac{ab}{2}+\frac{ac}{2}+\frac{abc}{4}\end{aligned}$$

and

$$\begin{aligned}& (a*b)*c\\ &=\left(a+b+\frac{ab}{2}\right)+c\\ &=a+b+\frac{ab}{2}+c+\frac{\left(a+b+\frac{ab}{2}\right)c}{2}\\ &=a+b+c+\frac{ab}{2}+\frac{ac}{2}+\frac{bc}{2}+\frac{abc}{4}\\ &=a*(b*c)\quad\text{for all}\ a,b,c\in S\end{aligned}$$

$\therefore\ *$ is associative on S.

Identity: If $a*e=e*a=a$, then

$$a+e+\frac{ae}{2}=e+a+\frac{ae}{2}=a$$

$$\therefore\ e+\frac{ae}{2}=0$$

$$\therefore\ e\left(1+\frac{a}{2}\right)=0\ \text{ for all }\ a\in S$$

$$\therefore\ e=0$$

Check: $a*0=a+0+\frac{0}{2}=a$ ✓

$0*a=0+a+\frac{0}{2}=a$ ✓

$\therefore$ 0 is the identity in $\{S,*\}$.

Inverse: If $a*x=x*a=0$ for all $a\in S$,

$$a+x+\frac{ax}{2}=x+a+\frac{xa}{2}=0$$

$$\therefore\ 2a+2x+ax=2x+2a+xa=0$$

$$\therefore\ (2+a)x=-2a$$

$$x=\frac{-2a}{a+2}$$

and as $a\neq-2,\ x\in S$

$\therefore$ every element $a\in S$ has a unique inverse $a^{-1}=\dfrac{-2a}{a+2}$.

Commutative:

$$\begin{aligned}b*a&=b+a+\frac{ba}{2}\\ &=a+b+\frac{ab}{2}\quad\{+\ \text{and}\ \times\ \text{are commutative on}\ \mathbb{R}\}\\ &=a*b\end{aligned}$$

$\therefore\ *$ is commutative on S.

Hence $\{S,*\}$ forms an Abelian group.

b **i**

$$\begin{aligned}2*x*5&=11\\ \therefore\ 2*5*x&=11\quad\{\text{commutative property}\}\\ \therefore\ (2+5+\tfrac{10}{2})*x&=11\\ \therefore\ 12*x&=11\\ \therefore\ 12+x+\frac{12x}{2}&=11\end{aligned}$$

$\therefore\ 7x + 12 = 11$

$\therefore\ x = -\frac{1}{7}$

ii $x * 3 * 8 = 12$

$\therefore\ x * (3 + 8 + \frac{24}{2}) = 12$

$\therefore\ x * 23 = 12$

$\therefore\ x + 23 + \dfrac{23x}{2} = 12$

$\therefore\ 2x + 46 + 23x = 24$

$\therefore\ 25x = -22$

$\therefore\ x = -\frac{22}{25}$

11 Since $*$ is associative we do not need brackets. We shall also write $a * b$ as ab.

We need to prove that:

(1) the left identity is also the right identity, so it will be the identity in G

(2) the left inverse is also the right inverse and so will be the inverse required for the group axiom.

Proof:

(1) We know that $ea = a$ {left identity}

Suppose $ae = y$, say for some $y \in G$

$\therefore\ a_L^{-1}ae = a_L^{-1}y$

$\therefore\ ee = a_L^{-1}y$

$\therefore\ e = a_L^{-1}y$ {e is the left identity}

$\therefore\ a_L^{-1}y = a_L^{-1}a$

$\therefore\ \underbrace{(a_L^{-1})^{-1}a_L^{-1}}y = (a_L^{-1})^{-1}a_L^{-1}a$

$\therefore\ ey = ea$

$\therefore\ y = a$

Hence $ae = a$ {as $ae = y$}

$\therefore$ e is also the right identity.

(2) For any $a \in G$, $a_L^{-1}a = e$ {given}

Suppose $a\,a_L^{-1} = d$, say where $d \neq e$

$\therefore\ \underbrace{a\,a_L^{-1}}a = da$

$\therefore\ ae = da$

$\therefore\ a = da$

$\therefore\ a_L^{-1}a = a_L^{-1}da$

$\therefore\ e = a_L^{-1}da$

As $a_L^{-1} \in G$, there exists a left inverse $(a_L^{-1})^{-1}$ for all a_L^{-1} in G.

Thus $(a_L^{-1})^{-1}e = (a_L^{-1})^{-1}a_L^{-1}da$

$\therefore\ (a_L^{-1})^{-1} = eda$

$\therefore\ (a_L^{-1})^{-1} = da$

$\therefore\ (a_L^{-1})^{-1} = a$ {as $a = da$}

Thus $a_L^{-1}a = e = aa_L^{-1}$

Hence a_L^{-1} is also the right inverse for a

$\therefore$ a_L^{-1} is the inverse of a for all $a \in G$.

So, $\{G, *\}$ has $*$ which is closed and associative. It contains an identity e where $a * e = e * a = a$ for all $a \in G$ and each element a has a unique inverse a^{-1} where $a * a^{-1} = a^{-1} * a = e$

$\therefore$ $\{G, *\}$ is a group.

12 As G is closed and associative under $*$ and e is the identity we only have to show that every element $a \in G$ has a **unique inverse**. Since e appears once in **each column**, for each $b \in G$ there exists a unique $a \in G$ such that $a * b = e$.

Thus, each element of G has a left inverse in G

$\therefore$ by question **11**, each element of g has an inverse in G

$\therefore$ $\{G, *\}$ is a group.

13 S is the set of all subsets of $\mathbb{U}$.

Closure: If $A, B \in S$ then $A \Delta B \in S$ as $A\Delta B = (A \backslash B) \cup (B \backslash A)$.

Associative: Proved in **Exercise D.2**, question **5**.

Identity: For each set $A \in \mathbb{U}$, $A\Delta\varnothing = \varnothing\Delta A = A$

$\therefore$ the empty set $\varnothing$ is the identity.

Inverse: Since $A \Delta A = \varnothing$ for all sets $A \in S$, each element is its own inverse

$\therefore\ A^{-1} = A$.

EXERCISE E.2

1 **a**

$*$	2	4	6	8
2	4	8	2	6
4	8	6	4	2
6	2	4	6	8
8	6	2	8	4

b 6 is the identity as $6 * a = a * 6 = a$ for all $a \in G$.

c *Closure*: Every entry in the Cayley table is an element of G.

$\therefore$ G is closed under $*$.

Associative: Since multiplication is associative, $*$ is associative.

Identity: The identity is 6.

Inverse: As $2 * 8 = 6$, $2^{-1} = 8$ and $8^{-1} = 2$

As $4 * 4 = 6$, $4^{-1} = 4$

As $6 * 6 = 6$, $6^{-1} = 6$

d $2^2 = 4$, $2^3 = 4 * 2 = 8$, $2^4 = 8 * 2 = 6$

$\therefore$ 2 has order 4

$4^2 = 6$, $\therefore$ 4 has order 2

$6^1 = 6$, $\therefore$ 6 has order 1

$8^2 = 4$, $8^3 = 8 * 4 = 2$, $8^4 = 2 * 8 = 6$

$\therefore$ 8 has order 4

e $y = 2 * x * 4 = 2 * 4 * x$ {$*$ is commutative}

$\therefore\ y = 8 * x$

When $x = 2$, $y = 6$, $x = 4$, $y = 2$,

$x = 6$, $y = 8$, $x = 8$, $y = 4$

$\therefore$ solutions are: (2, 6), (4, 2), (6, 8), (8, 4).

2 **a** The Cayley table is:

$*$	1	3	5	7
1	1	3	5	7
3	3	1	7	5
5	5	7	1	3
7	7	5	3	1

Closure: Each element in the table is an element of A

$\therefore$ A is closed under $*$.

Associative: Since multiplication is associative, $*$ is associative.

Identity: The identity is 1 as $a * 1 = 1 * a = a$ for all $a \in A$.

Inverse: $1 \times 1 = 1$ $\therefore\ 1^{-1} = 1$

$3 \times 3 = 1$ $\therefore\ 3^{-1} = 3$

$5 \times 5 = 1$ $\therefore\ 5^{-1} = 5$

$7 \times 7 = 1$ $\therefore\ 7^{-1} = 7$

Each element is its own inverse.

b $*$ is commutative as the Cayley table is symmetric about the leading diagonal
$\therefore$ $\{A, *\}$ is an Abelian group.

c $1^1 = 1$ $\therefore$ 1 has order 1
$3^2 = 1,\ 5^2 = 1,\ 7^2 = 1$
$\therefore$ 3, 5, and 7 each have order 2

d $H = \{1, 3\}$

$*$	1	3
1	1	3
3	3	1

has closure, associativity, identity 1, and inverses as $1^{-1} = 1$ and $3^{-1} = 3$
$\therefore$ $\{H, *\}$ is a group.

3 **a** $U_n = \{\alpha \mid \alpha^n = 1,\ \alpha \in \mathbb{C}\}$
If $\alpha^n = 1$ then
$\alpha = \text{cis}\left(\frac{2\pi}{n}\right)$
and $\alpha^i = \text{cis}\left(\frac{2\pi i}{n}\right)$ for $i = 1, 2, 3, 4, 5, ..., n$
and $\alpha^n = \text{cis}(2\pi) = 1$
$\therefore$ $U_n = \{1, \alpha, \alpha^2, \alpha^3, ..., \alpha^{n-1}\}$

Closure: We note that $\alpha^{i+n} = \alpha^i\alpha^n = \alpha^i$.
Hence $\alpha^i\alpha^j = \alpha^{i+j(\text{mod}\,n)} \in U_n$.
Thus U_n is closed under $\times$.

Associative: As $\times$ is associative on $\mathbb{C}$, $\times$ is associative on U_n, a subset of $\mathbb{C}$.

Identity: 1 is the identity as $\alpha^i \times 1 = 1 \times \alpha^i = \alpha^i$ for all $i = 0, 1, 2, 3, ..., n-1$.

Inverse: Since $\alpha^n = 1$, and $\alpha^i\alpha^{n-i} = \alpha^n = 1$ then $(\alpha^i)^{-1} = \alpha^{n-i}$ for all $i = 0, 1, 2, 3, ..., n-1$.
So, each element α^i has a unique inverse α^{n-i}.

Thus $\{U_n, \times\}$ is a group.

b $U_4 = \{1, i, -1, -i\}$

c $1^1 = 1$, $\therefore$ 1 has order 1
$i^4 = (i^2)^2 = (-1)^2 = 1$, $\therefore$ i has order 4
$(-1)^2 = 1$, $\therefore$ -1 has order 2
$(-i)^4 = i^4 = 1$, $\therefore$ $-i$ has order 4

4 As G has order m, m is the **smallest** positive integer such that $x^m = e$.
($\Leftarrow$) If n is a multiple of m, then $n = km$ where $k \in \mathbb{Z}$
$\therefore$ $x^n = x^{km} = (x^m)^k = e^k = e$
($\Rightarrow$) If $x^n = e$ for $n \in \mathbb{Z}$, we suppose that n is not a multiple of m.
Then $n = km + r$ {Division algorithm}
where $0 < r < m$
Then $e = x^n$
$= x^{km+r}$
$= x^{km}x^r$
$= (x^m)^k x^r$
$= e^k x^r$
$= ex^r$
$= x^r,\ 0 < r < m$
This is a contradiction as m is the smallest positive integer such that $x^m = e$
$\therefore$ the supposition is false
$\therefore$ $x^n = e \Rightarrow n$ is a multiple of m.
Hence $x^n = e \Leftrightarrow n$ is a multiple of m.

5 **a** ($\Rightarrow$) Suppose g has finite order m
$\therefore$ $g^m = e$ where m is the smallest positive integer for which this is true.
We now multiply on the left by g^{-1}, m times
$\therefore$ $(g^{-1})^m g^m = (g^{-1})^m e$
$\therefore$ $(g^m)^{-1} g^m = (g^{-1})^m$
{By **Theorem 8**, $(g^{-1})^m = (g^m)^{-1}$}
$\therefore$ $e = (g^{-1})^m$
and $(g^{-1})^i = g^m g^{-i}$ $\{g^m = e\}$
$= g^{m-i}$
$\neq e$ $\{i = 1, 2, 3, ..., m-1\}$
$\therefore$ g^{-1} has order m.
Thus g^{-1} has finite order and $\left|g^{-1}\right| = |g|$.
($\Leftarrow$) Similarly $\left|g^{-1}\right| = m \Rightarrow |g| = m$
{by reversing the roles of g and g^{-1}}

b Proved in **a**.

c Suppose $(fg)^n = e$
$\Rightarrow \underbrace{(fg)(fg)(fg)....(fg)}_{n \text{ of these}} = e$
Multiplying on the left by f^{-1} and on the right by g^{-1}
$\Rightarrow f^{-1}f\underbrace{(gf)(gf)(gf)....(gf)}_{n-1 \text{ of these}}gg^{-1} = f^{-1}eg^{-1}$
$\Rightarrow e(gf)^{n-1}e = f^{-1}g^{-1}$
$\Rightarrow (gf)^{n-1} = f^{-1}g^{-1}$
$\Rightarrow (gf)(gf)^{n-1} = gff^{-1}g^{-1}$
$\Rightarrow (gf)^n = geg^{-1}$
$= gg^{-1} = e$
By reversing the roles of f and g, we can show that $(gf)^n = e \Rightarrow (fg)^n = e$.
$\therefore$ $(fg)^n = e \Leftrightarrow (gf)^n = e$ and so $|fg| = |gf|$.
$\therefore$ if $|fg| = m$, then $|gf| = m$.

EXERCISE F.1

1 **a** $\begin{pmatrix} 1 & 2 & 3 & 4 \\ 3 & 4 & 2 & 1 \end{pmatrix}$ **b** $\begin{pmatrix} 1 & 2 & 3 & 4 \\ 4 & 1 & 2 & 3 \end{pmatrix}$

c $\begin{pmatrix} 1 & 2 & 3 & 4 \\ 1 & 2 & 3 & 4 \end{pmatrix}$ **d** $\begin{pmatrix} 1 & 2 & 3 & 4 \\ 2 & 4 & 1 & 3 \end{pmatrix}$

2 **a** $\begin{pmatrix} 1 & 2 & 3 & 4 \\ 2 & 4 & 1 & 3 \end{pmatrix}$ **b** $\begin{pmatrix} 1 & 2 & 3 & 4 \\ 2 & 1 & 4 & 3 \end{pmatrix}$

c $\left[\begin{pmatrix} 1 & 2 & 3 & 4 \\ 3 & 4 & 2 & 1 \end{pmatrix}\begin{pmatrix} 1 & 2 & 3 & 4 \\ 2 & 4 & 1 & 3 \end{pmatrix}\right]^{-1}$
$= \begin{pmatrix} 1 & 2 & 3 & 4 \\ 4 & 1 & 3 & 2 \end{pmatrix}^{-1}$
$= \begin{pmatrix} 1 & 2 & 3 & 4 \\ 2 & 4 & 3 & 1 \end{pmatrix}$

3 $(qp) * (p^{-1}q^{-1})$ and $(p^{-1}q^{-1}) * (qp)$

$= qpp^{-1}q^{-1}$	$= p^{-1}q^{-1}qp$
$= qeq^{-1}$	$= p^{-1}ep$
$= qq^{-1}$	$= p^{-1}p$
$= e$	$= e$

Thus $(qp) * (p^{-1}q^{-1}) = (p^{-1}q^{-1})(qp) = e$
$\therefore$ qp and $p^{-1}q^{-1}$ are inverses.
Hence $(qp)^{-1} = p^{-1}q^{-1}$

4 **a** $pq = r$

$\therefore\ pqq^{-1} = rq^{-1}$

$\therefore\ pe = rq^{-1}$

$\therefore\ p = rq^{-1}$

$$\therefore\ p = \begin{pmatrix}1&2&3&4\\2&1&4&3\end{pmatrix}\begin{pmatrix}1&2&3&4\\3&1&2&4\end{pmatrix}^{-1} = \begin{pmatrix}1&2&3&4\\2&1&4&3\end{pmatrix}\begin{pmatrix}1&2&3&4\\2&3&1&4\end{pmatrix} = \begin{pmatrix}1&2&3&4\\1&4&2&3\end{pmatrix}$$

b $$p = \begin{pmatrix}1&2&3&4\\2&4&1&3\end{pmatrix}\begin{pmatrix}1&2&3&4\\2&3&1&4\end{pmatrix}^{-1} = \begin{pmatrix}1&2&3&4\\2&4&1&3\end{pmatrix}\begin{pmatrix}1&2&3&4\\3&1&2&4\end{pmatrix} = \begin{pmatrix}1&2&3&4\\1&2&4&3\end{pmatrix}$$

5 **a** $S = \{A, B, C, D\}$

	A	B	C	D
A	A	B	C	D
B	B	C	D	A
C	C	D	A	B
D	D	A	B	C

Closure: All elements of the table are from S $\therefore$ S is closed under composition of permutations.

Associative: On checking, the operation is associative.

Identity: The identity is A.

Inverse: The inverse of A is A, B is D, C is C, and D is B.

Hence S is a group under permutation composition.

b $S = \{A, B, C, D\}$

	A	B	C	D
A	A	B	C	D
B	B	A	D	C
C	C	D	A	B
D	D	C	B	A

Closure: All elements of the table are from S $\therefore$ S is closed under the composition of purmutations.

Associative: On checking, the operation is associative.

Identity: The identity is A.

Inverse: Each element is its own inverse. That is, $A^{-1} = A$, $B^{-1} = B$, $C^{-1} = C$, $D^{-1} = D$.

Hence S under the operation is a group.

6 **a** $p = \begin{pmatrix}1&2&3&4\\4&3&2&1\end{pmatrix} \neq e$

$$p^2 = \begin{pmatrix}1&2&3&4\\4&3&2&1\end{pmatrix}\begin{pmatrix}1&2&3&4\\4&3&2&1\end{pmatrix} = \begin{pmatrix}1&2&3&4\\1&2&3&4\end{pmatrix} = e$$

$\therefore$ p has order 2.

b $q = \begin{pmatrix}1&2&3&4\\4&1&3&2\end{pmatrix} \neq e$

$$q^2 = \begin{pmatrix}1&2&3&4\\4&1&3&2\end{pmatrix}\begin{pmatrix}1&2&3&4\\4&1&3&2\end{pmatrix} = \begin{pmatrix}1&2&3&4\\2&4&3&1\end{pmatrix}$$

$\therefore\ q^2 \neq e$

$$q^3 = q^2q = \begin{pmatrix}1&2&3&4\\2&4&3&1\end{pmatrix}\begin{pmatrix}1&2&3&4\\4&1&3&2\end{pmatrix} = \begin{pmatrix}1&2&3&4\\1&2&3&4\end{pmatrix} = e$$

$\therefore$ q has order 3.

c $r = \begin{pmatrix}1&2&3&4&5\\4&5&1&3&2\end{pmatrix} \neq e$

$$r^2 = \begin{pmatrix}1&2&3&4&5\\4&5&1&3&2\end{pmatrix}\begin{pmatrix}1&2&3&4&5\\4&5&1&3&2\end{pmatrix} = \begin{pmatrix}1&2&3&4&5\\3&2&4&1&5\end{pmatrix} \neq e$$

$$r^3 = r^2r = \begin{pmatrix}1&2&3&4&5\\3&2&4&1&5\end{pmatrix}\begin{pmatrix}1&2&3&4&5\\4&5&1&3&2\end{pmatrix} = \begin{pmatrix}1&2&3&4&5\\1&5&3&4&2\end{pmatrix} \neq e$$

$$r^4 = r^3r = \begin{pmatrix}1&2&3&4&5\\1&5&3&4&2\end{pmatrix}\begin{pmatrix}1&2&3&4&5\\4&5&1&3&2\end{pmatrix} = \begin{pmatrix}1&2&3&4&5\\4&2&1&3&5\end{pmatrix} \neq e$$

$$r^5 = r^4r = \begin{pmatrix}1&2&3&4&5\\4&2&1&3&5\end{pmatrix}\begin{pmatrix}1&2&3&4&5\\4&5&1&3&2\end{pmatrix} = \begin{pmatrix}1&2&3&4&5\\3&5&4&1&2\end{pmatrix} \neq e$$

$$r^6 = r^5r = \begin{pmatrix}1&2&3&4&5\\3&5&4&1&2\end{pmatrix}\begin{pmatrix}1&2&3&4&5\\4&5&1&3&2\end{pmatrix} = \begin{pmatrix}1&2&3&4&5\\1&2&3&4&5\end{pmatrix} = e$$

$\therefore$ r has order 6.

d $e^1 = e$ $\therefore$ e has order 1.

e $s = \begin{pmatrix}1&2&3&4\\4&1&2&3\end{pmatrix} \neq e$

$$s^2 = \begin{pmatrix}1&2&3&4\\4&1&2&3\end{pmatrix}\begin{pmatrix}1&2&3&4\\4&1&2&3\end{pmatrix} = \begin{pmatrix}1&2&3&4\\3&4&1&2\end{pmatrix} \neq e$$

$$s^3 = s^2s = \begin{pmatrix}1&2&3&4\\3&4&1&2\end{pmatrix}\begin{pmatrix}1&2&3&4\\4&1&2&3\end{pmatrix} = \begin{pmatrix}1&2&3&4\\2&3&4&1\end{pmatrix} \neq e$$

$$s^4 = s^3s = \begin{pmatrix}1&2&3&4\\2&3&4&1\end{pmatrix}\begin{pmatrix}1&2&3&4\\4&1&2&3\end{pmatrix} = \begin{pmatrix}1&2&3&4\\1&2&3&4\end{pmatrix} = e$$

$\therefore$ s has order 4.

7 The elements of S_4 are:

$$\begin{pmatrix}1&2&3&4\\1&2&3&4\end{pmatrix}\begin{pmatrix}1&2&3&4\\1&2&4&3\end{pmatrix}\begin{pmatrix}1&2&3&4\\1&3&2&4\end{pmatrix}$$

$$\begin{pmatrix}1&2&3&4\\1&3&4&2\end{pmatrix}\begin{pmatrix}1&2&3&4\\1&4&2&3\end{pmatrix}\begin{pmatrix}1&2&3&4\\1&4&3&2\end{pmatrix}$$

$$\begin{pmatrix}1&2&3&4\\2&1&3&4\end{pmatrix}\begin{pmatrix}1&2&3&4\\2&1&4&3\end{pmatrix}\begin{pmatrix}1&2&3&4\\2&3&1&4\end{pmatrix}$$

$$\begin{pmatrix}1&2&3&4\\2&3&4&1\end{pmatrix}\begin{pmatrix}1&2&3&4\\2&4&1&3\end{pmatrix}\begin{pmatrix}1&2&3&4\\2&4&3&1\end{pmatrix}$$

$$\begin{pmatrix}1&2&3&4\\3&1&2&4\end{pmatrix}\begin{pmatrix}1&2&3&4\\3&1&4&2\end{pmatrix}\begin{pmatrix}1&2&3&4\\3&2&1&4\end{pmatrix}$$

$\begin{pmatrix}1&2&3&4\\3&2&4&1\end{pmatrix}\begin{pmatrix}1&2&3&4\\3&4&1&2\end{pmatrix}\begin{pmatrix}1&2&3&4\\3&4&2&1\end{pmatrix}$

$\begin{pmatrix}1&2&3&4\\4&1&2&3\end{pmatrix}\begin{pmatrix}1&2&3&4\\4&1&3&2\end{pmatrix}\begin{pmatrix}1&2&3&4\\4&2&1&3\end{pmatrix}$

$\begin{pmatrix}1&2&3&4\\4&2&3&1\end{pmatrix}\begin{pmatrix}1&2&3&4\\4&3&1&2\end{pmatrix}\begin{pmatrix}1&2&3&4\\4&3&2&1\end{pmatrix}$

8 $S = \{1, 2\}$

a Let $e = \begin{pmatrix}1&2\\1&2\end{pmatrix}$, $a = \begin{pmatrix}1&2\\2&1\end{pmatrix}$

These are the elements of S_2.

b Cayley table:

$*$	e	a
e	e	a
a	a	e

Closure: The table contains only elements a and e which are in S.
$\therefore$ S is closed under the operation.

Associative:

$(a*a)*a = e*a = a$ and $a*(a*a) = a*e = a$ ✓

$(a*a)*e = e*e = e$ and $a*(a*e) = a*a = e$ ✓

$(a*e)*a = a*a = e$ and $a*(e*a) = a*a = e$ ✓

$(e*a)*a = a*a = e$ and $e*(a*a) = e*e = e$ ✓

$(e*e)*a = e*a = a$ and $e*(e*a) = e*a = a$ ✓

$(e*a)*e = a*e = a$ and $e*(a*e) = e*a = a$ ✓

$(a*e)*e = a*e = a$ and $a*(e*e) = a*e = a$ ✓

$(e*e)*e = e*e = e$ and $e*(e*e) = e*e = e$ ✓

All 8 possibilities are checked.
So, the operation is associative on S_2.

Identity: The identity is e.

Inverse: $e^{-1} = e$ and $a^{-1} = a$

$\therefore$ S_2 is a group under composition of permutations.

c There is symmetry about the main diagonal
$\therefore$ the operation is commutative
$\therefore$ S_2 is Abelian.

EXERCISE F.2

1 **a** (1 4 3 2) **b** (1 6 4)(2 5 3) **c** (1 5)(2 3) **d** (1 6 2 3)(4 5) **e** (1 3) **f** (2 3 5)(4 6) **g** (1 6)(2 4)(3 5)

2 **a** (1 2 3 4)(1 5) = (1 5 2 3 4)

b (1 3)(1 2)(1 5) = (1 5 2 3)

c (1 2 3)(1 4 2 3) = (1 4 3 2)

d (1 6)(1 5)(1 4)(1 3)(1 2) = (1 2 3 4 5 6)

3 **a** $(1\ 3\ 2\ 4\ 5)^{-1}$
$= (5\ 4\ 2\ 3\ 1)$
$= (1\ 5\ 4\ 2\ 3)$
{writing 1 in position (1)}

b $[(1\ 3\ 2)(4\ 5)]^{-1}$
$= (4\ 5)^{-1}(1\ 3\ 2)^{-1}$ $\{(pq)^{-1} = q^{-1}p^{-1}\}$
$= (5\ 4)(2\ 3\ 1)$
$= (1\ 2\ 3)(4\ 5)$

c $[(1\ 3)(2\ 4\ 5)]^{-1}$
$= (2\ 4\ 5)^{-1}(1\ 3)^{-1}$
$= (5\ 4\ 2)(3\ 1)$
$= (1\ 3)(2\ 5\ 4)$

d $[(1\ 2\ 3)(1\ 4\ 5)]^{-1}$
$= (1\ 4\ 5)^{-1}(1\ 2\ 3)^{-1}$
$= (5\ 4\ 1)(3\ 2\ 1)$
$= (1\ 3\ 2\ 5\ 4)$

e $[(1\ 3)(1\ 4)(1\ 5)]^{-1}$
$= (1\ 5)^{-1}(1\ 4)^{-1}(1\ 3)^{-1}$
$= (5\ 1)(4\ 1)(3\ 1)$
$= (1\ 3\ 4\ 5)(2)$
$= (1\ 3\ 4\ 5)$

f $[(1\ 2\ 3)(1\ 5)]^{-1}$
$= (1\ 5)^{-1}(1\ 2\ 3)^{-1}$
$= (5\ 1)(3\ 2\ 1)$
$= (1\ 3\ 2\ 5)(4)$
$= (1\ 3\ 2\ 5)$

4 $p = (1\ 3\ 2\ 4)$, $q = (1\ 2\ 3\ 5)$

a $p^{-1} = (4\ 2\ 3\ 1)$
$= (1\ 4\ 2\ 3)$

b $q^{-1} = (5\ 3\ 2\ 1)$
$= (1\ 5\ 3\ 2)$

c $(pq)^{-1} = q^{-1}p^{-1}$
$= (1\ 5\ 3\ 2)(1\ 4\ 2\ 3)$
$= (1\ 4)(2)(3\ 5)$
$= (1\ 4)(3\ 5)$

d $q^{-1}pq = (1\ 5\ 3\ 2)(1\ 3\ 2\ 4)(1\ 2\ 3\ 5)$
$= (1\ 4\ 5\ 2)(3)$
$= (1\ 4\ 5\ 2)$

e $p^{-2}q^{-1} = (1\ 4\ 2\ 3)(1\ 4\ 2\ 3)(1\ 5\ 3\ 2)$
$= (1\ 5\ 4\ 3)(2)$
$= (1\ 5\ 4\ 3)$

f $pr^{-1} = q$
$\therefore$ $p^{-1}pr^{-1} = p^{-1}q$
$\therefore$ $er^{-1} = p^{-1}q$
$\therefore$ $(r^{-1})^{-1} = (p^{-1}q)^{-1}$
$\therefore$ $r = q^{-1}p$
$\therefore$ $r = (1\ 5\ 3\ 2)(1\ 3\ 2\ 4)$
$= (1\ 2\ 4\ 5\ 3)$

5 **a** Let $p = (1\ 4\ 3\ 2)$
$\therefore$ $p^2 = (1\ 4\ 3\ 2)(1\ 4\ 3\ 2)$
$= (1\ 3)(2\ 4)$
$p^3 = p^2p = (1\ 3)(2\ 4)(1\ 4\ 3\ 2)$
$= (1\ 2\ 3\ 4)$
$p^4 = p^3p = (1\ 2\ 3\ 4)(1\ 4\ 3\ 2)$
$= (1)(2)(3)(4)$
$= e$ $\therefore$ p has order 4

b Let $p = (1\ 2)(1\ 3\ 4)$
$= (1\ 3\ 4\ 2)$
$p^2 = (1\ 3\ 4\ 2)(1\ 3\ 4\ 2)$
$= (1\ 4)(2\ 3)$

$p^3 = p^2p = (1\ 4)(2\ 3)(1\ 3\ 4\ 2)$
$= (1\ 2\ 4\ 3)$
$p^4 = p^3p = (1\ 2\ 4\ 3)(1\ 3\ 4\ 2)$
$= (1)(2)(3)(4)$
$= e$ $\therefore$ p has order 4

c Let $p = (1\ 2\ 3)(2\ 3\ 4)$
$= (1\ 2)(3\ 4)$
$p^2 = (1\ 2)(3\ 4)(1\ 2)(3\ 4)$
$= (1)(2)(3)(4)$
$= e$ $\therefore$ p has order 2

6 **a** Let $p = (1\ 2\ 3\ 4)(1\ 5)$
$= (1\ 5\ 2\ 3\ 4)$ {from **2 a**}
$\therefore$ p has order 5 {**Theorem 12**}

b Let $p = (1\ 3)(1\ 2)(1\ 5)$
$= (1\ 5\ 2\ 3)$ {from **2 b**}
$\therefore$ p has order 4 {**Theorem 12**}

c Let $p = (1\ 2\ 3)(1\ 4\ 2\ 3)$
$= (1\ 4\ 3\ 2)$ {from **2 c**}
$\therefore$ p has order 4 {**Theorem 12**}

d Let $p = (1\ 6)(1\ 5)(1\ 4)(1\ 3)(1\ 2)$
$= (1\ 2\ 3\ 4\ 5\ 6)$ {from **2 d**}
$\therefore$ p has order 6 {**Theorem 12**}

7 **a** $p = (1\ 3\ 2\ 4)$ is a cycle of length 4
$\therefore$ $p^4 = e$
$p^5 = p^4p = ep = p$
$p^6 = p^4p^2 = ep^2 = p^2$
$\therefore$ the Cayley table is:

	e	p	p^2	p^3
e	e	p	p^2	p^3
p	p	p^2	p^3	e
p^2	p^2	p^3	e	p
p^3	p^3	e	p	p^2

Closure: As all elements of the table are in G, G is closed under the operation.

Associative: As permutations are functions and functions are associative under composition, then, permutations are associative under composition.

Identity: $e = \begin{pmatrix} 1 & 2 & 3 & 4 \\ 1 & 2 & 3 & 4 \end{pmatrix}$ is the identity as $ep^i = p^ie = p^i$ for $i = 0, 1, 2, 3$.

Inverse: $e^{-1} = e$, p and p^3 are inverses {as $pp^3 = p^3p = e$}
and $p^2p^2 = e$
$\therefore$ p^2 is its own inverse.
$\therefore$ $e^{-1} = e$, $p^{-1} = p^3$, $(p^2)^{-1} = p^2$, and $(p^3)^{-1} = p$.

Hence, G is a group.

b **i** $q = p^2$
$= (1\ 3\ 2\ 4)(1\ 3\ 2\ 4)$
$= (1\ 2)(3\ 4)$

ii $p^{159} = (p^4)^{39}p^3$
$= e^{39}p^3$
$= ep^3$
$= p^3$
$= p^2p$
$= (1\ 2)(3\ 4)(1\ 3\ 2\ 4)$
$= (1\ 4\ 2\ 3)$

iii $q^{159} = (p^2)^{159}$
$= p^{318}$
$= (p^4)^{79}p^2$
$= e^{79}p^2$
$= ep^2$
$= p^2$
$= (1\ 2)(3\ 4)$

iv $p^{508} = (p^4)^{127}$
$= e^{127}$
$= e$

8 **a** order $=$ LCM $(3, 4, 5) = 60$
b order $=$ LCM $(2, 4, 5) = 20$
c order $=$ LCM $(2, 4, 8) = 8$
d order $=$ LCM $(7, 3, 6) = 42$

EXERCISE F.3

1 **a** e is an anticlockwise rotation through $0°$ about O.
r is an anticlockwise rotation through $90°$ about O.
r^2 is an anticlockwise rotation through $180°$ about O.
r^3 is an anticlockwise rotation through $270°$ about O.

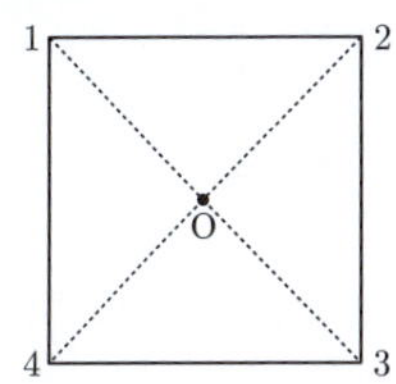

We notice that $r^4 = e$, and so the order of rotational symmetry is 4.

b It has 4 lines of symmetry 1 - 3, A - C, 2 - 4, D - B.

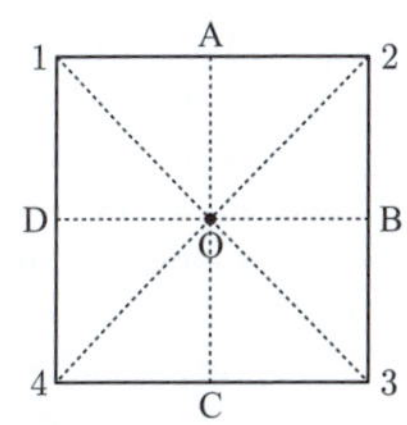

c There are 8 symmetries of the square
$e = (1)(2)(3)(4)$
$r = (1\ 4\ 3\ 2)$
$r^2 = (1\ 3)(2\ 4)$
$r^3 = (1\ 2\ 3\ 4)$
$l_{13} = (2\ 4)$
$l_{24} = (1\ 3)$
$l_{AC} = (1\ 2)(3\ 4)$
$l_{DB} = (1\ 4)(2\ 3)$

d $|e| = 1$, $|r| = \left|r^3\right| = 4$, $\left|r^2\right| = 2$, $|l_{13}| = |l_{24}| = 2$, $|l_{AC}| = |l_{DB}| = 2$

e $|D_4| = 8$ as there are 8 symmetries.

f $|S_4| = 4! = 24$

2 **a** The 4 symmetries are:
e, an anticlockwise rotation about O through $0°$;
r, an anticlockwise rotation about O through $180°$;

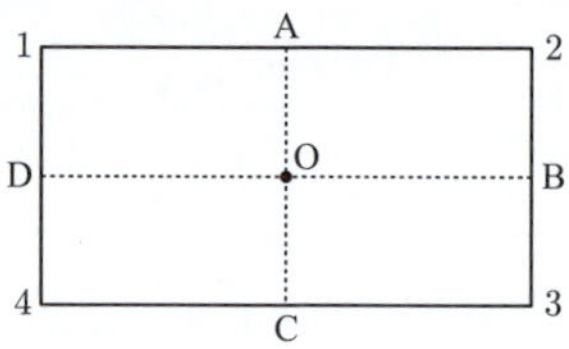

$R_1 = l_{AC}$, a reflection in the line [AC];
$R_2 = l_{DB}$, a reflection in the line [DB].

$$e = (1)(2)(3)(4)$$
$$r = (1\ 3)(2\ 4)$$
$$R_1 = (1\ 2)(3\ 4)$$
$$R_2 = (1\ 4)(2\ 3)$$

b $|e| = 1$, $|r| = |R_1| = |R_2| = 2$

We notice that $r^2 = R_1^{\,2} = R_2^{\,2} = e$

c The Cayley table is:

	e	r	R_1	R_2
e	e	r	R_1	R_2
r	r	e	R_2	R_1
R_1	R_1	R_2	e	r
R_2	R_2	R_1	r	e

Closure: Each element of the Cayley table is in $G = \{e, r, R_1, R_2\}$
$\therefore$ G is closed under the operation.

Associative: Follows from the associativity of composition of permuations (which follows from the associativity of bijective functions on a set).

Identity: $ea = ae = a$ for all $a \in G$
$\therefore$ e is the identity.

Inverse: $e^{-1} = e$, $r^{-1} = r$, $R_1^{\,-1} = R_1$, $R_2^{\,-1} = R_2$.

d G is an Abelian group as the Cayley table is symmetric about the main diagonal.

3 **a** e is an anticlockwise rotation about O through $0°$.
$e = (1)(2)(3)(4)(5)$
r_1 is an anticlockwise rotation about O through $72°$ and $r_1 = (1\ 5\ 4\ 3\ 2)$.
r_2 is an anticlockwise rotation about O through $144°$ and $r_2 = (1\ 4\ 2\ 5\ 3)$.

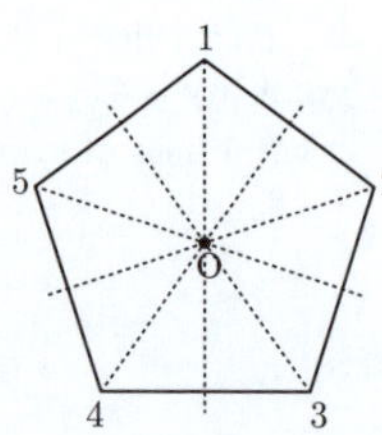

r_3 is an anticlockwise rotation about O through $216°$ and $r_3 = (1\ 3\ 5\ 2\ 4)$.
r_4 is an anticlockwise rotation about O through $288°$ and $r_4 = (1\ 2\ 3\ 4\ 5)$.

R_1 is a reflection in line O_1 and $R_1 = (2\ 5)(3\ 4)$.
R_2 is a reflection in line O_2 and $R_2 = (1\ 3)(4\ 5)$.
R_3 is a reflection in line O_3 and $R_3 = (1\ 5)(2\ 4)$.
R_4 is a reflection in line O_4 and $R_4 = (1\ 2)(3\ 5)$.
R_5 is a reflection in line O_5 and $R_5 = (1\ 4)(2\ 3)$.

b $|e| = 1$, $|r_1| = |r_2| = |r_3| = |r_4| = 5$,
$|R_1| = |R_2| = |R_3| = |R_4| = |R_5| = 2$

c $|D_5| = 10$

d $|S_5| = 5! = 120$

4 The statement is true.
We need to find an element in S_n which is not in D_n.
For $n > 3$ there is no symmetry of the figure which fixes two vertices and permutes the remaining vertices in a cycle.
For example, $(1)(2)(3\ 4\ 5\ 6\ ....\ n)$ is in S_n, but not in D_n.

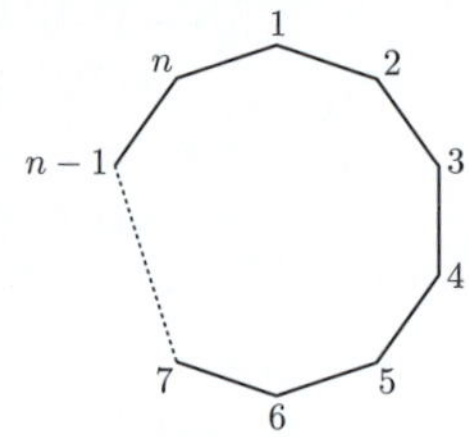

EXERCISE G

1 **a** $\{\mathbb{R}\setminus\{0\}, \times\}$ is a group with identity $e = 1$.
$\mathbb{R}^+ \subseteq \mathbb{R}\setminus\{0\}$ and $\mathbb{R}^+$ is non-empty as $1 \in \mathbb{R}^+$.
Suppose $a, b \in \mathbb{R}^+$.
$b^{-1} = \dfrac{1}{b}$ {as $b \times \dfrac{1}{b} = 1$, the identity}
and $ab^{-1} = a \times \dfrac{1}{b} = \dfrac{a}{b} \in \mathbb{R}^+$.
Hence, by the subgroup test, $H < G$.

b $\{\mathbb{R}^+, \times\}$ is a group with identity 1.
$\mathbb{Q}^+ \subseteq \mathbb{R}^+$ and $\mathbb{Q}^+$ is non-empty as $1 \in \mathbb{Q}^+$.
Suppose $\dfrac{a}{b}, \dfrac{c}{d} \in \mathbb{Q}^+$ then $a, b, c, d \in \mathbb{Z}^+$.

Now $\left(\dfrac{c}{d}\right)^{-1} = \dfrac{d}{c}$ $\left\{\text{since } \dfrac{c}{d} \times \dfrac{d}{c} = 1\right\}$

and so $\dfrac{a}{b} \times \left(\dfrac{c}{d}\right)^{-1} = \dfrac{a}{b}\dfrac{d}{c} = \dfrac{ad}{bc} > 0$
and $bc \neq 0$ {as $b \neq 0$, $c \neq 0$}.

Thus $\dfrac{a}{b} \times \left(\dfrac{c}{d}\right)^{-1} \in \mathbb{Q}^+$.
Hence, by the subgroup test, $H < G$.

c $\{\mathbb{Z}\setminus\{0\}, \times\}$ is not a group with identity 1.
For example, 2 has no inverse in $\mathbb{Z}\setminus\{0\}$.
$\therefore$ $H \not< G$.

d $\{\mathbb{Z}_6, +_6\}$ is a finite group with identity 0.
H has Cayley table:

$+_6$	0	2	4
0	0	2	4
2	2	4	0
4	4	0	2

Clearly, H is closed under $+_6$
$\therefore$ $a +_6 b \in H$ for all $a, b \in H$
$\therefore$ by the subgroup test for finite groups, $H < G$.

e $\{\mathbb{Z}, +\}$ is a group with identity 0.
$H = \{$all multiples of 4$\}$
If $4m, 4n \in H$, then $(4n)^{-1} = -4n$ and
$4m + (4n)^{-1} = 4m - 4n$
$= 4(m - n) \in H$ {as $m - n \in \mathbb{Z}$}
$\therefore$ by the subgroup test, $H < G$.

f From **e**, $\{G, +\}$ is a group with identity 0.
For $8m, 8n \in H$, $(8n)^{-1} = -8n$ and
$8m + (8n)^{-1} = 8m - 8n$
$= 8(m - n) \in H$ {as $m - n \in \mathbb{Z}$}
$\therefore$ by the subgroup test, $H < G$.

g D_3 has Cayley table:
$\therefore$ D_3 is a finite group with identity e.
$H \subseteq D_3$ and H is non-empty.

$*$	e	r	r^2	x	y	z
e	e	r	r^2	x	y	z
r	r	r^2	e	z	x	y
r^2	r^2	e	r	y	z	x
x	x	y	z	e	r	r^2
y	y	z	x	r^2	e	r
z	z	x	y	r	r^2	e

The Cayley table for H, where $r^3 = e$, is:

$*$	e	r	r^2
e	e	r	r^2
r	r	r^2	e
r^2	r^2	e	r

H is closed under $* =$ composition of transformations.
$\therefore$ by the subgroup test for finite groups, $H < G$.

h $H = \{e, a, b, c\}$ has Cayley table:

$*$	e	a	b	c
e	e	a	b	c
a	a	e	c	b
b	b	c	e	a
c	c	b	a	e

For example, $c * b = (1\ 4)(2\ 3) * (1\ 3)(2\ 4)$
$= (1\ 2)(3\ 4)$
$= a$

S_4 is a finite group with identity e and $H \subseteq S_4$, and H is non-empty.
H is closed under $*$
$\therefore$ by the subgroup test for finite groups, $H < G$.
{**Note:** H is the group of symmetries of a rectangle.}

i $G = \{\mathbb{C} \setminus \{0\}, \times\}$ is a group with identity 1.
$H = U_n \subseteq \mathbb{C} \setminus \{0\}$ and U_n is non-empty as $1 \in U_n$.
Since $zz^{n-1} = z^{n-1}z = 1$, $z^{-1} = z^{n-1}$ for all z in U_n.
Now suppose $z_1, z_2 \in U_n$, then

$(z_1 {z_2}^{-1})^n = (z_1 {z_2}^{n-1})^n$
$= {z_1}^n ({z_2}^n)^{n-1}$
$= 1 \times 1^{n-1} = 1$

$\therefore$ $z_1 {z_2}^{-1} \in U_n$
Thus, by the subgroup test, $H < G$.

j $\{\mathbb{C}, +\}$ is a group with identity 0.
$H \subseteq \mathbb{C}$ and H is non-empty as for example $1 + \sqrt{5}i \in H$.
Consider $a_1 + ib_1\sqrt{5}$ and $a_2 + ib_2\sqrt{5} \in H$
Now $a_2 + ib_2\sqrt{5} + (-a_2 - ib_2\sqrt{5}) = 0$
$\therefore$ $(a_2 + ib_2\sqrt{5})^{-1} = -a_2 - ib_2\sqrt{5}$
and so $(a_1 + ib_1\sqrt{5}) + (a_2 + ib_2\sqrt{5})^{-1}$
$= a_1 + ib_1\sqrt{5} - a_2 - ib_2\sqrt{5}$
$= (a_1 - a_2) + i(b_1 - b_2)\sqrt{5}$
which is $\in H$ as $a_1 - a_2, b_1 - b_2 \in \mathbb{R}$
Thus, by the subgroup test, $H < G$.

2 **a** *Closure*: Suppose $(a, b), (c, d) \in S$ then
$(a, b) * (c, d) = (a + c, (-1)^c b + d) \in S$
as $a + c \in \mathbb{Z}$ and $(-1)^c b + d \in \mathbb{Z}$
$\therefore$ S is closed under $*$.

Associative: Let $(p, q), (r, s), (t, u)$ be in S.
$\therefore$ $[(p, q) * (r, s)] * (t, u)$
$= (p + r, (-1)^r q + s) * (t, u)$
$= (p + r + t, (-1)^t((-1)^r q + s) + u)$
$= (p + r + t, (-1)^{t+r} q + (-1)^t s + u)$
and $(p, q) * [(r, s) * (t, u)]$
$= (p, q) * (r + t, (-1)^t s + u)$
$= (p + r + t, (-1)^{r+t} q + (-1)^t s + u)$
$= [(p, q) * (r, s)] * (t, u)$
$\therefore$ $*$ is associative on S.

Identity: Suppose $(x, y) * (a, b) = (a, b)$
$\therefore$ $(x + a, (-1)^a y + b) = (a, b)$
$\therefore$ $x + a = a$ and $(-1)^a y + b = b$
$\therefore$ $x = 0$ and $(-1)^a y = 0$ for all a
$\therefore$ $x = 0$ and $y = 0$
Thus $(0, 0) * (a, b) = (a, b)$ for all $(a, b) \in S$.
Also $(a, b) * (0, 0)$
$= (a + 0, (-1)^0 b + 0)$
$= (a, b)$ for all $(a, b) \in S$
$\therefore$ $(0, 0)$ is the identity.

Inverse: Suppose $(x, y) * (a, b) = (0, 0)$
$\therefore$ $(x + a, (-1)^a y + b) = (0, 0)$
$\therefore$ $x + a = 0$ and $(-1)^a y + b = 0$
$\therefore$ $x = -a$ and $y = \dfrac{-b}{(-1)^a}$
$\therefore$ $y = (-1)^{1-a} b$
$\therefore$ each element (a, b) has a unique inverse $(-a, (-1)^{1-a} b)$.

Check: $(a, b) * (-a, (-1)^{1-a} b)$
$= (a - a, (-1)^a b + (-1)^{1-a} b)$
$= (0, (-1)^{-a}(b + (-1)b))$
$= (0, 0)$

Thus $\{S, *\}$ is a group.

b $\{S, *\}$ is not Abelian as, for example,

$(1, 3) * (1, 0)$	and $(1, 0) * (1, 3)$
$= (1 + 1, (-1)^1 3 + 0)$	$= (1 + 1, (-1)^1 0 + 3)$
$= (2, -3)$	$= (2, 3)$

$\therefore$ $(1, 3) * (1, 0) \neq (1, 0) * (1, 3)$
$\therefore$ $*$ is not commutative
$\therefore$ $\{S, *\}$ is not Abelian.

c **i** Suppose $(a_1, 0), (a_2, 0) \in H_1$
then from **a** $(a_2, 0)^{-1} = (-a_2, (-1)^{1-a_2} 0)$
$= (-a_2, 0)$
$\therefore$ $(a_1, 0) * (a_2, 0)^{-1}$
$= (a_1, 0) * (-a_2, 0)$
$= (a_1 - a_2, (-1)^{-a_2}(0) + 0)$
$= (a_1 - a_2, 0)$
where $a_1 - a_2 \in \mathbb{Z}$ since $a_1, a_2 \in \mathbb{Z}$.
$\therefore$ by the subgroup test, $H_1 < S$.

ii Suppose $(0, b_1), (0, b_2) \in H_2$
$\therefore$ $(0, b_2)^{-1} = (-0, (-1)^{1-0} b_2)$
$= (0, -b_2)$
$\therefore$ $(0, b_1) * (0, b_2)^{-1}$
$= (0, b_1) * (0, -b_2)$
$= (0 + 0, (-1)^0 b_1 + (-b_2))$
$= (0, b_1 - b_2)$
where $b_1 - b_2 \in \mathbb{Z}$ since $b_1, b_2 \in \mathbb{Z}$.
$\therefore$ by the subgroup test, $H_2 < S$.

iii As the identity $(0, 0) \notin H_3$, $H_3 \not< S$.

3 $\{H_1, *\}$ and $\{H_2, *\}$ are subgroups of $\{G, *\}$
$e \in H_1$ and $e \in H_2$ $\therefore$ $e \in H_1 \cap H_2$
$\therefore$ $H_1 \cap H_2 \neq \varnothing$
$\therefore$ $H_1 \cap H_2$ is non-empty.

Since $H_1 \subseteq G$ and $H_2 \subseteq G$ then $H_1 \cap H_2 \subseteq G$.
Suppose $a, b \in H_1 \cap H_2$
then $a * b^{-1} \in H_1$ as $H_1 < G$
and $a * b^{-1} \in H_2$ as $H_2 < G$
$\therefore\ a * b^{-1} \in H_1 \cap H_2$
Thus, by the subgroup test, $H_1 \cap H_2 < G$.

4 $\{G, *\}$ is a group with identity e.
Since $e * a = a * e = a$ then $e \in H$
$\therefore\ H \neq \varnothing$ and $H \subseteq G$.
Suppose $x, y \in H$ $\therefore\ y * a = a * y$
Now $y^{-1} * y * a = y^{-1} * a * y$
$\therefore\ e * a = y^{-1} * a * y$
$\therefore\ a = y^{-1} * a * y$
and so $a * y^{-1} = y^{-1} * a * y * y^{-1}$
$\therefore\ a * y^{-1} = y^{-1} * a * e$
$\therefore\ a * y^{-1} = y^{-1} * a$ .... (1)
$\therefore\ y^{-1} \in H$
Consider $x * y^{-1}$
Now $x * y^{-1} * a = x * a * y^{-1}$ {from (1)}
$= a * x * y^{-1}$ $\{x \in H\}$
$\therefore\ (x * y^{-1}) * a = a * (x * y^{-1})$
$\therefore\ x * y^{-1} \in H$
Thus for all $x, y \in H$, $x * y^{-1} \in H$.
Hence, by the subgroup test, $H < G$.

5 G is Abelian and $H < G$
$S = \{x \mid x \in G,\ x^2 \in H\}$
To prove: $S < G$
Proof: If e is the identity of G, $e^2 = e \in S$.
Thus $S \neq \varnothing$.
By definition, $S \subseteq G$.
Suppose $x, y \in S$, then $y^2 \in H$
$\therefore\ (y^2)^{-1} \in H$ {as $H < G$}
$\therefore\ y^{-2} \in H$
$\therefore\ (y^{-1})^2 \in H$
$\therefore\ y^{-1} \in S$ {as $y^{-1} \in G$ and $(y^{-1})^2 \in H$}
Consider $xy^{-1} \in G$ {G is closed}
Now $(xy^{-1})^2 = xy^{-1}xy^{-1}$
$= xxy^{-1}y^{-1}$ {G is Abelian}
$= x^2(y^{-1})^2$
which is in H as $x^2, (y^{-1})^2 \in H$ and H is a group.
$\therefore\ xy^{-1} \in S$
Thus, by the subgroup test, $S < G$.

6 **a** $\alpha = \frac{1}{\sqrt{2}} + \frac{1}{\sqrt{2}}i = \text{cis}\left(\frac{\pi}{4}\right)$
$\alpha^2 = \text{cis}\left(\frac{\pi}{2}\right) = i$
$\alpha^3 = \text{cis}\left(\frac{3\pi}{4}\right) = -\frac{1}{\sqrt{2}} + \frac{1}{\sqrt{2}}i$
$\alpha^4 = \text{cis}\,(\pi) = -1$
$\alpha^5 = \text{cis}\left(\frac{5\pi}{4}\right) = -\frac{1}{\sqrt{2}} - \frac{1}{\sqrt{2}}i$
$\alpha^6 = \text{cis}\left(\frac{3\pi}{4}\right) = -i$
$\alpha^7 = \text{cis}\left(\frac{7\pi}{4}\right) = \frac{1}{\sqrt{2}} - \frac{1}{\sqrt{2}}i$
$\alpha^8 = \text{cis}\,2\pi = 1$

Let $H = \{1, \alpha, \alpha^2, \alpha^3, \alpha^4, \alpha^5, \alpha^6, \alpha^7\}$
$\therefore\ \{H, \times\} \subseteq \{\mathbb{C}\setminus\{0\}, \times\}$, a group
1 is the identity of H and $\mathbb{C}\setminus\{0\}$
We notice that $\begin{cases} \alpha^i\alpha^{8-i} = \alpha^8 = 1 \\ \alpha^{8-i}\alpha^i = \alpha^8 = 1 \end{cases}$
$\therefore\ (\alpha^i)^{-1} = \alpha^{8-i}$
and if $\alpha^i, \alpha^j \in H$,
$\alpha^i(\alpha^j)^{-1} = \alpha^i\alpha^{8-j}$
$= \alpha^{8-j+i}$
$\in H$ $\{\alpha^9 = \alpha^8\alpha = \alpha,\ \alpha^{10} = \alpha^2,$ etc.$\}$
Thus, by the subgroup test, $H < G$
$\therefore\ H$ is a group.

b As $\alpha^8 = 1$ where 8 is the smallest integer where $\alpha^n = 1$, the order of H is 8.

c **i** $G = \{\mathbb{C}\setminus\{0\}, \times\}$
ii If $\beta = \text{cis}\left(\frac{\pi}{8}\right)$, then let $G = \{1, \beta, \beta^2, ..., \beta^{15}\}$
$\beta^2 = \alpha$
$\therefore\ H < G$

7 For $a, b \in H$, $a = x^n$ and $b = x^m$ for some $n, m \in \mathbb{Z}^+$
$\therefore\ ab = x^n x^m = x^{n+m}$
Thus $ab \in H$ {as $n + m \in \mathbb{Z}^+$}
$\therefore\ H$ is closed under $*$.
So, by the subgroup test for finite groups, $H < G$
$\therefore\ \{H, *\}$ is a group.

EXERCISE H

1 $\mathbb{Z}_{12} = \{0, 1, 2, 3, 4, 5, 6, 7, 8, 9, 10, 11\}$
a $1+1+1+1+1+1+1+1+1+1+1+1 = 12 \equiv 0$
$\therefore\ |1| = 12$
$2+2+2+2+2+2 = 12 \equiv 0$
$\therefore\ |2| = 6$
$3+3+3+3 = 12 \equiv 0$
$\therefore\ |3| = 4$
$4+4+4 = 12 \equiv 0$
$\therefore\ |4| = 3$
$5+5+5+5+5+5+5+5+5+5+5+5 = 60 \equiv 0$
$\therefore\ |5| = 12$
etc. with $|6| = 2$
$|7| = 12$
$|8| = 3$
$|9| = 4$
$|10| = 6$
$|11| = 12$

b **i** $\{\langle 4\rangle, +_{12}\}$ where $\langle 4\rangle = \{0, 4, 8\}$
ii $\{\langle 3\rangle, +_{12}\}$ where $\langle 3\rangle = \{0, 3, 6, 9\}$

c Let $\langle 2\rangle = \{0, 2, 4, 6, 8, 10\}$.
$\langle 2\rangle$ is a non-empty subset of $\mathbb{Z}_{12}$, and since $\{\mathbb{Z}_{12}, +_{12}\}$ is a finite group we need to show $\langle 2\rangle$ is closed under $+_{12}$.

$+_{12}$	0	2	4	6	8	10
0	0	2	4	6	8	10
2	2	4	6	8	10	0
4	4	6	8	10	0	2
6	6	8	10	0	2	4
8	8	10	0	2	4	6
10	10	0	2	4	6	8

As each element in the Cayley table $\in \langle 2\rangle$
$\langle 2\rangle < \{\mathbb{Z}_{12}, +_{12}\}$.

d **i** As 1, 5, 7, and 11 are the elements of order 12, they are the generators of $\{\mathbb{Z}_{12}, +_{12}\}$.

ii As 2 and 10 are the elements of order 6, they are the generators of $\{\langle 2\rangle, +_{12}\}$.

2 **a** $\{\{2, 4, 6, 8\}, \times_{10}\}$

i As $2^1 = 2,\ 2^2 = 4,\ 2^3 = 8,\ 2^4 = 6$ the group is cyclic.

$\therefore\ \langle 2\rangle = \{\{2, 4, 6, 8\}, \times_{10}\}$ is cyclic.

ii 2 and 8 are generators.

b $\{\{1, 3, 5, 7\}, \times_8\}$

i As $3^2 = 1$
$5^2 = 1$
$7^2 = 1$
the group is not cyclic.

ii The subgroups are:
$\{\{1\}, \times_8\}$
$\{\{1, 3\}, \times_8\}$
$\{\{1, 5\}, \times_8\}$
$\{\{1, 7\}, \times_8\}$ (proper subgroups)
$\{\{1, 3, 5, 7\}, \times_8\}$

3 **a** $n = 3$, $G = \{1, 2\}$ under $\times_3$.
The identity 1 is not a generator.
$2^2 = 1$
$\therefore\ G = \langle 2\rangle$ and 2 is the unique generator.

b $n = 5$, $G = \{1, 2, 3, 4\}$ under $\times_5$.
The identity 1 is not a generator.
$2^1 = 2,\ 2^2 = 4,\ 2^3 = 3,\ 2^4 = 1$ ✓
$3^1 = 3,\ 3^2 = 4,\ 3^3 = 2,\ 3^4 = 1$ ✓
$4^1 = 4,\ 4^2 = 1$ ✗
$\therefore$ 2 and 3 are generators of G
$\therefore\ G = \langle 2\rangle = \langle 3\rangle$.

c $n = 7$, $G = \{1, 2, 3, 4, 5, 6\}$ under $\times_7$.
The identity 1 is not a generator.
$2^1 = 2,\ 2^2 = 4,\ 2^3 = 1$ ✗
$3^1 = 3,\ 3^2 = 2,\ 3^3 = 6,\ 3^4 = 4,\ 3^5 = 5,\ 3^6 = 1$ ✓
$4^1 = 4,\ 4^2 = 2,\ 4^3 = 1$ ✗
$5^1 = 5,\ 5^2 = 4,\ 5^3 = 6,\ 5^4 = 2,\ 5^5 = 3,\ 5^6 = 1$ ✓
$6^1 = 6,\ 6^2 = 1$ ✗
$|1| = 1,\ |2| = |4| = 3,\ |3| = |5| = 6,\ |6| = 2$
$\therefore$ 3 and 5 are generators of G
$\therefore\ G = \langle 3\rangle = \langle 5\rangle$.

d $n = 11$, $G = \{1, 2, 3, 4, 5, 6, 7, 8, 9, 10\}$ under $\times_{11}$.
The identity 1 is not a generator.
$2^1 = 2,\ 2^2 = 4,\ 2^3 = 8,\ 2^4 = 5,\ 2^5 = 10,\ 2^6 = 9,$
$2^7 = 7,\ 2^8 = 3,\ 2^9 = 6,\ 2^{10} = 1$ ✓
$3^1 = 3,\ 3^2 = 9,\ 3^3 = 5,\ 3^4 = 4,\ 3^5 = 1$ ✗
$4^1 = 4,\ 4^2 = 5,\ 4^3 = 9,\ 4^4 = 3,\ 4^5 = 1$ ✗
$5^1 = 5,\ 5^2 = 3,\ 5^3 = 4,\ 5^4 = 9,\ 5^5 = 1$ ✗
$6^1 = 6,\ 6^2 = 3,\ 6^3 = 7,\ 6^4 = 9,\ 6^5 = 10,\ 6^6 = 5,$
$6^7 = 8,\ 6^8 = 4,\ 6^9 = 2,\ 6^{10} = 1$ ✓
$7^1 = 7,\ 7^2 = 5,\ 7^3 = 2,\ 7^4 = 3,\ 7^5 = 10,\ 7^6 = 4,$
$7^7 = 6,\ 7^8 = 9,\ 7^9 = 8,\ 7^{10} = 1$ ✓
$8^1 = 8,\ 8^2 = 9,\ 8^3 = 6,\ 8^4 = 4,\ 8^5 = 10,\ 8^6 = 3,$
$8^7 = 2,\ 8^8 = 5,\ 8^9 = 7,\ 8^{10} = 1$ ✓
$9^1 = 9,\ 9^2 = 4,\ 9^3 = 3,\ 9^4 = 5,\ 9^5 = 1$ ✗
$10^1 = 10,\ 10^2 = 1$ ✗
$\therefore\ |1| = 1,\ |10| = 2,\ |3| = |4| = |5| = |9| = 5,$
$|2| = |6| = |7| = |8| = 10$
$\therefore$ 2, 6, 7, and 8 are generators of G and
$G = \langle 2\rangle = \langle 6\rangle = \langle 7\rangle = \langle 8\rangle$.

4 G is a finite cyclic group under $*$
g is a generator of G
$\therefore\ G = \langle g\rangle$ and $|g| = |G| = n$
Since G is a group, $g \in G \Rightarrow g^{-1} \in G$
and $\left|g^{-1}\right| = |g|$ {**Exercise E.2**, question **5 b**}
$\therefore\ \left|g^{-1}\right| = n$
$\therefore\ g^{-1}, (g^{-1})^2, (g^{-1})^3, (g^{-1})^4, ..., (g^{-1})^{n-1}, (g^{-1})^n$ are n distinct elements of G with $(g^{-1})^n = e$.
But G has only n distinct elements
$\therefore\ G = \left\langle g^{-1}\right\rangle$.

5 Let r be an anticlockwise rotation through $\dfrac{2\pi}{n}$ radians.
$r^i =$ a rotation through $\dfrac{2\pi i}{n}$ radians for
$i = 0, 1, 2, 3, 4, 5, ..., n - 1$
Thus, $R = \{e, r, r^2, r^3, r^4, ..., r^{n-1}\}$
Note: $r^n =$ a rotation through 2π radians
$= e$
By **Exercise G** question **7**, R is a group.
So, R is a cyclic subgroup of D_n, and r is a generator of R.

6 $\{G, *\}$ is an Abelian group of order 6.

a The identity is e and it contains α where $\alpha^2 = e$ and β where $\beta^3 = e$.
$\therefore\ G = \{e, \alpha, \beta, \beta^2, \alpha\beta, \alpha\beta^2\}$

b The Cayley table is:

$*$	e	α	β	β^2	$\alpha\beta$	$\alpha\beta^2$
e	e	α	β	β^2	$\alpha\beta$	$\alpha\beta^2$
α	α	e	$\alpha\beta$	$\alpha\beta^2$	β	β^2
β	β	$\alpha\beta$	β^2	e	$\alpha\beta^2$	α
β^2	β^2	$\alpha\beta^2$	e	β	α	$\alpha\beta$
$\alpha\beta$	$\alpha\beta$	β	$\alpha\beta^2$	α	β^2	e
$\alpha\beta^2$	$\alpha\beta^2$	β^2	α	$\alpha\beta$	e	β

c $|e| = 1$
$\alpha^2 = 1, \quad \therefore\ |\alpha| = 2$
$\beta^3 = 1, \quad \therefore\ |\beta| = 3$
$(\beta^2)^1 = \beta^2, \quad (\beta^2)^2 = \beta^4 = \beta$
$(\beta^2)^3 = (\beta^3)^2 = e^2 = e$
$\therefore\ \left|\beta^2\right| = 3$
$(\alpha\beta)^1 = \alpha\beta, \quad (\alpha\beta)^2 = \alpha^2\beta^2 = e\beta^2 = \beta^2$
$\therefore\ (\alpha\beta)^3 = \alpha^3\beta^3 = \alpha\alpha^2\beta^3 = \alpha ee = \alpha$
$(\alpha\beta)^4 = (\alpha\beta)^3(\alpha\beta) = \alpha\alpha\beta = \alpha^2\beta = e\beta = \beta$
$(\alpha\beta)^5 = (\alpha\beta)^4(\alpha\beta) = \beta\alpha\beta = \alpha\beta^2$
$(\alpha\beta)^6 = \left[(\alpha\beta)^3\right]^2 = \alpha^2 = e$
$\therefore\ |\alpha\beta| = 6$
From the table, $(\alpha\beta)(\alpha\beta^2) = e$
$\therefore\ \alpha\beta^2 = (\alpha\beta)^{-1}$
$\therefore\ \left|\alpha\beta^2\right| = |\alpha\beta| = 6$

d Since G is a finite group of order 6 and $|\alpha\beta| = 6$ and $\left|\alpha\beta^2\right| = 6$, G must be cyclic with generators $\alpha\beta$ and $\alpha\beta^2$
$\therefore\ G = \langle\alpha\beta\rangle = \left\langle\alpha\beta^2\right\rangle$.

e From the Cayley table $\{\{e, \beta, \beta^2\}, *\}$ is a unique subgroup of order 3.

$*$	e	β	β^2
e	e	β	β^2
β	β	β^2	e
β^2	β^2	e	β

f Since $\alpha\beta * \beta = \alpha\beta^2 \notin S$ where $S = \{e, \alpha, \beta, \alpha\beta\}$, S is not closed under $*$.
$\therefore \; S \not< G$.

7 **a** $i^1 = i, \; i^2 = -1, \; i^3 = -i, \; i^4 = 1$
$\therefore$ the order of $\langle i \rangle$ is 4.

b Let $\alpha = \frac{1}{2} + i\frac{\sqrt{3}}{2} = \text{cis}\left(\frac{\pi}{3}\right)$
then $\alpha^6 = \left[\text{cis}\left(\frac{\pi}{3}\right)\right]^6 = \text{cis}\,(2\pi) = 1$, and 6 is the least positive integer n such that $\alpha^n = 1$.
$\therefore$ the order of $\left\langle \frac{1}{2} + i\frac{\sqrt{3}}{2} \right\rangle$ is 6.

c Let $\alpha = \frac{\sqrt{3}}{2} + \frac{i}{2} = \text{cis}\left(\frac{\pi}{6}\right)$
then $\alpha^{12} = \left[\text{cis}\left(\frac{\pi}{6}\right)\right]^{12} = \text{cis}\,(2\pi) = 1$, and 12 is the least positive integer n such that $\alpha^n = 1$.
$\therefore$ the order of $\left\langle \frac{\sqrt{3}}{2} + \frac{i}{2} \right\rangle$ is 12.

d Let $\alpha = \sqrt{3} + i = 2\left(\frac{\sqrt{3}}{2} + \frac{i}{2}\right) = 2\,\text{cis}\left(\frac{\pi}{6}\right)$
$\alpha^n = 2^n\,\text{cis}\left(\frac{n\pi}{6}\right)$ and there is no $n \in \mathbb{Z}^+$ such that $\alpha^n = 1$.
$\therefore \; \left\langle \sqrt{3} + i \right\rangle$ has infinite order.

8 $G = \langle g \rangle$ is a cyclic group of order 12 with generator g.

a $\left\langle g^4 \right\rangle = \{g^4, g^8, g^{12} = e\}$
$\therefore \; \left\langle g^4 \right\rangle \subseteq G$

The Cayley table is:

$*$	e	g^4	g^8
e	e	g^4	g^8
g^4	g^4	g^8	e
g^8	g^8	e	g^4

and $a, b \in \left\langle g^4 \right\rangle \Rightarrow ab \in \left\langle g^4 \right\rangle$.
So, by the subgroup test for finite groups $\left\langle g^4 \right\rangle < G$, and $\left|\left\langle g^4 \right\rangle\right| = 3$.

b **i** $(g^2)^6 = 1$ and 6 is the least positive integer n such that $(g^2)^n = 1$.
$\therefore$ order of $\left\langle g^2 \right\rangle = 6$.

ii $(g^3)^4 = 1$ likewise
$\therefore$ order of $\left\langle g^3 \right\rangle = 4$.

iii $(g^6)^2 = 1$ likewise
$\therefore$ order of $\left\langle g^6 \right\rangle$ is 2.

EXERCISE I

1 **a** $f(x) = x^2$
$\therefore \; f(x+y) = (x+y)^2$
$= x^2 + 2xy + y^2$
$= f(x) + f(y) + 2xy$
$\therefore \; f(x+y) \neq f(x) + f(y)$ for all $x, y \in \{\mathbb{R}, +\}$
$\therefore \; f$ is not a homomorphism.

b $f(x+y) = 7(x+y)$
$= 7x + 7y$
$= f(x) + f(y)$ for all $x, y \in \{\mathbb{R}, +\}$
$\therefore \; f$ is a homomorphism.

i 0 is the identity in $\{\mathbb{R}, +\}$
$\therefore$ Ker$(f) = \{a \mid f(a) = 0\}$
$= \{a \mid 7a = 0\}$
$= \{0\}$
and $R(f) = \{f(a) \mid a \in \mathbb{R}\}$
$= \{7a \mid a \in \mathbb{R}\}$
$= \mathbb{R}$

ii Now since Ker$(f) = \{0\}$, the identity of the domain group, then f is one-to-one.
Since $R(f) = \mathbb{R}$, f is onto.
Thus f is an isomorphism.

c $f(x) = x^2$
$f(xy) = (xy)^2$
$= x^2y^2$ $\quad\{\times$ is commutative$\}$
$= f(x)f(y)$
$\therefore \; f$ is a homomorphism.

i The identity in $\{\mathbb{R}\setminus\{0\}, \times\}$ is 1
$\therefore$ Ker$(f) = \{a \mid f(a) = 1\}$
$= \{a \mid a^2 = 1\}$
$= \{1, -1\}$
and $R(f) = \{f(a) \mid a \in \mathbb{R}\setminus\{0\}\}$
$= \{a^2 \mid a \in \mathbb{R}\setminus\{0\}\}$
$= \mathbb{R}^+$

ii As $R(f) = \mathbb{R}^+ \neq \mathbb{R}\setminus\{0\}$, f is not onto.
$\therefore \; f$ is not an isomorphism.

d $f(x) = x^2$
$f(xy) = (xy)^2$
$= x^2y^2$ $\quad\{\times$ is commutative$\}$
$= f(x)f(y)$
$\therefore \; f$ is a homomorphism.

i The identity in $\{\mathbb{R}^+, \times\}$ is 1
$\therefore$ Ker$(f) = \{a \mid a \in \mathbb{R}^+$ and $f(a) = 1\}$
$= \{a \mid a \in \mathbb{R}^+$ and $a^2 = 1\}$
$= \{1\}$ $\quad\{-1 \notin \mathbb{R}^+\}$
and $R(f) = \{f(a) \mid a \in \mathbb{R}^+\}$
$= \{a^2 \mid a \in \mathbb{R}^+\}$
$= \mathbb{R}^+$

ii Ker$(f) = \{1\}$, the identity of the domain group $\{\mathbb{R}^+, \times\}$ $\quad\therefore \; f$ is one-to-one.
Also, $R(f) = \mathbb{R}^+$, $\quad\therefore \; f$ is onto.
Thus, as f is a homomorphism which is one-to-one and onto, f is an isomorphism.

e $f(x) = e^x$ $\qquad f(x+y) = e^{x+y}$
$= e^x e^y$
$= f(x)f(y)$
$\therefore \; f$ is a homomorphism.

i The identity of $\{\mathbb{R}, \times\}$ is 1
$\therefore$ Ker$(f) = \{a \mid a \in \mathbb{R}, \; f(a) = 1\}$
$= \{a \mid a \in \mathbb{R}, \; e^a = 1\}$
$= \{0\}$

and $R(f) = \{f(a) \mid a \in \mathbb{R}\}$
$= \{e^a \mid a \in \mathbb{R}\}$
$= \mathbb{R}^+$

ii $\text{Ker}(f) = \{0\}$, the identity of the domain group $\{\mathbb{R}, +\}$
$\therefore$ f is one-to-one.
Also, $R(f) = \mathbb{R}^+$, $\therefore$ f is onto.
Thus, as f is a homomorphism which is one-to-one and onto, f is an isomorphism.

f $f(x) = 5x$

$$\begin{aligned} f(x+y) &= 5(x+y) \\ &= 5x + 5y \\ &= f(x) + f(y) \end{aligned}$$

$\therefore$ f is a homomorphism.

i The identity of $\{5\mathbb{Z}, +\}$ is 0

$$\begin{aligned} \therefore\ \text{Ker}(f) &= \{a \mid a \in \mathbb{Z},\ f(a) = 0\} \\ &= \{a \mid a \in \mathbb{Z},\ 5a = 0\} \\ &= \{0\} \end{aligned}$$

and $R(f) = \{f(a) \mid a \in \mathbb{Z}\}$
$= \{5a \mid a \in \mathbb{Z}\}$
$= 5\mathbb{Z}$, all integer multiples of 5

ii $\text{Ker}(f) = \{0\}$, the identity of $\{\mathbb{Z}, +\}$
$\therefore$ f is one-to-one.
Also, $R(f) = 5\mathbb{Z}$, $\therefore$ f is onto.
Thus, as f is a homomorphism which is one-to-one and onto, f is an isomorphism.

g $f(x) = x \pmod 5$

$$\begin{aligned} f(a) +_5 f(b) &= a \pmod 5 +_5 b \pmod 5 \\ &= [a \pmod 5 + b \pmod 5] \pmod 5 \\ &= a + b \pmod 5 \\ &= f(a+b) \end{aligned}$$

f is a homomorphism.

i The identity of $\{\mathbb{Z}_5, +_5\}$ is 0

$$\begin{aligned} \therefore\ \text{Ker}(f) &= \{a \mid a \in \mathbb{Z},\ f(a) = 0\} \\ &= \{a \mid a \in \mathbb{Z},\ a \pmod 5 = 0\} \\ &= \{0, 5, 10, 15, ...\} \end{aligned}$$

and $R(f) = \{f(a) \mid a \in \mathbb{Z}\}$
$= \{a \pmod 5 \mid a \in \mathbb{Z}\}$
$= \mathbb{Z}_5$

ii As $\text{Ker}(f) \neq \{0\}$, f is not one-to-one.
Hence f is not an isomorphism.

h

$$\begin{aligned} f(\alpha\beta) &= (\alpha\beta)^2 \\ &= \alpha\beta\alpha\beta \\ &\neq \alpha\alpha\beta\beta \qquad \{\text{as } S_3 \text{ is not Abelian}\} \\ \therefore\ &\neq \alpha^2\beta^2 \\ \therefore\ &\neq f(\alpha)f(\beta) \end{aligned}$$

$\therefore$ f is not a homomorphism.

2 $f : \{P, +\} \to \{P, +\}$ and $f(p(x)) = p'(x)$

a For $p(x), q(x) \in P$

$$\begin{aligned} & f(p(x) + q(x)) \\ &= \frac{d}{dx}(p(x) + q(x)) \\ &= p'(x) + q'(x) \qquad \{\text{rule of differentiation}\} \\ &= f(p(x)) + f(q(x)) \end{aligned}$$

$\therefore$ f is a homomorphism.

b The identity of $\{P, +\}$ is the zero polynomial, 0 {which is actually $0x + 0$, etc.}

i

$$\begin{aligned} \text{Ker}(f) &= \{p(x) \mid p(x) \in P \text{ and } f(p(x)) = 0\} \\ &= \{p(x) \mid p(x) \in P \text{ and } p'(x) = 0\} \\ &= \{c \text{ where } c \in \mathbb{R}\} \end{aligned}$$

which is the set of all constant functions on $\mathbb{R}$.

ii

$$\begin{aligned} R(f) &= \{f(p(x)) \mid p(x) \in P\} \\ &= \{p'(x) \mid p(x) \in P\} \end{aligned}$$

Each polynomial $q(x)$ in P has an antiderivative $p(x) = \int q(x)\,dx$, where $p(x) \in P$ and $p'(x) = q(x)$.
$\therefore$ $R(f) = P$.

c $\text{Ker}(f) < \{P, +\}$, the domain group
$\therefore$ $\{\{p(x) = c,\ c \in \mathbb{R}\}, +\}$ forms an additive subgroup of $\{P, +\}$.
This is a proper subgroup of $\{P, +\}$ as the only improper subgroups are $\{\{0\}, +\}$ and $\{P, +\}$.

d f is *not* an isomorphism since, for example $f(3) = f(4) = 0$
for polynomials $p(x) = 3$ and $q(x) = 4 \in \{P, +\}$.

3 $f(G) = \{f(g) \mid g \in G\}$ for $f : \{(G, *) \to (H, \circ)\}$

a $f(G) = \{f(g) \mid g \in G\} = R(f)$
$\therefore$ $f(G)$ is the range of f.
By **Theorem 23**, $R(f) < H$ and $\therefore$ $f(G) < H$.
$\therefore$ $f(G)$ is a group.

b G is Abelian
$\Rightarrow$ $a * b = b * a$ for all $a, b \in G$
$\Rightarrow$ $f(a * b) = f(b * a)$
$\Rightarrow$ $f(a) \circ f(b) = f(b) \circ f(a)$ {as f is a homomorphism}
$\Rightarrow$ $f(G)$ is Abelian

c If $G = \langle g \rangle$ is cyclic with generator g then

$$\begin{aligned} G &= \{g^n \mid n \in \mathbb{Z}\} \\ &= \{e_G, g, g^2, g^3, ..., g^{-1}, g^{-2}, g^{-3}, ...\} \end{aligned}$$

and

$$\begin{aligned} f(G) &= \{f(e_G), f(g), f(g^2), ...\} \\ &= \{f(g^n) \mid n \in \mathbb{Z}\} \\ &= \{f(\underbrace{g * g * g * .... * g}_{n \text{ times}}) \mid n \in \mathbb{Z}\} \\ &= \{\underbrace{f(g) \circ f(g) \circ f(g) \circ .... \circ f(g)}_{n \text{ times}} \mid n \in \mathbb{Z}\} \\ &= \{(f(g))^n \mid n \in \mathbb{Z}\} \\ &= \langle f(g) \rangle \end{aligned}$$

$\therefore$ $f(G)$ is cyclic with generator $f(g) \in H$.

d The statement is false in general.
As $|g| = m$, $g^m = e_G$

$$\begin{aligned} &\Rightarrow f(g^m) = f(e_G) \\ &\Rightarrow f(\underbrace{g * g * g * .... * g}_{m \text{ times}}) = e_H \\ &\Rightarrow \underbrace{f(g) \circ f(g) \circ f(g) \circ .... \circ f(g)}_{m \text{ times}} = e_H \\ &\Rightarrow (f(g))^m = e_H \\ &\Rightarrow \text{order of } f(g) \text{ is a factor of } m \end{aligned}$$

Also $f : \{G, *\} \to \{G, *\}$ where $f(g) = e_G$ is a homomorphism and $|f(g)| = |e_G| = 1$ for all $g \in G$.
But $|g| \neq 1$ for all $g \in G$ if $G \neq \{e_G\}$.

EXERCISE J

1 $f : \{\{0, 1, 2\}, +_3\} \to \{\{1, 2, 4\}, \times_7\}$

$+_3$	0	1	2
0	0	1	2
1	1	2	0
2	2	0	1

$0 \mapsto 1$
$1 \mapsto 2$
$2 \mapsto 4$

$\times_7$	1	2	4
1	1	2	4
2	2	4	1
4	4	1	2

f is a bijection and for all $a, b \in G$, $f(a +_3 b) = f(a) \times_7 f(b)$.
As the Cayley tables have the same structure,
$\{\{0, 1, 2\}, +_3\} \cong \{\{1, 2, 4\}, \times_7\}$

2 If $\alpha = -\frac{1}{2} + i\frac{\sqrt{3}}{2} = \text{cis}\left(\frac{2\pi}{3}\right)$,
$\alpha^2 = \text{cis}\left(\frac{4\pi}{3}\right) = -\frac{1}{2} - i\frac{\sqrt{3}}{2}$ and
$\alpha^3 = \text{cis}\, 2\pi = 1$

Thus $G = \{\{1, \alpha, \alpha^2\}, \times\}$

$\times$	1	α	α^2
1	1	α	α^2
α	α	α^2	1
α^2	α^2	1	α

$1 \mapsto 1$
$\alpha \mapsto 2$
$\alpha^2 \mapsto 4$

$\times_7$	1	2	4
1	1	2	4
2	2	4	1
4	4	1	2

f is a bijection and for all $a, b \in G$, $f(ab) = f(a) \times_7 f(b)$.
As the Cayley tables have the same structure,
$\{\{1, \alpha, \alpha^2\}, \times\} \cong \{\{1, 2, 4\}, \times_7\}$

3 $G = \{0, 1, 2, 3, 4\}$ under $+_5$
$H = \{1, \alpha, \alpha^2, \alpha^3, \alpha^4\}$ where $\alpha = \text{cis}\left(\frac{2\pi}{5}\right)$ under $\times$

$+_5$	0	1	2	3	4
0	0	1	2	3	4
1	1	2	3	4	0
2	2	3	4	0	1
3	3	4	0	1	2
4	4	0	1	2	3

$\times$	1	α	α^2	α^3	α^4
1	1	α	α^2	α^3	α^4
α	α	α^2	α^3	α^4	1
α^2	α^2	α^3	α^4	1	α
α^3	α^3	α^4	1	α	α^2
α^4	α^4	1	α	α^2	α^3

$0 \mapsto 1,\ 1 \mapsto \alpha,\ 2 \mapsto \alpha^2,\ 3 \mapsto \alpha^3,\ 4 \mapsto \alpha^4$
f is a bijection and for all $a, b \in G$, $f(a +_5 b) = f(a)f(b)$
$\therefore\ \{G, +_5\} \cong \{H, \times\}$

4 **a** In G, $2^1 = 2,\ 2^2 = 4,\ 2^3 = 8,\ 2^4 = 6$
$\therefore\ G = \langle 2 \rangle$
In H, $1^2 = 1,\ 3^2 = 1,\ 5^2 = 1,\ 7^2 = 1$
$\therefore$ no element has order 4
$\therefore$ H is not cyclic.
Hence $G \ncong H$.

b Suppose $f(x) = nx$ for all $x \in G$, $n \in \mathbb{Z}^+$.
Then $f(x) = f(y) \Leftrightarrow nx = ny$
$\Leftrightarrow x = y \quad \{n \neq 0\}$
$\therefore$ f is one-to-one.
The range of f, $R(f) = \{nx \mid x \in \mathbb{Z}\}$
$= n\mathbb{Z}$
$\therefore$ f is onto.
As f is one-to-one and onto it is a bijection.
For $x, y \in \mathbb{Z}$, $f(x + y) = n(x + y)$
$= nx + ny$
$= f(x) + f(y)$
Thus f is a bijective homomorphism
$\therefore$ f is an isomorphism.
Thus $G \cong H$.

c $|G| = 6$ and $|S_6| = 6! = 720$.
As G and S_6 have different orders their elements cannot be put into a one-to-one correspondence.
$\therefore\ G \ncong H$.

d $G = \{\mathbb{Z}_6, +_6\}$ is cyclic as $G = \langle 1 \rangle$.
But, H is not cyclic. $\therefore\ G \ncong H$.

e $G = \langle i \rangle = \{1, i, -1, -i\}$ is a finite cyclic group.
$H = \langle 1 + i \rangle = \left\langle \sqrt{2}\ \text{cis}\left(\frac{\pi}{4}\right) \right\rangle$ and
$\left[\sqrt{2}\ \text{cis}\left(\frac{\pi}{4}\right)\right]^n = 2^{\frac{n}{2}}\ \text{cis}\left(\frac{n\pi}{4}\right)$
$\neq 1$ for any $n \in \mathbb{Z}^+$
$\therefore$ $\langle 1 + i \rangle$ has infinite order
$\therefore$ H is an infinite group
$\therefore$ $G \ncong H$ {as G is finite, H infinite}.
Their elements cannot be put in a one-to-one correspondence.

5 We define f by $f : \{\mathbb{R}^+, \times\} \to \{\mathbb{R}, +\},\ f(x) = \ln x$

(1) If $f(x) = f(y)$ then $\ln x = \ln y$
$\therefore\ \ln x - \ln y = 0$
$\therefore\ \ln\left(\frac{x}{y}\right) = 0$
$\therefore\ \frac{x}{y} = e^0 = 1$
$\therefore\ x = y$
$\therefore$ f is one-to-one.

(2) 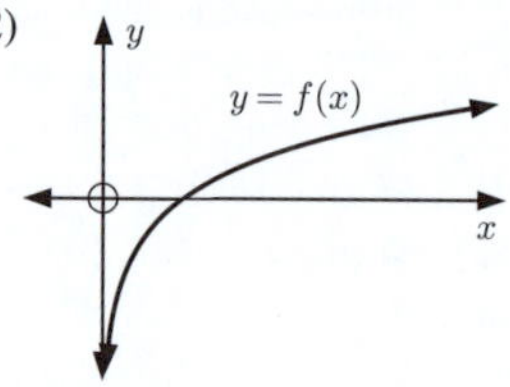

No horizontal line cuts the increasing function f in more than one place.
$\therefore$ f is onto.

(3) Also for all $a, b \in \mathbb{R}^+$, $f(ab) = \ln(ab)$
$= \ln a + \ln b$
$= f(a) + f(b)$

From (1), (2), and (3), f is a bijective homomorphism.
$\therefore$ f is an isomorphism.
$\therefore$ $\{\mathbb{R}^+, \times\} \cong \{\mathbb{R}, +\}$

6 Let $f : \{G, *\} \to \{H, \circ\}$ be an isomorphism and suppose $g \in G$ has finite order $m \in \mathbb{Z}^+$.

$\therefore\ g^m = e_G$
$\therefore\ f(g^m) = f(e_G) = e_H$
$\therefore\ f(\underbrace{g * g * g * \dots * g}_{m \text{ of these}}) = e_H$
$\therefore\ \underbrace{f(g) \circ f(g) \circ f(g) \circ \dots \circ f(g)}_{m \text{ of these}} = e_H$
$\therefore\ [f(g)]^m = e_H$

Now suppose $f(g)$ has order $n < m$, $n \in \mathbb{Z}^+$
$\therefore\ [f(g)]^n = e_H$
$\therefore\ f(g) \circ f(g) \circ f(g) \circ \dots \circ f(g) = e_H$
$\therefore\ f(g * g * g * \dots * g) = e_H$
$\therefore\ f(g^n) = e_H$
$\therefore\ g^n = e_G$
{as f is an isomorphism, e_G is unique and maps to e_H}

This is a contradiction as $|g| = m$ and $n < m$, $n \in \mathbb{Z}^+$
$\therefore\ |g| = |f(g)|$
Thus, the order of g is unchanged by the isomorphism.

7 $\{G, *\}$ and $\{H, \circ\}$ are isomorphic.

To prove: $\{G, *\}$ is Abelian $\Leftrightarrow$ $\{H, \circ\}$ is Abelian

$(\Rightarrow)$ $\{G, *\}$ Abelian

$\Rightarrow$ $a * b = b * a$ for all $a, b \in G$

$\Rightarrow$ $f(a * b) = f(b * a)$

$\Rightarrow$ $f(a) \circ f(b) = f(b) \circ f(a)$

for all $f(a), f(b) \in R(f) = H$

$\Rightarrow$ H is Abelian.

$(\Leftarrow)$ Since f^{-1} is an isomorphism from H to G, $\{H, \circ\}$ Abelian would imply that $\{G, *\}$ is Abelian.

Thus $\{G, *\}$ is Abelian $\Leftrightarrow$ $\{H, \circ\}$ is Abelian.

8 **a** The Cayley table for D_3 is:

$\circ$	e	r	r^2	x	y	z
e	e	r	r^2	x	y	z
r	r	r^2	e	z	x	y
r^2	r^2	e	r	y	z	x
x	x	y	z	e	r	r^2
y	y	z	x	r^2	e	r
z	z	x	y	r	r^2	e

and for S_3 is:

$*$	e	α	α^2	β	γ	δ
e	e	α	α^2	β	γ	δ
α	α	α^2	e	δ	β	γ
α^2	α^2	e	α	γ	δ	β
β	β	γ	δ	e	α	α^2
γ	γ	δ	β	α^2	e	α
δ	δ	β	γ	α	α^2	e

$e \mapsto e$, $r \mapsto \alpha$, $r^2 \mapsto \alpha^2$, $x \mapsto \beta$, $y \mapsto \gamma$, $z \mapsto \delta$

f is a bijection.

And, as the Cayley tables have the same structure $D_3 \cong S_3$.

b For $n \in \mathbb{Z}^+$, $n > 3$ the finite groups D_n and S_n have different orders.

The order of $D_n = 2n$

The order of $S_n = n!$

$\therefore$ they cannot be isomorphic.

$\therefore$ $D_n \ncong S_n$ for $n > 3$.

9 $\{G_1, *\}$ and $\{G_2, \triangle\}$ are isomorphic.

$\{G_2, \triangle\}$ and $\{G_3, \square\}$ are also isomorphic.

$\therefore$ $G_1 \cong G_2$ and $G_2 \cong G_3$

To prove: $G_1 \cong G_3$

Proof:

If $G_1 \cong G_2$, there exists an isomorphism $f : G_1 \to G_2$.

If $G_2 \cong G_3$, there exists an isomorphism $g : G_2 \to G_3$.

f and g are one-to-one, onto, and are homomorphisms.

Consider $g \circ f : G_1 \to G_3$

$$\begin{aligned}\therefore\ (g \circ f)(a * b) &= g(f(a * b)) \\ &= g(f(a) \triangle f(b)) \\ &= g(f(a)) \square g(f(b)) \\ &= (g \circ f)(a) \square (g \circ f)(b)\end{aligned}$$

$\therefore$ $g \circ f$ is a homomorphism from G_1 to G_3.

Now as f and g are one-to-one and onto, $g \circ f$ is one-to-one and onto.

$\therefore$ $g \circ f$ is an isomorphism from G_1 to G_3.

$\therefore$ $G_1 \cong G_3$

10 **a** R is the group of symmetries of a rectangle and has Cayley table:

	e	r	R_1	R_2
e	e	r	R_1	R_2
r	r	e	R_2	R_1
R_1	R_1	R_2	e	r
R_2	R_2	R_1	r	e

Let $e = \begin{pmatrix} 1 & 2 & 3 & 4 \\ 1 & 2 & 3 & 4 \end{pmatrix}$, $a = (1\ 2)(3\ 4)$

$b = (1\ 3)(2\ 4)$, $c = (1\ 4)(2\ 3)$

G has Cayley table:

$*$	e	a	b	c
e	e	a	b	c
a	a	e	c	b
b	b	c	e	a
c	c	b	a	e

For example,

$$\begin{aligned}&a * c \\ &= (1\ 2)(3\ 4) * (1\ 4)(2\ 3) \\ &= (1\ 3)(2\ 4) \\ &= b\end{aligned}$$

For example, $c * b = (1\ 4)(2\ 3) * (1\ 3)(2\ 4)$

$= (1\ 2)(3\ 4)$

$= a$

$e \mapsto e$, $r \mapsto a$, $R_1 \mapsto b$, $R_2 \mapsto c$

f is one-to-one and onto and is a homomorphism since the two tables have the same structure.

$\therefore$ f is an isomorphism.

$\therefore$ $R \cong G$

b The **Klein 4-group** V_4 has Cayley table:

	i	p	q	r
i	i	p	q	r
p	p	i	r	q
q	q	r	i	p
r	r	q	p	i

G		**Klein 4**
e	$\mapsto$	i
a	$\mapsto$	p
b	$\mapsto$	q
c	$\mapsto$	r

$\therefore$ $G \leftrightarrow$ Klein 4 is one-to-one, onto, and a homomorphism since the tables have the same structure.

$\therefore$ $G \cong$ Klein 4

c $R \cong G$ and $G \cong$ Klein 4

$\therefore$ from question **9**, $R \cong$ Klein 4

EXERCISE K

1 $\{H = \{0, 3\}, +_6\} < \{\mathbb{Z}_6, +_6\}$

a The left cosets of H in $\mathbb{Z}_6$ are:

$$\begin{aligned}0H &= \{0 +_6 h \mid h \in H\} \\ &= \{0 +_6 0,\ 0 +_6 3\} \\ &= \{0, 3\} \\ &= H\end{aligned}$$

$$\begin{aligned}1H &= \{1 +_6 h \mid h \in H\} \\ &= \{1 +_6 0,\ 1 +_6 3\} \\ &= \{1, 4\}\end{aligned}$$

$$\begin{aligned}2H &= \{2 +_6 0,\ 2 +_6 3\} \\ &= \{2, 5\}\end{aligned}$$

$$\begin{aligned}3H &= \{3 +_6 0,\ 3 +_6 3\} \\ &= \{3, 0\} \\ &= H\end{aligned}$$

$$\begin{aligned}4H &= \{4 +_6 0,\ 4 +_6 3\} \\ &= \{4, 1\} \\ &= \{1, 4\}\end{aligned}$$

$$\begin{aligned}5H &= \{5 +_6 0,\ 5 +_6 3\} \\ &= \{5, 2\} \\ &= \{2, 5\}\end{aligned}$$

Hence H has three distinct left cosets in $\mathbb{Z}_6$:

$\{0, 3\} = H$, $\{2, 5\}$, and $\{1, 4\}$

b As $\{\mathbb{Z}_6, +_6\}$ is Abelian, the right cosets are the same as the corresponding left cosets.

$\therefore$ H has three distinct right cosets in $\mathbb{Z}_6$:

$\{0, 3\} = H$, $\{2, 5\}$, and $\{1, 4\}$

2 $H = \{0, 4, 8\}$ under $+_{12} < \{\mathbb{Z}_{12}, +_{12}\}$

a The left cosets of H in $\mathbb{Z}_{12}$ are:

$$\begin{aligned}
0H &= \{0 +_{12} 0,\ 0 +_{12} 4,\ 0 +_{12} 8\} \\
&= \{0, 4, 8\} \\
&= H \\
1H &= \{1 +_{12} 0,\ 1 +_{12} 4,\ 1 +_{12} 8\} \\
&= \{1, 5, 9\} \\
2H &= \{2 +_{12} 0,\ 2 +_{12} 4,\ 2 +_{12} 8\} \\
&= \{2, 6, 10\} \\
3H &= \{3 +_{12} 0,\ 3 +_{12} 4,\ 3 +_{12} 8\} \\
&= \{3, 7, 11\} \\
4H &= \{4 +_{12} 0,\ 4 +_{12} 4,\ 4 +_{12} 8\} \\
&= \{4, 8, 0\} \\
&= H \\
5H &= \{5 +_{12} 0,\ 5 +_{12} 4,\ 5 +_{12} 8\} \\
&= \{5, 9, 1\} \\
6H &= \{6 +_{12} 0,\ 6 +_{12} 4,\ 6 +_{12} 8\} \\
&= \{6, 10, 2\} \\
7H &= \{7 +_{12} 0,\ 7 +_{12} 4,\ 7 +_{12} 8\} \\
&= \{7, 11, 3\} \\
8H &= \{8 +_{12} 0,\ 8 +_{12} 4,\ 8 +_{12} 8\} \\
&= \{8, 0, 4\} \\
&= H \\
9H &= \{9 +_{12} 0,\ 9 +_{12} 4,\ 9 +_{12} 8\} \\
&= \{9, 1, 5\} \\
10H &= \{10 +_{12} 0,\ 10 +_{12} 4,\ 10 +_{12} 8\} \\
&= \{10, 2, 6\} \\
11H &= \{11 +_{12} 0,\ 11 +_{12} 4,\ 11 +_{12} 8\} \\
&= \{11, 3, 7\}
\end{aligned}$$

$\therefore$ H has four distinct left cosets in $\mathbb{Z}_{12}$:

$\{0, 4, 8\} = H$, $\{1, 5, 9\}$, $\{2, 6, 10\}$, $\{3, 7, 11\}$.

b As $\{\mathbb{Z}_{12}, +_{12}\}$ is Abelian, the right cosets of H are the same as the left cosets.

Proof: For $g \in \mathbb{Z}_{12}$,

$$\begin{aligned}
g +_{12} H &= \{g +_{12} h \mid h \in H\} \\
&= \{h +_{12} g \mid h \in H\} \\
&= H +_{12} g
\end{aligned}$$

$\therefore$ each left coset of H is also a right coset.

3 **a** As $|S_4| \div |R| = 4! \div 4 = 6$, there are 6 left cosets of R in S_4.

b 1, the subgroup R itself.

c
$$\begin{aligned}
&(1\,2\,3)(1\,2\,3\,4)^{-1} \\
&= (1\,2\,3)(4\,3\,2\,1) \\
&= (1\,4)(2)(3) \notin R
\end{aligned}$$

$\therefore$ by **Theorem 26** parts **3** and **4**,

$(1\,2\,3)R \cap (1\,2\,3\,4)R = \varnothing$, so they have no elements in common.

d
$$\begin{aligned}
(12)R &= \{(1\,2)e,\ (1\,2)(1\,2)(3\,4), \\
&\qquad (1\,2)(1\,3)(2\,4),\ (1\,2)(1\,4)(2\,3)\} \\
&= \{(1\,2),\ (3\,4),\ (1\,3\,2\,4),\ (1\,4\,2\,3)\}
\end{aligned}$$

$$\begin{aligned}
R(12) &= \{e(1\,2),\ (1\,2)(3\,4)(1\,2), \\
&\qquad (1\,3)(2\,4)(1\,2),\ (1\,4)(2\,3)(1\,2)\} \\
&= \{(1\,2),\ (3\,4),\ (1\,4\,2\,3),\ (1\,3\,2\,4)\}
\end{aligned}$$

$\therefore$ $(1\,2)R = R(1\,2)$

4 If $gH = Hg$ for all $g \in G$.

As G is a group and $g \in G$, g^{-1} exists.

Multiplying on the left by g^{-1} gives $g^{-1}gH = g^{-1}Hg$

$\therefore$ $eH = g^{-1}Hg$

$\therefore$ $H = g^{-1}Hg$

Multiplying on the right by g^{-1} gives $Hg^{-1} = g^{-1}Hgg^{-1}$

$\therefore$ $Hg^{-1} = g^{-1}He$

$\therefore$ $Hg^{-1} = g^{-1}H$

5 $|S_5| = 5! = 120$ and 120 *is not* a multiple of 7.

$\therefore$ by Lagrange's theorem, S_5 can have no subgroup of order 7.

6 For any $m \in \mathbb{Z}^+$, $m \leqslant n$, consider the symmetric group S_m of degree m.

Elements of S_m permute the values 1, 2, ..., m, whilst keeping $m+1, m+2, ..., n$ fixed. $\therefore$ $S_m \subseteq S_n$.

$\therefore$ $S_m < S_n$ and $|S_m| = m!$

7 Given: $|G| = 36$ and G has identity e.

Consider $g \in G$

$|g|$ is a factor of $|G| = 36$

$\therefore$ $|g| = 1, 2, 3, 4, 6, 9, 12, 18$, or 36

{Corollary of Lagrange's theorem}

Since $g^7 = e$, $7 = k\,|g|$ for some $k \in \mathbb{Z}^+$

$\therefore$ $|g| = 1$

$\therefore$ $g = e$

Thus e is the unique element of G such that $g^7 = e$.

8 Let G be a group with prime order.

That is G is a group with $|G| = p$.

As $p \geqslant 2$, $|G| \geqslant 2$

$\therefore$ there exists an element $g \in G$ which is not the identity element e of G.

Let $m = |g| > 1$, $m \in \mathbb{Z}^+$.

The group $\langle g \rangle = \{e, g, g^2, g^3, ..., g^{m-1}\}$ is a subgroup of G and $|\langle g \rangle| = |g| = m > 1$

$\therefore$ by Lagrange's theorem $|G|$ is a multiple of $|\langle g \rangle|$

$\therefore$ p is a multiple of m where $m > 1$.

But p is prime and so $m = p$.

Since $|\langle g \rangle| = |G| = p$ is finite, $G = \langle g \rangle$

$\therefore$ G is cyclic with generator g.

REVIEW SET A

1 $A = \{a, b, c, d, e, f\}$, $B = \{c, e, g, h\}$

a $A \cup B = \{a, b, c, d, e, f, g, h\}$

b $A \backslash B = \{a, b, d, f\}$

c
$$\begin{aligned}
A \Delta B &= (A \backslash B) \cup (B \backslash A) \\
&= \{a, b, d, f\} \cup \{g, h\} \\
&= \{a, b, d, f, g, h\}
\end{aligned}$$

Check:

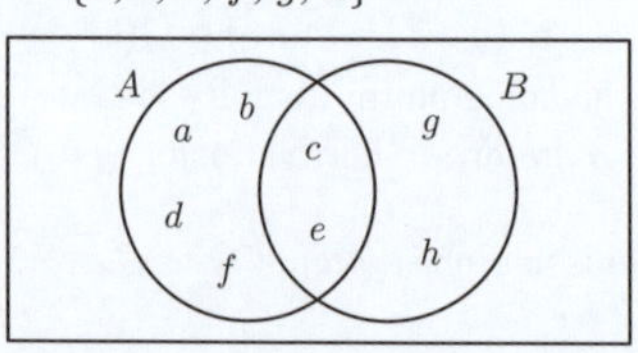

✓

2 $A = \{1, 2, 3\}, \quad B = \{2, 4\}$
$A \times B = \{(1, 2), (1, 4), (2, 2), (2, 4), (3, 2), (3, 4)\}$

3 **To prove:** $(A \cap B) \times (C \cap D) = (A \times C) \cap (B \times D)$

Proof:

$(\Rightarrow)$ Let $(x, y) \in (A \cap B) \times (C \cap D)$ then $x \in A$ and $x \in B$ and $y \in C$ and $y \in D$
$\therefore \ (x, y) \in A \times C$ and $(x, y) \in B \times D$
$\therefore \ (x, y) \in (A \times C) \cap (B \times D)$
$\therefore \ (A \cap B) \times (C \cap D) \subseteq (A \times C) \cap (B \times D)$ .... (1)

$(\Leftarrow)$ Let $(x, y) \in (A \times C) \cap (B \times D)$
$\therefore \ (x, y) \in A \times C$ and $(x, y) \in B \times D$
$\therefore \ x \in A$ and $y \in C$ and $x \in B$ and $y \in D$
$\therefore \ x \in A \cap B$ and $y \in C \cap D$
$\therefore \ (x, y) \in (A \cap B) \times (C \cap D)$
$\therefore \ (A \times C) \cap (B \times D) \subseteq (A \cap B) \times (C \cap D)$ .... (2)

From (1) and (2), $(A \cap B) \times (C \cap D) = (A \times C) \cap (B \times D)$

4 **a** $\{(0, 0), (1, 1), (2, 2), (3, 3), (4, 4), (5, 5), (0, 1), (1, 0), (0, 2), (2, 0), (1, 2), (2, 1), (1, 3), (3, 1), (2, 3), (3, 2), (2, 4), (4, 2), (3, 5), (5, 3), (4, 5), (5, 4), (3, 4), (4, 3)\}$

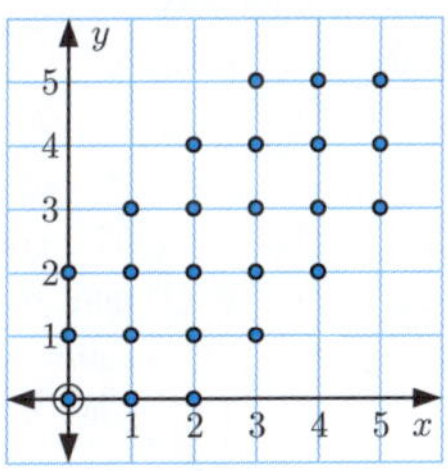

b $xRy \Leftrightarrow |x - y| < 3$

i $|x - x| = 0 < 3$
$\therefore \ xRx$ for all $x \in \{0, 1, 2, 3, 4, 5\}$.
$\therefore \ R$ is reflexive.

ii If $xRy \Rightarrow |x - y| < 3$
$\Rightarrow |y - x| < 3 \quad \{\text{as } |x - y| = |y - x|\}$
$\Rightarrow yRx$
$\therefore \ R$ is symmetric.

iii As $0R2$ and $2R4 \not\Rightarrow 0R4$, R is not transitive.

c R *is not* an equivalence relation as R is not transitive.

5 **a** $f : \mathbb{R} \to \mathbb{R}$,
$f(x) = 2x^3 + 3x - 1$
$f'(x) = 6x^2 + 3$
> 0 for all $x \in \mathbb{R}$
$\therefore \ f(x)$ is increasing for all $x > \mathbb{R}$.

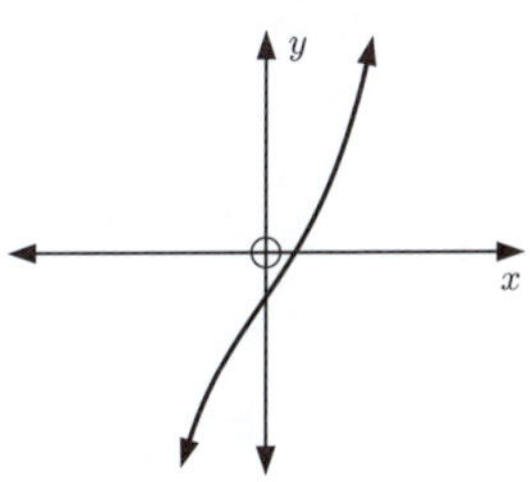

i $f(x)$ is an injection as no horizontal line cuts the graph more than once.

ii The range of f is $\mathbb{R}$, so f is a surjection.

b $f : \mathbb{Z} \to \mathbb{Z}^+, \ f(x) = x^2$

i f is not one-to-one as, for example, the horizontal line $y = 1$, cuts the graph twice. Thus f is not an injection.

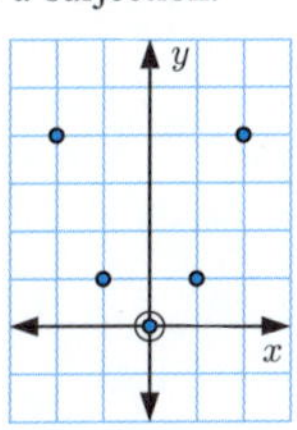

ii f is not onto as, for example, $2 \in \mathbb{Z}^+$, but there is no $x \in \mathbb{Z}$ such that $x^2 = 2$.
$\therefore \ f$ is not a surjection.

c $f : \mathbb{C} \to \mathbb{R}^+ \cup \{0\}, \ f(z) = |z|$

i f is not one-to-one as, for example, $f(i) = |i| = 1$ and $f(-i) = |-i| = 1$
$\therefore \ f$ is not an injection.

ii For any $r \in \mathbb{R}^+ \cup \{0\}$, $z = ri$ has $|z| = r$
$\therefore \ f$ is a surjection.

d $f : \mathbb{Z}^+ \to \mathbb{R}^+, \ f(x) = \sqrt{x}$

i f is one-to-one as $\sqrt{x_1} = \sqrt{x_2} \Leftrightarrow x_1 = x_2$ for all $x_1, x_2 \in \mathbb{Z}^+$
$\therefore \ f$ is an injection.

ii f is not onto as, for example $\pi \in \mathbb{R}^+$, but there is no $x \in \mathbb{Z}^+$ such that $\sqrt{x} = \pi$.
$\therefore \ f$ is not a surjection.

6 **a** **i** $3 * 4 = 2$ **ii** $2 * (1 * 3) = 2 * 3 = 4$ **iii** $(2 * 1) * 3 = 3 * 3 = 3$

b It is not a Latin square as row 4 contains the element 1 twice. This indicates that $\{S, *\}$ is not a group.

7 **a** **i** $gf = \begin{pmatrix} 1 & 2 & 3 & 4 \\ 2 & 3 & 1 & 4 \end{pmatrix}\begin{pmatrix} 1 & 2 & 3 & 4 \\ 1 & 3 & 4 & 2 \end{pmatrix} = \begin{pmatrix} 1 & 2 & 3 & 4 \\ 2 & 1 & 4 & 3 \end{pmatrix}$

ii $fg = \begin{pmatrix} 1 & 2 & 3 & 4 \\ 1 & 3 & 4 & 2 \end{pmatrix}\begin{pmatrix} 1 & 2 & 3 & 4 \\ 2 & 3 & 1 & 4 \end{pmatrix} = \begin{pmatrix} 1 & 2 & 3 & 4 \\ 3 & 4 & 1 & 2 \end{pmatrix}$

iii $f^{-1} = \begin{pmatrix} 1 & 2 & 3 & 4 \\ 1 & 4 & 2 & 3 \end{pmatrix}$

iv $g^{-1} = \begin{pmatrix} 1 & 2 & 3 & 4 \\ 3 & 1 & 2 & 4 \end{pmatrix}$

b $f^2 = \begin{pmatrix} 1 & 2 & 3 & 4 \\ 1 & 3 & 4 & 2 \end{pmatrix}\begin{pmatrix} 1 & 2 & 3 & 4 \\ 1 & 3 & 4 & 2 \end{pmatrix} = \begin{pmatrix} 1 & 2 & 3 & 4 \\ 1 & 4 & 2 & 3 \end{pmatrix}$

$f^3 = f^2 f = \begin{pmatrix} 1 & 2 & 3 & 4 \\ 1 & 4 & 2 & 3 \end{pmatrix}\begin{pmatrix} 1 & 2 & 3 & 4 \\ 1 & 3 & 4 & 2 \end{pmatrix}$
$= \begin{pmatrix} 1 & 2 & 3 & 4 \\ 1 & 2 & 3 & 4 \end{pmatrix}$

As $f^3 = e$, $f^{3k} = (f^3)^k = e^k = e$, $k \in \mathbb{Z}^+$
$\therefore \ n = 3k, \ k \in \mathbb{Z}^+$

8 **a** $n = 3, \ G = \{0, 1, 2\}$ under $+_3$
$1^1 = 1, \ 1^2 = 1 +_3 1 = 2, \ 1^3 = 0$
$2^1 = 2, \ 2^2 = 2 +_3 2 = 1, \ 2^3 = 0$
$\therefore$ 1 and 2 are its generators.

b $n = 5, \ G = \{0, 1, 2, 3, 4\}$ under $+_5$
$1^1 = 1, \ 1^2 = 2, \ 1^3 = 3, \ 1^4 = 4, \ 1^5 = 0$
$2^1 = 2, \ 2^2 = 4, \ 2^3 = 1, \ 2^4 = 3, \ 2^5 = 0$
$3^1 = 3, \ 3^2 = 1, \ 3^3 = 4, \ 3^4 = 2, \ 3^5 = 0$
$4^1 = 4, \ 4^2 = 3, \ 4^3 = 2, \ 4^4 = 1, \ 4^5 = 0$
$\therefore$ 1, 2, 3, and 4 are all generators.

c $n = 6, \ G = \{0, 1, 2, 3, 4, 5\}$ under $+_6$
$1^1 = 1, \ 1^2 = 2, \ 1^3 = 3, \ 1^4 = 4, \ 1^5 = 5, \ 1^6 = 0$
$2^1 = 2, \ 2^2 = 4, \ 2^3 = 0$
$3^1 = 3, \ 3^2 = 0$
$4^1 = 4, \ 4^2 = 2, \ 4^3 = 0$
$5^1 = 5, \ 5^2 = 4, \ 5^3 = 3, \ 5^4 = 2, \ 5^5 = 1, \ 5^6 = 0$
$\therefore$ 1 and 5 are its generators.

9 Notice that:

$xRy \Leftrightarrow p^a(x) = y$

$\Leftrightarrow x$ and y belong to the same cycle c_i in p.

a (1) xRx since x lies in the same cycle as x.

$\therefore$ R is reflexive.

(2) $xRy \Rightarrow x$ and y belong to the same cycle

$\Rightarrow y$ and x belong to the same cycle

$\Rightarrow yRx$

$\therefore$ R is symmetric.

(3) $xRy \Rightarrow x$ and y belong to the same cycle.

$yRz \Rightarrow y$ and z belong to the same cycle.

So, if xRy and yRz then x, y, and z belong to the same cycle.

$\Rightarrow xRz$

$\therefore$ R is transitive.

Thus, from (1), (2), and (3), R is an equivalence relation.

b The equivalence classes are the 'orbits' of the distinct, disjoint cycles of p.

c $p = (1\,2)(3\,6\,8)(4\,5)(7)$ is the product of disjoint cycles.

The equivalence classes are:

$\{1, 2\}$, $\{3, 6, 8\}$, $\{4, 5\}$, and $\{7\}$

10 **a**

$*$	1	2
1	1	2
2	1	2

$S = \{1, 2\}$ is **closed** under $*$ as $a*b \in S$ whenever $a, b \in S$, as the table contains only members of S.

$(1*1)*1$	$(1*1)*2$	$(1*2)*1$	$(2*1)*1$
$= 1*1$	$= 1*2$	$= 2*1$	$= 1*1$
$= 1$	$= 2$	$= 1$	$= 1$
and	and	and	and
$1*(1*1)$	$1*(1*2)$	$1*(2*1)$	$2*(1*1)$
$= 1*1$	$= 1*2$	$= 1*1$	$= 2*1$
$= 1$ ✓	$= 2$ ✓	$= 1$ ✓	$= 1$ ✓
$(2*2)*1$	$(2*1)*2$	$(1*2)*2$	$(2*2)*2$
$= 2*1$	$= 1*2$	$= 2*2$	$= 2*2$
$= 1$	$= 2$	$= 2$	$= 2$
and	and	and	and
$2*(2*1)$	$2*(1*2)$	$1*(2*2)$	$2*(2*2)$
$= 2*1$	$= 2*2$	$= 1*2$	$= 2*2$
$= 1$ ✓	$= 2$ ✓	$= 2$ ✓	$= 2$ ✓

Thus $(a*b)*c = a*(b*c)$ for all $a, b \in S$

$\therefore$ $*$ is **associative** on S.

Hence, $\{S, *\}$ is a semi-group.

But the Cayley table is not a Latin square.

$\therefore$ $\{S, *\}$ is not a group.

b

$*$	1	2
1	1	2
2	2	1

$S = \{1, 2\}$ under $*$ is either $\mathbb{Z}_3$ under $\times_3$ or isomorphic to it, and $\{\mathbb{Z}_3, \times_3\}$ is a group.

$\therefore$ $\{S, *\}$ is closed and $*$ is associative

$\therefore$ $\{S, *\}$ is a semi-group.

$\{S, *\}$ is a group.

c

$*$	1	2	3
1	1	2	3
2	2	3	1
3	3	1	2

$S = \{1, 2, 3\}$ under $*$

S is isomorphic to $U_3 = \{1, \alpha, \alpha^2\}$ where $\alpha = \text{cis}\left(\frac{2\pi}{3}\right)$ under $\times$ of complex numbers.

$1 \mapsto 1$

$2 \mapsto \alpha$

$3 \mapsto \alpha^2$

$\times$	1	α	α^2
1	1	α	α^2
α	α	α^2	1
α^2	α^2	1	α

Now S is closed and $*$ is associative on S, as $\times$ is associative on U_3.

$\therefore$ $\{S, *\}$ is a semi-group.

In fact $\{S, *\}$ is a group as $\{U_3, \times\}$ is a group.

d $S = \{1, 2, 3\}$ under $*$

S is clearly closed as the Cayley table contains only elements of S.

$*$	1	2	3
1	1	2	3
2	3	2	3
3	3	2	3

On checking all $3 \times 3 \times 3 = 27$ possibilities $*$ is associative on S.

$\therefore$ $\{S, *\}$ is a semi-group.

However, the Cayley table is not a Latin square.

$\therefore$ $\{S, *\}$ cannot be a group.

11 $\{G, *\}$ is a group with identity e.

$\{G', \circ\}$ is a group with identity e'.

$(a, a')(b, b') = (a*b, a' \circ b')$

a *Closure*: If $(a, a') \in S$ and $(b, b') \in S$

$a \in G$, $a' \in G'$ and $b \in G$, $b' \in G'$

$\therefore$ $a, b \in G$ and $a', b' \in G'$

$\therefore$ $a*b \in G$ and $a' \circ b' \in G'$

$\therefore$ $(a*b, a' \circ b') \in S$

$\therefore$ S is closed.

Associative: $*$ is associative on G

$\circ$ is associative on G'

$\therefore$ 'product' is associative on $S = G \times G'$.

Identity: The identity is (e, e').

Inverse: $(a^{-1}, (a')^{-1})$ is the inverse of (a, a').

b $S_1 = \{(g, e') \mid g \in G\}$

Let (g_1, e') and (g_2, e') be in S

$\therefore$ $(g_2, e')^{-1} = ({g_2}^{-1}, e') \in S$ $\quad\{(e')^{-1} = e'\}$

and $(g_1, e')({g_2}^{-1}, e') = (g_1{g_2}^{-1}, e')$

Now $g_1{g_2}^{-1} \in G$ $\quad\{$as $G < G\}$

$\therefore$ $(g_1{g_2}^{-1}, e') \in S_1$

$\therefore$ $S_1 < S$

12 If $|G| = 1, 2, 3,$ or 5, then $|G| = 1$ or a prime

$\therefore$ G is cyclic

$\therefore$ G is Abelian

If $|G| = 4$, then G is either cyclic or isomorphic to the Klein 4-group and therefore is Abelian.

Hence, a non-Abelian group must have order 6 or more.

$\therefore$ $|G| \geqslant 6$

13 $(a, b)*(c, d) = (a + c,\ 2^c b + d)$, $x \in \mathbb{Z}$, $y \in \mathbb{Q}$

a *Closure*: For all $(a, b), (c, d) \in G$,

$a, c \in \mathbb{Z}$ and $c, d \in \mathbb{Q}$

$\therefore$ $a + c \in \mathbb{Z}$ and $2^c b + d \in \mathbb{Q}$

$\therefore$ $(a, b)*(c, d) \in G$

$\therefore$ G is closed under $*$.

Associative: $[(a, b) * (c, d)] * (e, f)$
$= (a + c,\ 2^c b + d) * (e, f)$
$= (a + c + e,\ 2^e(2^c b + d) + f)$
$= (a + c + e,\ 2^{e+c} b + 2^e d + f)$
and $(a, b) * [(c, d) * (e, f)]$
$= (a, b) * (c + e,\ 2^e d + f)$
$= (a + c + e,\ 2^{c+e} b + 2^e d + f)$
$= (a + c + e,\ 2^{e+c} b + 2^e d + f)$
$= [(a, b) * (c, d)] * (e, f)$
$\therefore$ $*$ is associative on G.

Identity: Suppose $(a, b) * (x, y) = (a, b)$
$\therefore\ (a + x,\ 2^x b + y) = (a, b)$
$\therefore\ a + x = a$ and $2^x b + y = b$
$\therefore\ x = 0$ and $b + y = b$
$\therefore\ x = 0,\ y = 0$
Check: $(0, 0) * (a, b)$
$= (0 + a,\ 2^a(0) + b)$
$= (a, b)$ ✓
$\therefore\ (a, b) * (0, 0) = (a, b)$ and
$(0, 0) * (a, b) = (a, b)$ for all $(a, b) \in G$

Inverse: Suppose $(a, b) * (x, y) = (0, 0)$
$\therefore\ (a + x,\ 2^x b + y) = (0, 0)$
$\therefore\ a + x = 0$ and $2^x b + y = 0$
$\therefore\ x = -a$ and $y = -2^{-a} b$
$\therefore\ y = -\dfrac{b}{2^a}$
Each element has an inverse $\left(-a, -\dfrac{b}{2^a}\right)$
$\{2^a > 0$ for all $a \in \mathbb{Z},\ \therefore\ 2^a \neq 0\}$
Check: $\left(-a, -\dfrac{b}{2^a}\right) * (a, b)$
$= \left(-a + a,\ 2^a\left(-\dfrac{b}{2^a}\right) + b\right)$
$= (0,\ -b + b)$
$= (0, 0)$ ✓
$\therefore\ \{G, *\}$ is a group.

b $\{G, *\}$ is **not** Abelian, as for example,

$(2, 3) * (1, 0)$	and $(1, 0) * (2, 3)$
$= (2 + 1,\ 2^1(3) + 0)$	$= (1 + 2,\ 2^2(0) + 3)$
$= (3, 6)$	$= (3, 3)$

$\therefore\ (2, 3) * (1, 0) \neq (1, 0) * (2, 3)$

c **i** $H_1 = \{(a, 0) \mid a \in \mathbb{Z}\}$ is non-empty and
$(b, 0)^{-1} = \left(-b, -\dfrac{0}{2^b}\right) = (-b, 0)$
Thus $(a, 0) * (-b, 0) = (a - b,\ 2^{-b}(0) + 0)$
$= (a - b,\ 0) \in H_1$
$\therefore$ by the subgroup test, $H_1 < G$.

ii $H_2 = \{(0, b) \mid b \in \mathbb{Q}\}$ is non-empty and
$(0, a)^{-1} = \left(-0, -\dfrac{a}{2^0}\right) = (0, -a)$
Thus $(0, b) * (0, a)^{-1} = (0, b) * (0, -a)$
$= (0 + 0,\ 2^0 b - a)$
$= (0,\ b - a)$
$\in H_2$
$\therefore$ by the subgroup test, $H_2 < G$.

14 $G_1 : \{\mathbb{R} \setminus \{1\},\ *\},\ a * b = a + b - ab$
$G_2 : \{\mathbb{R}^+,\ \times\}$
$f : G_1 \to G_2$ under $f(a) = |a - 1|$.

a For $a, b \in \mathbb{R} \setminus \{1\}$,
$f(a * b) = f(a + b - ab)$
$= |a + b - ab - 1|$
$= |ab - a - b + 1|$ $\{|-x| = |x|\}$
$= |(a - 1)(b - 1)|$
$= |a - 1|\,|b - 1|$
$= f(a) \times f(b)$
$\therefore\ f$ is a homomorphism.

b **i**

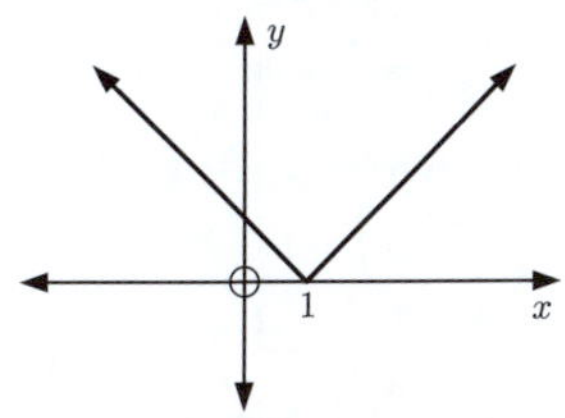

Graph of $f(x) = |x - 1|$

f is not one-to-one as the horizontal line through $(0, 1)$ meets the graph more than once.
$\therefore\ f$ is not an injection.

ii The range of $f(x)$ is $\mathbb{R}^+$.
$\{f(a) \neq 0$ since $a \neq 1\}$
$\therefore\ f$ is onto and so f is a surjection.

iii As f is not one-to-one, it is not a bijection.
$\therefore\ f$ is not an isomorphism.

15 **a** Suppose $p = (1\ 2\ 3),\ q = (1\ 4)$ for example.
In this case $pq = (1\ 2\ 3)(1\ 4) = (1\ 4\ 2\ 3)$
and $qp = (1\ 4)(1\ 2\ 3) = (1\ 2\ 3\ 4)$
Hence $pq \neq qp$

b $f(pq) = (pq)^2 = (pq)(pq) = pqpq$
and $f(p)f(q) = p^2 q^2 = ppqq$
$\therefore\ f(pq) = f(p)f(q)$
$\Leftrightarrow\ pqpq = ppqq$
$\Leftrightarrow\ p^{-1}pqpqq^{-1} = p^{-1}ppqqq^{-1}$ {multiply on left by p^{-1} and on right by q^{-1}}
$\Leftrightarrow\ pq = qp$
However, from **a**, $pq \neq qp$ for all $p, q \in S_4$
$\therefore\ f(pq) \neq f(p)f(q)$ for all $p, q \in S_4$
$\therefore\ f$ is not a homomorphism.

REVIEW SET B

1 $A = \{0, 3, 6, 9, 12\}$, $B = \{1, 2, 3, 4, 5, 6\}$
$C = \{2, 4, 6, 8, 10\}$
$\mathbb{U} = \{0, 1, 2, 3, 4, 5, 6, 7, 8, 9, 10, 11, 12, 13\}$

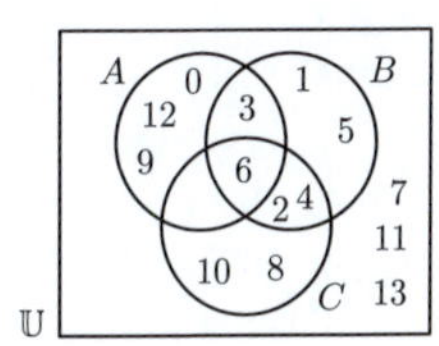

a $A \cap (B \cup C)$
$= \{0, 3, 6, 9, 12\} \cap \{1, 2, 3, 4, 5, 6, 8, 10\}$
$= \{3, 6\}$

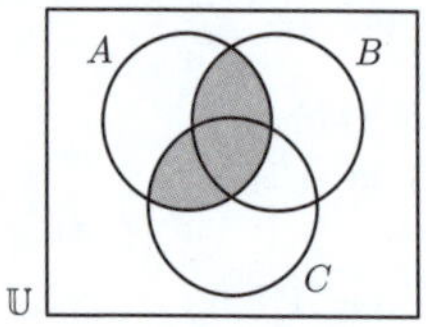

b $A \Delta (B \backslash C) = \{0, 3, 6, 9, 12\} \Delta \{1, 3, 5\}$
$= \{0, 6, 9, 12\} \cup \{1, 5\}$
$= \{0, 1, 5, 6, 9, 12\}$

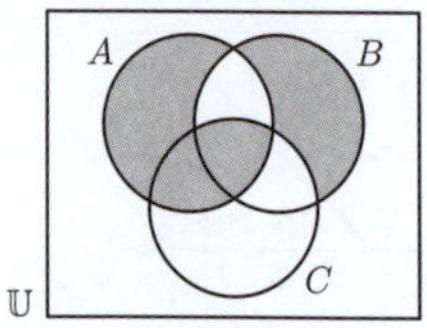

c $B' \cup C' = (B \cap C)'$ {De Morgan's law}
$= \{2, 4, 6\}'$
$= \{0, 1, 3, 5, 7, 8, 9, 10, 11, 12, 13\}$

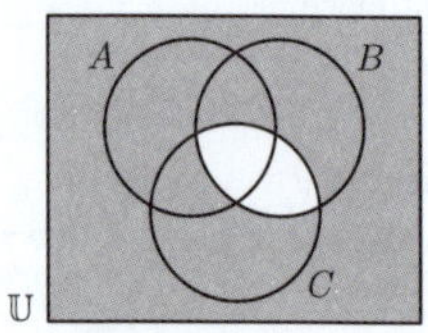

d $A \cup (B \Delta C)$
$= \{0, 3, 6, 9, 12\} \cup ((B \backslash C) \cup (C \backslash B))$
$= \{0, 3, 6, 9, 12\} \cup (\{1, 3, 5\} \cup \{8, 10\})$
$= \{0, 3, 6, 9, 12\} \cup \{1, 3, 5, 8, 10\}$
$= \{0, 1, 3, 5, 6, 8, 9, 10, 12\}$

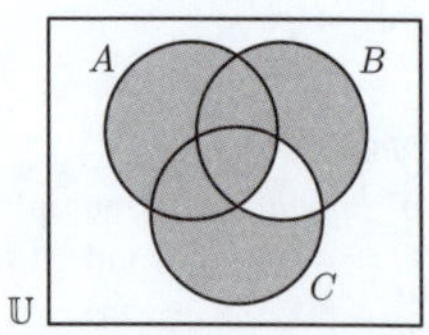

e $A' \cap (B' \Delta C')$
$= A' \cap (B \Delta C)$ {**Exercise A.5**, question **7**}
$= \{1, 2, 4, 5, 7, 8, 10, 11, 13\} \cap \underbrace{\{1, 3, 5, 8, 10\}}_{\text{from } \mathbf{d}}$
$= \{1, 5, 8, 10\}$

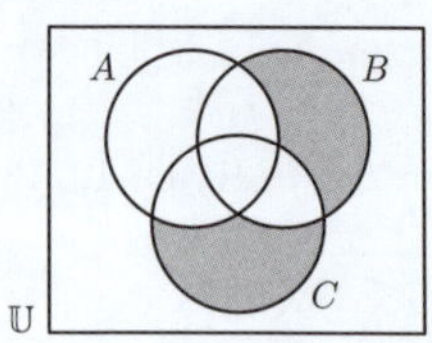

2 If $x \in (A \cap B)'$ then $x \notin A \cap B$
$\Rightarrow x \notin A$ or $x \notin B$
$\Rightarrow x \in A'$ or $x \in B'$
$\Rightarrow x \in A' \cup B'$
Thus $(A \cap B)' \subseteq A' \cup B'$ .... (1)

If $x \in A' \cup B'$ then $x \in A'$ or $x \in B'$
$\Rightarrow x \notin A$ or $x \notin B$
$\Rightarrow x \notin A \cap B$
$\Rightarrow x \in (A \cap B)'$
Thus $A' \cup B' \subseteq (A \cap B)'$ .... (2)
From (1) and (2), $(A \cap B)' = A' \cup B'$

3 **a** $xRy \Leftrightarrow x - y$ is divisible by 6
$\Leftrightarrow x - y = 6n$ for some $n \in \mathbb{Z}$
(1) As $x - x = 0 = 6 \times 0$, $0 \in \mathbb{Z}$ then xRx
$\therefore$ R is reflexive.
(2) If xRy then $x - y = 6n$ for some $n \in \mathbb{Z}$
$\Rightarrow y - x = 6(-n)$, $-n \in \mathbb{Z}$
$\Rightarrow yRx$
$\therefore$ R is symmetric.
(3) If xRy and yRz then $x - y = 6n$ and $y - z = 6m$, $n, m \in \mathbb{Z}$
$\Rightarrow x - y + y - z = 6n + 6m$
$\Rightarrow x - z = 6(n + m)$ where $n + m \in \mathbb{Z}$
$\Rightarrow xRz$
$\therefore$ R is transitive.
From (1), (2), and (3), R is an equivalence relation on $\mathbb{Z}$.

b Each integer belongs to exactly one equivalence class containing all integers which have the same remainder on division by 6. These equivalence classes are [0], [1], [2], [3], [4], and [5]. For example, [3] represents all integers of the form $6n + 3$. 27 is in this class as $27 = 6(4) + 3$.

4 **a** $(a, b)R(x, y) \Leftrightarrow |x| + |y| = |a| + |b|$ on $\mathbb{R} \times \mathbb{R}$.
(1) $|x| + |y| = |x| + |y|$
$\therefore$ $(x, y)R(x, y)$ for all $(x, y) \in \mathbb{R} \times \mathbb{R}$.
$\therefore$ R is reflexive.
(2) If $(a, b)R(x, y) \Rightarrow |x| + |y| = |a| + |b|$
$\Rightarrow |a| + |b| = |x| + |y|$
$\Rightarrow (x, y)R(a, b)$
$\therefore$ R is symmetric.
(3) If $(a, b)R(c, d)$ and $(c, d)R(e, f)$ then
$|c| + |d| = |a| + |b|$ and $|e| + |f| = |c| + |d|$
$\Rightarrow |e| + |f| = |a| + |b|$
$\Rightarrow (a, b)R(e, f)$
$\therefore$ R is transitive.
From (1), (2), and (3), R is an equivalence relation on $\mathbb{R} \times \mathbb{R}$.

b Each point (a, b) is an element of an equivalence class containing all points lying on a square with equation $|x| + |y| = c$, say, where $c = |a| + |b|$.

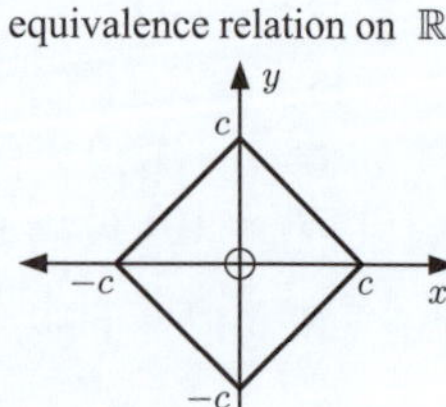

Note: When $a = b = 0$, $c = 0$ and $\{(0, 0)\}$ is an equivalence class, which has only one element.

5 **a** $\mathbb{R} \to \mathbb{R}$ and $f(x) = x^3 + 5$

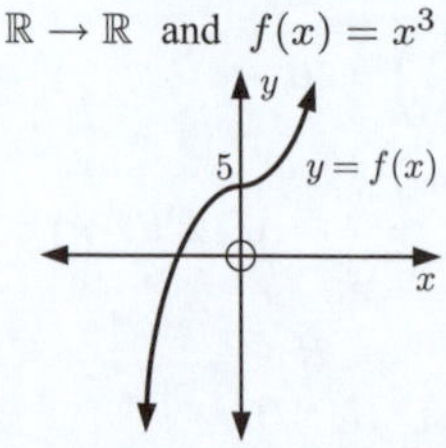

Any horizontal line cuts the function at most once
$\therefore$ f is one-to-one.
The range of f is $\mathbb{R}$
$\therefore$ f is onto.
$\therefore$ $f(x) = x^3 + 5$ is a bijection.

$\therefore\ f^{-1}$ exists and $x = y^3 + 5$

$\therefore\ y^3 = x - 5$

$\therefore\ y = \sqrt[3]{x-5}$

$\therefore\ f^{-1}(x) = \sqrt[3]{x-5}$

b $\mathbb{R}^+ \to \mathbb{R}$ and $f(x) = \ln x$

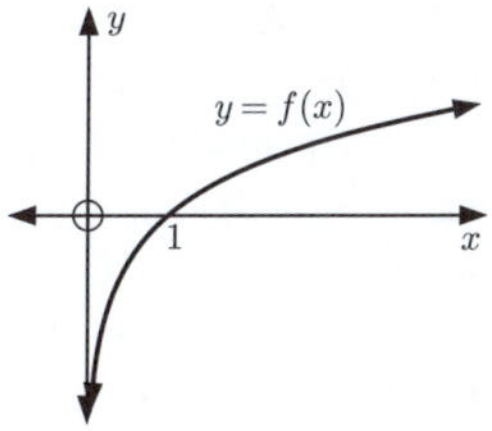

Any horizontal line cuts the curve at most once.
$\{f'(x) = \frac{1}{x} > 0$ for all $x \in \mathbb{R}^+\}$
$\therefore\ f(x)$ is increasing for all $x \in \mathbb{R}^+$.
Thus f is one-to-one and its range is $\mathbb{R} \Rightarrow f$ is onto.

$\therefore\ f(x) = \ln x$ is a bijection.

$\therefore\ f^{-1}$ exists and $x = \ln y$

$\therefore\ y = e^x$

$\therefore\ f^{-1}(x) = e^x$

c $f : \mathbb{Z} \to \mathbb{Z},\ f(x) = 2x$

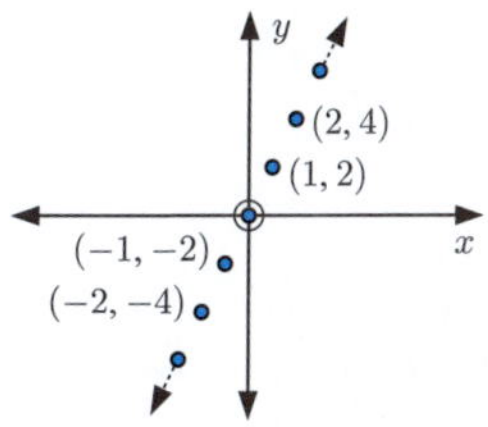

Range of $f = \{$even integers$\}$ $\neq \mathbb{Z}$
$\therefore\ f$ is not onto.
$\therefore\ f$ is not a bijection.

d $f : \mathbb{R} \to \mathbb{R},\ f(x) = 2x$

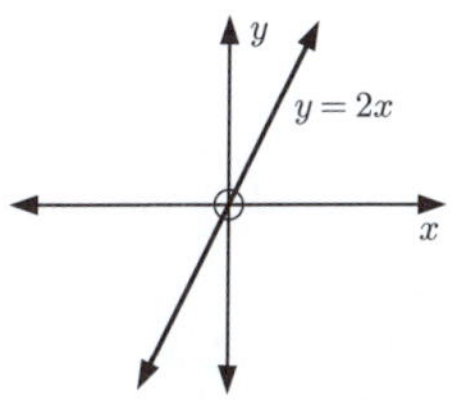

f is one-to-one
{no horizontal line cuts the graph more than once}
and range of $f = \mathbb{R}$
$\therefore\ f$ is onto.

Hence, f is a bijection with inverse $x = 2y$

$$\therefore\ y = f^{-1}(x) = \frac{x}{2}$$

e $f : \mathbb{R} \to [-1, 1],\ f(x) = \sin x$ is not a bijection as, for example, the horizontal line $y = 0$ cuts the graph more than once.

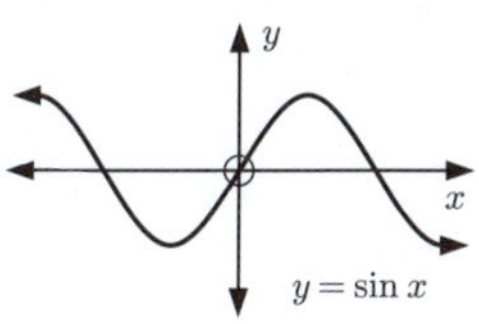

6 **a** $a * b = ab + 2$

i Not associative, as for example,

$(1 * 1) * 2$	and	$1 * (1 * 2)$
$= (1 + 2) * 2$		$= 1 * (2 + 2)$
$= 3 * 2$		$= 1 * 4$
$= 6 + 2 = 8$		$= 4 + 2 = 6$

and $8 \neq 6$

ii $b * a = ba + 2 = ab + 2 = a * b$
{as $\mathbb{R}$ under $\times$ is commutative}
$\therefore\ a * b = b * a$ for all $a, b \in \mathbb{R}$
$\therefore\ *$ is commutative on $\mathbb{R}$.

iii Suppose $a * x = x * a = a$ for all $a \in \mathbb{R}$

$\therefore\ ax + 2 = xa + 2 = a$

$\therefore\ x = \dfrac{a-2}{a}$ which is not unique for $a \in \mathbb{R}$

$\therefore$ no identity exists.

iv If there is no identity there cannot be inverses.

b $a * b = |a + b|$

i Not associative, as for example,

$(-1 * -1) * 2$	and	$-1 * (-1 * 2)$
$= \lvert -2 \rvert * 2$		$= -1 * \lvert 1 \rvert$
$= 2 * 2$		$= -1 * 1$
$= \lvert 4 \rvert = 4$		$= \lvert 0 \rvert = 0$

and $4 \neq 0$

ii $b * a = |b + a| = |a + b| = a * b$ for all $a, b \in \mathbb{R}$
$\therefore\ *$ is commutative on $\mathbb{R}$.

iii Suppose $a * x = a$ for all $a \in \mathbb{R}$

$\therefore\ |a + x| = a$ for all $a \in \mathbb{R}$

$\therefore\ a + x = \pm a$ for all $a \in \mathbb{R}$

$\therefore\ x = 0$ or $-2a$ for all $a \in \mathbb{R}$

As x is not unique, no identity exists.

iv As no identity exists, no inverses exist.

c $a * b = |ab|$

i

$(a * b) * c$	and	$a * (b * c)$
$= \lvert ab \rvert * c$		$= a * \lvert bc \rvert$
$= \lvert \lvert ab \rvert\, c \rvert$		$= \lvert a \lvert bc \rvert \rvert$
$= \lvert ab \rvert \lvert c \rvert$		$= \lvert a \rvert \lvert bc \rvert$
$= \lvert abc \rvert$		$= \lvert abc \rvert$

{using laws of modulus}

$\therefore\ (a * b) * c = a * (b * c)$ for all $a, b, c \in \mathbb{R}$
$\therefore\ *$ is associative on $\mathbb{R}$.

ii $b * a = |ba| = |ab| = a * b$ for all $a, b \in \mathbb{R}$
$\therefore\ *$ is commutative on $\mathbb{R}$.

iii Suppose $a * x = a$ for all $a \in \mathbb{R}$

$\therefore\ |ax| = a$ for all $a \in \mathbb{R}$

$\therefore\ ax = \pm a$

$\therefore\ x = \pm 1$

As x is not unique, no identity exists.

iv As no identity exists, no inverses can exist.

7 **a** $p = \begin{pmatrix} 1 & 2 & 3 & 4 \\ 3 & 1 & 2 & 4 \end{pmatrix}$

$= (1\ 3\ 2)$

$\therefore\ p$ has order 3 **{Theorem 12}**

b $q = \begin{pmatrix} 1 & 2 & 3 & 4 \\ 1 & 2 & 4 & 3 \end{pmatrix}$

$= (3\ 4)$

$\therefore\ q$ has order 2 **{Theorem 12}**

c $r = \begin{pmatrix} 1 & 2 & 3 & 4 \\ 2 & 1 & 4 & 3 \end{pmatrix}$

$= (1\ 2)(3\ 4)$

$\therefore$ order of r = LCM $(2, 2) = 2$ **{Theorem 13}**

8 $S = \{I, A, B, C, D\}$

Closure: S is closed under $*$ as the table contains only the elements of S.

Associative: $A * (B * C) = A * D = C$
whereas $(A * B) * C = D * C = A$
$\therefore\ *$ is not associative on S.

Identity: The identity is I as $X * I = I * X = X$ for all $X \in S$.

Inverse: As $A^2 = B^2 = C^2 = D^2 = I$, each element is its own inverse.
That is, $A^{-1} = A$, $B^{-1} = B$, $C^{-1} = C$, $D^{-1} = D$, and $I^{-1} = I$.

Thus only associativity fails.

9 **a** $S = \{1, 3, 5, 9, 11, 13\}$ under $\times_{14}$ has Cayley table:

$\times_{14}$	1	3	5	9	11	13
1	1	3	5	9	11	13
3	3	9	1	13	5	11
5	5	1	11	3	13	9
9	9	13	3	11	1	5
11	11	5	13	1	9	3
13	13	11	9	5	3	1

Closure: As the Cayley table contains only elements in S, S is closed under $\times_{14}$.

Associative: $\times_{14}$ is associative since $\times$ is associative.

Identity: 1 is the identity as $a \times_{14} 1 = 1 \times_{14} a = a$ for all $a \in S$.

Inverse: From the table,
$1^{-1} = 1$, $3^{-1} = 5$, $5^{-1} = 3$, $9^{-1} = 11$, $11^{-1} = 9$, $13^{-1} = 13$
Each element has a unique inverse.

Thus $\{S, \times_{14}\}$ is a group.

b $1^1 = 1$ $\quad 3^2 = 9 \quad 5^2 = 11 \quad 9^2 = 11 \quad 11^2 = 9$
$3^3 = 13 \quad 5^3 = 13 \quad 9^3 = 1 \quad 11^3 = 1$
$3^4 = 11 \quad 5^4 = 9$
$3^5 = 5 \quad 5^5 = 3$
$3^6 = 1 \quad 5^6 = 1$

$13^2 = 1$

$\therefore$ 1 has order 1, 13 has order 2, 9 and 11 have order 3, 3 and 5 have order 6.

c As 3 and 5 have order 6, the group is cyclic, and the generators are 3 and 5.

10 $a \in G$, $a \neq e$, has order 2 $\quad \therefore a^2 = e$
$\therefore a^{-1} = a$

Let $a, b \in G$ be two non-identity elements, and $b \neq a$
$\therefore a^2 = e$ and $b^2 = e$.

Now $a * b \in G$ $\quad \{\{G, *\}$ is a group $\therefore$ closure$\}$
and $a * b \neq e$, so $a * b$ has order 2.

$\therefore (a * b)^2 = e$
$\therefore (a * b) * (a * b) = e$
$\therefore a * b * a * b = e \quad$ {associativity}
$\therefore a^{-1} * (a * b * a * b) * b^{-1} = a^{-1} * e * b^{-1}$
$\therefore \underbrace{a^{-1} * a} * b * a * \underbrace{b * b^{-1}} = a^{-1} * b^{-1}$
$\therefore e * b * a * e = a * b \quad \{a^{-1} = a, \ b^{-1} = b\}$
$\therefore b * a = a * b$

$\therefore$ G is Abelian.

11 $\{A, +_m\}$ is a group, $A = \{0, 1, 2, ..., (m-1)\}$
$\{B, +_{m^2}\}$ is a group, $B = \{0, 1, 2, ..., (m^2-1)\}$
$(a, b) * (x, y) = (a + x \pmod m, \ b + y + mxb \pmod{m^2})$

Closure: For all $(a, b), (c, d) \in G$
$(a, b) * (c, d)$
$= (a + c \pmod m, \ b + d + mcb \pmod{m^2})$
Now $a + c \pmod m \in A$
and $b + d + mcb \pmod{m^2} \in B$
$\therefore (a, b) * (c, d) \in G$
$\therefore *$ is closed on G.

{We will now write $(p \pmod m), q \pmod{m^2})$ as (p, q).}

Associative: $[(a, b) * (x, y)] * (c, d)$
$= (a + x, \ b + y + mxb) * (c, d)$
$= (a + x + c,$
$\quad b + y + mxb + d + mc(b + y + mxb))$
$= (a + x + c,$
$\quad b + y + mxb + d + mcb + mcy + \cancel{m^2bcx})$ $\{\equiv 0\}$
$= (a + x + c, \ b + y + d + m[bx + bc + cy])$
and
$(a, b) * [(x, y) * (c, d)]$
$= (a, b) * (x + c, \ y + d + mcy)$
$= (a + x + c, \ b + y + d + mcy + m(x + c)b)$
$= (a + x + c, \ b + y + d + mcy + mxb + mbc)$
$= (a + x + c, \ b + y + d + m[bx + bc + cy])$
$= [(a, b) * (x, y)] * (c, d)$
$\therefore *$ is associative on G.

Identity: Suppose $(a, b) * (x, y) = (a, b)$
$\therefore (a + x, \ b + y + mxb) = (a, b)$
$\therefore a + x \pmod m = a$
$\therefore x = 0 \quad \{x \in A\}$
and $b + y + m(0)b \pmod{m^2} = b$
$\therefore y = 0 \quad \{y \in B\}$
Check: $(0, 0) * (a, b) = (0 + a, \ 0 + b + ma(0))$
$= (a, b)$
Thus $(a, b) * (0, 0) = (0, 0) * (a, b) = (a, b)$ for all $(a, b) \in G$. Hence $(0, 0)$ is the identity.

Inverse: Suppose $(a, b) * (x, y) = (0, 0)$
$\therefore (a + x, \ b + y + mxb) = (0, 0)$
$\therefore a + x \pmod m = 0$
$\therefore x = m - a \quad \{x \in A\}$
and $b + y + m(m - a)b \pmod{m^2} = 0$
$\therefore b + y - mab \pmod{m^2} = 0$
$\therefore y = b(ma - 1) \pmod{m^2}$
$\therefore (a, b)^{-1} = (m - a, \ b(ma - 1))$
Check: $(m - a, \ b(ma - 1)) * (a, b)$
$= (m - a + a,$
$\quad b(ma - 1) + b + mab(ma - 1))$
$= (m, \ \cancel{abm} - \cancel{b} + \cancel{b} + m^2a^2b - \cancel{abm})$
$= (m, \ m^2a^2b)$
$= (0, 0)$

$\therefore$ G is a group under $*$ $\quad$ {the 4 axioms are satisfied}

Consider elements $(0, 2)$ and $(1, 0)$, where $m = 5$:
$(0, 2) * (1, 0) = (0 + 1, \ 2 + 0 + 5(1)(2))$
$= (1, 12)$

and $(1, 0) * (0, 2) = (1+0,\ 0+2+5(0)(0))$
$= (1, 2)$

$\therefore$ $*$ is not Abelian.

The order of G = order of $A \times$ order of B
$= m \times m^2 = m^3$

12 $|G| = p$, an odd prime

Suppose $a \in G$, then $|a|$ is a factor of $|G| = p$
{Lagrange's theorem, Corollary 1}

$\therefore$ $|a| = 1$ or p .... (1)

Now if a is its own inverse, $a = a^{-1}$
$\therefore$ $aa = aa^{-1}$
$\therefore$ $a^2 = e$
$\therefore$ $a = e$ or $|a| = 2$
But $|a| \neq 2$ {from (1)}
$\therefore$ $a = e$

$\therefore$ e is the unique element of G which is its own inverse.

13 $\{H_1, *\}$ and $\{H_2, *\}$ are subgroups of $\{G, *\}$, where $\{G, *\}$ is a group.

Let $a \in H_1 \cup H_2$ and $b \in H_1 \cup H_2$

$\therefore$ $a \in H_1$ or H_2 and $b \in H_1$ or H_2

Suppose $a \in H_1$ and $b \in H_2$
$\Rightarrow$ $b^{-1} \in H_2$

Since b^{-1} is not necessarily $\in H_1$, we cannot say that $a * b^{-1} \in H_1$.

Likewise, since a is not necessarily $\in H_2$, we cannot say that $a * b^{-1} \in H_2$.

$\therefore$ $a, b \in H_1 \cup H_2 \not\Rightarrow a * b^{-1} \in H_1 \cup H_2$.

$\therefore$ $\{H_1 \cup H_2, *\}$ is not a subgroup of $\{G, *\}$.

For example, $\{\{0, 6\}, +_{12}\}$ and $\{\{0, 4, 8\}, +_{12}\}$ are subgroups of $\{\mathbb{Z}_{12}, +_{12}\}$, but $\{\{0, 4, 6, 8\}, +_{12}\}$ is not a subgroup as it is not closed.

14 $f : \{\mathbb{R}^+, \times\} \to \{\mathbb{R}^+, \times\}$ and $f(x) = x^2$
$g : \{\mathbb{R}^+, \times\} \to \{\mathbb{R}, +\}$ and $g(x) = \ln x$

a **i** $(f \circ g)(x) = f(g(x))$
$= f(\ln x)$
$= (\ln x)^2$

ii $(g \circ f)(x) = g(f(x))$
$= g(x^2)$
$= \ln(x^2)$

b The domain of f is $\mathbb{R}^+$ and the domain of g is $\mathbb{R}^+$.

$\therefore$ $(f \circ g)(x) = (\ln x)^2$ can only be defined for $x \in \mathbb{R}^+$ such that $g(x) \in \mathbb{R}^+$, the domain of f and for $\ln x > 0$, $x > 1$

$\therefore$ $x \in]1, \infty[$

$\therefore$ $f \circ g$ has domain $]1, \infty[$.

Thus $f \circ g$ is **not** an isomorphism from $\{\mathbb{R}^+, \times\}$ onto $\{\mathbb{R}^+, \times\}$.

c $g \circ f$ is an isomorphism.

15 **a** $f(a+b) = 3^{a+b}$
$= 3^a \times 3^b$
$= f(a) \times f(b)$

$\therefore$ f is a homomorphism.

i $\text{Ker}(f) = \{x \mid f(x) = 1\}$
$= \{x \mid 3^x = 1\}$
$= \{0\}$

$\therefore$ since the kernel contains only the identity in $\{\mathbb{R}, +\}$, f is one-to-one.

$R(f) = \{f(x) \mid x \in \{\mathbb{R}, +\}\}$
$= \{3^x \mid x \in \mathbb{R}\}$
$= \mathbb{R}^+$

$\therefore$ f is onto.

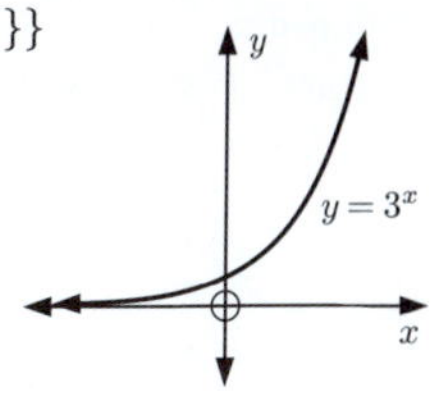

ii Since f is one-to-one and onto and a homomorphism, f is an isomorphism.

b G is a finite group of order 4.

$G = \{e, p, p^2, p^3\}$
$= \{e, (1\ 4\ 3\ 2), (1\ 3)(2\ 4), (1\ 2\ 3\ 4)\}$

$f(p^m p^n) = f(p^{m+n})$
$= i^{m+n}$
$= i^m \times i^n$
$= f(p^m) \times f(p^n)$

$\therefore$ f is a homomorphism.

i $\text{Ker}(f) = \{p^m \mid f(p^m) = i^m = 1,$ the identity in $\{\mathbb{C} \setminus \{0\}, \times\}\}$
$= \{p^0 = e\}$

$\therefore$ the kernel contains only the identity in G.

$\therefore$ f is one-to-one.

$R(f) = \{f(e), f(p), f(p^2), f(p^3)\} = \{i^0, i, i^2, i^3\}$
$= \{1, i, -1, -i\}$

$R(f) \neq \mathbb{C} \setminus \{0\}$

$\therefore$ f is not onto.

ii Since f is one-to-one but not onto, f is not an isomorphism. Note that we could deduce this directly from the fact that G is finite but $\{\mathbb{C} \setminus \{0\}, \times\}$ is an infinite group.

REVIEW SET C

1 $P(A) = \{\varnothing, \{1\}, \{2\}, \{1, 2\}\}$

a Under $\cap$

$\cap$	$\varnothing$	$\{1\}$	$\{2\}$	$\{1, 2\}$
$\varnothing$	$\varnothing$	$\varnothing$	$\varnothing$	$\varnothing$
$\{1\}$	$\varnothing$	$\{1\}$	$\varnothing$	$\{1\}$
$\{2\}$	$\varnothing$	$\varnothing$	$\{2\}$	$\{2\}$
$\{1, 2\}$	$\varnothing$	$\{1\}$	$\{2\}$	$\{1, 2\}$

Although $\{1, 2\}$ is the identity e, not all elements of $P(A)$ have an inverse.

For example, $\{2\} \cap a \neq \{1, 2\}$ for any $a \in P(A)$.

$\therefore$ $P(A)$ cannot be a group under $\cap$.

b Under $\cup$

$\cup$	$\varnothing$	$\{1\}$	$\{2\}$	$\{1, 2\}$
$\varnothing$	$\varnothing$	$\{1\}$	$\{2\}$	$\{1, 2\}$
$\{1\}$	$\{1\}$	$\{1\}$	$\{1, 2\}$	$\{1, 2\}$
$\{2\}$	$\{2\}$	$\{1, 2\}$	$\{2\}$	$\{1, 2\}$
$\{1, 2\}$	$\{1, 2\}$	$\{1, 2\}$	$\{1, 2\}$	$\{1, 2\}$

$\varnothing$ is the identity under $\cup$.

However, not all elements of $P(A)$ have an inverse.

For example, $\{2\} \cup a \neq \varnothing$ for any $a \in P(A)$, hence $\{2\}$ has no inverse.

$\therefore$ $P(A)$ cannot be a group under $\cup$.

2 **To prove:** $(A\backslash B) \times C = (A \times C) \backslash (B \times C)$

Proof: $(\Rightarrow)$ Let $(x, y) \in (A\backslash B) \times C$

then $x \in A\backslash B$ and $y \in C$

$\therefore$ $x \in A$ and $x \notin B$ and $y \in C$

$\therefore$ $x \in A$ and $y \in C$, but $(x, y) \notin B \times C$

$\therefore$ $(x, y) \in (A \times C) \backslash (B \times C)$

$\therefore$ $(A\backslash B) \times C \subseteq (A \times C) \backslash (B \times C)$ .... (1)

$(\Leftarrow)$ Let $(x, y) \in (A \times C) \backslash (B \times C)$

$\therefore$ $(x, y) \in A \times C$, but $(x, y) \notin B \times C$

$\therefore$ $x \in A$ and $y \in C$, but $x \notin B$

$\therefore$ $x \in A\backslash B$ and $y \in C$

$\therefore$ $(x, y) \in (A\backslash B) \times C$

$\therefore$ $(A \times C) \backslash (B \times C) \subseteq (A\backslash B) \times C$ .... (2)

From (1) and (2), $(A\backslash B) \times C = (A \times C) \backslash (B \times C)$.

3 $(a, b)R(x, y) \Leftrightarrow bx^2 = a^2y$ for $(a, b), (x, y) \in \mathbb{R} \backslash \{0\} \times \mathbb{R}$.

a (1) $(x, y)R(x, y)$ is true as $yx^2 = x^2y$ $\{\times$ is commutative on $\mathbb{R}\}$

$\therefore$ R is reflexive.

(2) If $(a, b)R(x, y)$

$\Rightarrow bx^2 = a^2y$

$\Rightarrow ya^2 = x^2b$ $\{\times$ is commutative on $\mathbb{R}\}$

$\Rightarrow (x, y)R(a, b)$

$\therefore$ R is symmetric.

(3) If $(a, b)R(c, d)$ **and** $(c, d)R(e, f)$

$\Rightarrow bc^2 = a^2d$ and $de^2 = c^2f$

$$\Rightarrow \frac{c^2}{d} = \underbrace{\frac{a^2}{b} = \frac{e^2}{f}}$$

$\Rightarrow be^2 = a^2f$

$\Rightarrow (a, b)R(e, f)$

$\therefore$ R is transitive.

From (1), (2), and (3), R is an equivalence relation on $\mathbb{R} \backslash \{0\} \times \mathbb{R}$.

b $y = \left(\frac{b}{a^2}\right)x^2$ where $a \neq 0$, $b \neq 0$

$\therefore$ points lie on the parabola $y = kx^2$ with $(0, 0)$ removed and $k > 0$.

So, each point (a, b) belongs to one of these parabolas.

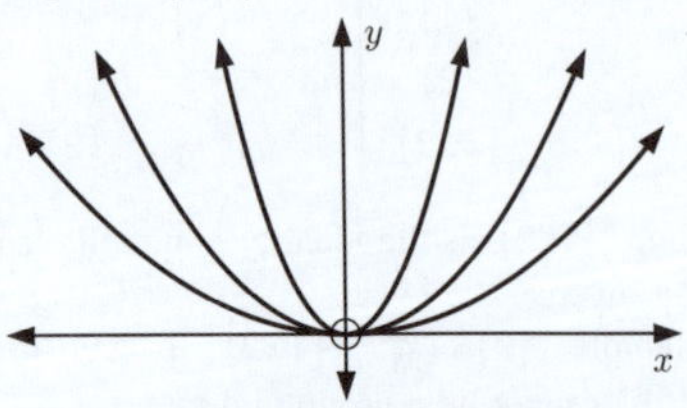

4 This is not correct. The argument assumes that xRy for some $y \in S$. x may not be related to any other element in the set.

5 **a** $f_1(x) = x$, $f_2(x) = -x$, $f_3(x) = \frac{1}{x}$, $f_4(x) = -\frac{1}{x}$

$(f_1 \circ f_1)(x) = f_1(f_1(x)) = f_1(x)$

$(f_1 \circ f_2)(x) = f_1(f_2(x)) = f_2(x)$

$(f_1 \circ f_3)(x) = f_1(f_3(x)) = f_3(x)$

$(f_1 \circ f_4)(x) = f_1(f_4(x)) = f_4(x)$

$(f_2 \circ f_1)(x) = f_2(f_1(x)) = f_2(x)$

$(f_2 \circ f_2)(x) = f_2(f_2(x)) = f_2(-x) = -(-x) = x = f_1(x)$

$(f_2 \circ f_3)(x) = f_2(f_3(x)) = f_2\left(\frac{1}{x}\right) = -\frac{1}{x} = f_4(x)$

$(f_2 \circ f_4)(x) = f_2(f_4(x)) = f_2\left(-\frac{1}{x}\right) = -\left(-\frac{1}{x}\right) = \frac{1}{x} = f_3(x)$

$(f_3 \circ f_1)(x) = f_3(f_1(x)) = f_3(x)$

$(f_3 \circ f_2)(x) = f_3(f_2(x)) = f_3(-x) = \frac{1}{-x} = f_4(x)$

$(f_3 \circ f_3)(x) = f_3(f_3(x)) = f_3\left(\frac{1}{x}\right) = \frac{1}{\frac{1}{x}} = x = f_1(x)$

$(f_3 \circ f_4)(x) = f_3(f_4(x)) = f_3\left(-\frac{1}{x}\right) = \frac{1}{-\frac{1}{x}} = -x = f_2(x)$

$(f_4 \circ f_1)(x) = f_4(f_1(x)) = f_4(x)$

$(f_4 \circ f_2)(x) = f_4(f_2(x)) = f_4(-x) = \frac{-1}{-x} = \frac{1}{x} = f_3(x)$

$(f_4 \circ f_3)(x) = f_4(f_3(x)) = f_4\left(\frac{1}{x}\right) = \frac{-1}{\frac{1}{x}} = -x = f_2(x)$

$(f_4 \circ f_4)(x) = f_4(f_4(x)) = f_4\left(-\frac{1}{x}\right) = \frac{-1}{-\frac{1}{x}} = x = f_1(x)$

$\therefore$ the Cayley table is:

$\circ$	f_1	f_2	f_3	f_4
f_1	f_1	f_2	f_3	f_4
f_2	f_2	f_1	f_4	f_3
f_3	f_3	f_4	f_1	f_2
f_4	f_4	f_3	f_2	f_1

Closure: In the table every entry is either f_1, f_2, f_3, or f_4. $\therefore$ G is closed under $\circ$.

Associative: Since function composition is associative, $\circ$ is associative in G.

Identity: f_1 is the identity as $f_1 \circ f_i = f_i \circ f_1 = f_i$ $(i = 1, 2, 3, 4)$.

Inverse: As $f_i \circ f_i = f_1$ for $i = 1, 2, 3, 4$

$f_i^{-1} = f_i$

$\therefore$ each element is its own inverse.

Thus $\{G, \circ\}$ is a group.

b $|f_1| = 1$ and $|f_2| = |f_3| = |f_4| = 2$

c G is not cyclic as no element has order 4 (the order of G).

d The Klein 4-group has Cayley table:

$*$	i	p	q	r
i	i	p	q	r
p	p	i	r	q
q	q	r	i	p
r	r	q	p	i

$f_1 \mapsto i$

$f_2 \mapsto p$

$f_3 \mapsto q$

$f_4 \mapsto r$

$\therefore$ $G \to$ Klein 4 is one-to-one, onto, and a homomorphism since the tables have the same structure.

$\therefore$ $G \cong$ Klein 4

6 **a** $a \circ b = \dfrac{1}{ab}$ on $S = \mathbb{R} \setminus \{0\}$

i $(1 \circ 2) \circ 3 = \frac{1}{2} \circ 3 = \dfrac{1}{1\frac{1}{2}} = \frac{2}{3}$

and $1 \circ (2 \circ 3) = 1 \circ \frac{1}{6} = \dfrac{1}{\frac{1}{6}} = 6$

$\therefore$ in general $(a \circ b) \circ c \neq a \circ (b \circ c)$

$\therefore$ $\circ$ is not associative.

ii $b \circ a = \dfrac{1}{ba} = \dfrac{1}{ab} = a \circ b$ for all $a, b \in S$

$\therefore$ $\circ$ is commutative on S.

iii Suppose $a \circ x = x \circ a = a$ for some $x \in S$

$\Rightarrow \dfrac{1}{ax} = \dfrac{1}{xa} = a$

$\Rightarrow x = \dfrac{1}{a^2}$ which is not unique.

$\therefore$ no identity exists.

iv As no identity exists, inverses cannot be found.

b $a \circ b = (a+2)(b+3)$

i $(1 \circ 0) \circ 1 = (3 \times 3) \circ 1 = 9 \circ 1 = 11 \times 4 = 44$

$1 \circ (0 \circ 1) = 1 \circ (2 \times 4) = 1 \circ 8 = 3 \times 11 = 33$

$\therefore$ in general $(a \circ b) \circ c \neq a \circ (b \circ c)$ for all $a, b, c \in \mathbb{R}$.

$\therefore$ $\circ$ is not associative on $\mathbb{R}$.

ii $1 \circ 0 = 3 \times 3 = 9$ and

$0 \circ 1 = 2 \times 4 = 8$

$\therefore$ in general $a \circ b \neq b \circ a$ for all $a, b \in \mathbb{R}$

$\therefore$ $\circ$ does not commute in $\mathbb{R}$.

iii Suppose $a \circ x = a$ for all $a \in \mathbb{R}$

$\therefore$ $(a+2)(x+3) = a$

$\therefore$ $x = -3 + \dfrac{a}{a+2}$

$\therefore$ x is not unique.

$\therefore$ no identity exists in $\mathbb{R}$.

iv As no identity exists, inverses cannot be found.

c $a \circ b = a + b + 3ab$

i $(a \circ b) \circ c$
$= (a + b + 3ab) \circ c$
$= a + b + 3ab + c + 3(a + b + 3ab)c$
$= a + b + c + 3ab + 3ac + 3bc + 9abc$

and $a \circ (b \circ c)$
$= a \circ (b + c + 3bc)$
$= a + b + c + 3bc + 3a(b + c + 3bc)$
$= a + b + c + 3bc + 3ab + 3ac + 9abc$
$= (a \circ b) \circ c$ for all $a, b, c \in \mathbb{R}$

$\therefore$ $\circ$ is associative on $\mathbb{R}$.

ii $b \circ a$
$= b + a + 3ba$
$= a + b + 3ab$ $\{\times$ and $+$ are commutative on $S\}$
$= a \circ b$

$\therefore$ $a \circ b = b \circ a$ for all $a, b \in \mathbb{R}$

$\therefore$ $\circ$ is commutative on $\mathbb{R}$.

iii If $a \circ e = e \circ a = a$ for all $a \in \mathbb{R}$

$a + e + 3ae = e + a + 3ea = a$

$\Rightarrow e + 3ae = 0$

$\Rightarrow e(1 + 3a) = 0$ for all $a \in \mathbb{R}$

$\Rightarrow e = 0$

Check: $a \circ 0 = a + 0 + 3a(0) = a$ ✓

$0 \circ a = 0 + a + 3(0)a = a$ ✓

Thus the identity is 0.

iv If $a \circ x = x \circ a = 0$ for all $a \in \mathbb{R}$

then $a + x + 3ax = x + a + 3xa = 0$

$\Rightarrow x(1 + 3a) = -a$

$\Rightarrow x = -\dfrac{a}{3a+1}, \quad a \neq -\frac{1}{3}$

$\therefore$ $-\dfrac{a}{3a+1}$ is the inverse of a for all $a \in \mathbb{R} \setminus \{-\frac{1}{3}\}$.

7 **a** $\begin{pmatrix} 1 & 2 & 3 \\ 1 & 2 & 3 \end{pmatrix} = (1)(2)(3)$, $\begin{pmatrix} 1 & 2 & 3 \\ 1 & 3 & 2 \end{pmatrix} = (2\ 3)$

$\begin{pmatrix} 1 & 2 & 3 \\ 2 & 1 & 3 \end{pmatrix} = (1\ 2)$, $\begin{pmatrix} 1 & 2 & 3 \\ 2 & 3 & 1 \end{pmatrix} = (1\ 2\ 3)$

$\begin{pmatrix} 1 & 2 & 3 \\ 3 & 1 & 2 \end{pmatrix} = (1\ 3\ 2)$, $\begin{pmatrix} 1 & 2 & 3 \\ 3 & 2 & 1 \end{pmatrix} = (1\ 3)$

b **i** If $p = (2\ 3)$ then $p^{-1} = (3\ 2) = (2\ 3)$

ii If $p = (1\ 3\ 2)$ then $p^{-1} = (2\ 3\ 1) = (1\ 2\ 3)$

c $(1)(2)(3) = e$ has order 1

Using **Theorem 12**: $(1\ 2)$ has order 2
$(1\ 3)$ has order 2
$(2\ 3)$ has order 2
$(1\ 2\ 3)$ has order 3
$(1\ 3\ 2)$ has order 3

8 **a**

$\times_{16}$	1	7	9	15
1	1	7	9	15
7	7	1	15	9
9	9	15	1	7
15	15	9	7	1

Closure: S is closed under $\times_{16}$ as every element in the table is in S.

Associative: $\times_{16}$ is associative as $\times$ in $\mathbb{R}$ is associative.

Identity: The identity is 1 as $a \times_{16} 1 = 1 \times_{16} a = a$ for all $a \in S$.

Inverse: As $7^2 = 1$, $9^2 = 1$, $15^2 = 1$, and $1^2 = 1$
$1^{-1} = 1$, $7^{-1} = 7$, $9^{-1} = 9$, and $15^{-1} = 15$.

Thus $\{S, \times_{16}\}$ is a group.

b 1 has order 1; 7, 9, and 15 have order 2

c The group is not cyclic as $|S| = 4$ and no element of S has order 4.

9 $(a, b) * (c, d) = (ac,\ bc + d)$

a $[(a, b) * (c, d)] * (e, f)$
$= (ac,\ bc + d) * (e, f)$
$= (ace,\ (bc + d)e + f)$
$= (ace,\ bce + de + f)$

Also $(a, b) * [(c, d) * (e, f)]$
$= (a, b) * (ce,\ de + f)$
$= (ace,\ bce + de + f)$
$= [(a, b) * (c, d)] * (e, f)$

$\therefore$ $*$ is associative on S.

b $*$ is not commutative.
For example,

$$\begin{aligned}&(0, 1) * (2, 1) \\ &= (0, 2+1) \\ &= (0, 3)\end{aligned} \quad \text{and} \quad \begin{aligned}&(2, 1) * (0, 1) \\ &= (0, 0+1) \\ &= (0, 1) \\ &\neq (0, 3)\end{aligned}$$

c Consider $(a, b) * (x, y) = (a, b)$ for all $(a, b) \in \mathbb{R} \times \mathbb{R}$.

$\therefore \quad (ax,\ bx + y) = (a, b)$

$\Rightarrow \quad ax = a$ and $bx + y = b$

$\Rightarrow \quad x = 1$ and $b + y = b$

$\therefore \quad y = 0$

Also $(1, 0) * (a, b) = (1 \times a,\ 0(a) + b) = (a, b)$

$\therefore \quad (1, 0)$ is the identity.

d Suppose $(a, b) * (p, q) = (1, 0)$, the identity

$\therefore \quad (ap,\ bp + q) = (1, 0)$

$\therefore \quad ap = 1$ and $bp + q = 0$

$\therefore \quad p = \dfrac{1}{a}$ and $q = -b\left(\dfrac{1}{a}\right)$, $a \neq 0$

Thus $(0, b)$ has no inverse.

e $\{S, *\}$ is not a group since not every element has an inverse.

10 $\{G, *\}$ is a finite group of order n, with identity e.
Suppose $a \in G$,

$\therefore \quad |a| \mid |G|$ {Lagrange's theorem Corollary}

$\therefore \quad |a| \mid n$

$\therefore \quad n = |a|\, s$ for some $s \in \mathbb{Z}^+$

$\therefore \quad a^n = a^{|a|s}$

$\therefore \quad a^n = \left(a^{|a|}\right)^s$

$\therefore \quad a^n = e^s$

$\therefore \quad a^n = e$

11 $\{G, *\}$ is a finite group of order n, identity e.
H is a non-empty subset of G and $a * b \in H$ for all $a, b \in H$.

To prove: $\{H, *\}$ is a group.

Closure: As $a*b \in H$ for all $a, b \in H$ then H is closed under $*$.

Associative: $\{G, *\}$ is a group and $*$ is associative in G. As $H \subseteq G$, $*$ is associative in H also.

Identity: From question **10**, as G has order n, $a^n = e$ for all $a \in G$.
Now for $a \in H$, $a^n \in H$ {closure in H}
$\therefore$ the identity $e \in H$.

Inverse: If $a \in H$, a has a finite order since $a \in G$ and G is a finite group.
If $|a| = m$ then $a^m = e$, $m \in \mathbb{Z}^+$
$\therefore \quad a * a^{m-1} = a^{m-1} * a = e$
So, if $a \neq e$, $a^{-1} = a^{m-1}$ $\{m - 1 \in \mathbb{Z}^+\}$
and $a^{m-1} \in H$. {closure}
Thus, each element in H has in inverse in H.

Hence $\{H, *\}$ is a group (and therefore a subgroup of $\{G, *\}$).

12 $S = \{1, -1\} \subseteq G = \{1, -1, i, -i\}$

$\times$	1	-1
1	1	-1
-1	-1	1

S is a non-empty subset of the finite set G and $\{S, *\}$ is closed.

$\therefore$ by the subgroup test for finite groups, $S < G$.

$T = \{i, -1\}$ does not contain the identity $1 \in G$, which is unique in G.

$\therefore$ as T has no identity $\{T, \times\}$ cannot be a group.

$\therefore \quad T \not< G$

13 Let $G = \langle g \rangle = \{e, g, g^2, g^3, ..., g^{m-1}\}$ where $m \in \mathbb{Z}^+$.
$e = g^0 = g^m$ So, G is a cyclic group of order m.
$\{\mathbb{Z}_m, +_m\} = \{\{0, 1, 2, 3, ..., m-1\}, +_m\}$
Consider $f : \{G, *\} \to \{\mathbb{Z}_m, +_m\}$
where $g^i \mapsto i$ for $i = 0, 1, 2, 3, ..., m - 1$.
Now f is one-to-one and onto and

$$\begin{aligned} f(g^i * g^j) &= f(g^{i+j}), \quad i, j \in \mathbb{Z}_m \\ &= f(g^{i+j\,(\text{mod}\, m)}) \quad \{g^m = e\} \\ &= i + j\ (\text{mod}\, m) \\ &= f(g^i) +_m f(g^j) \end{aligned}$$

$\therefore \quad f$ is an isomorphism.

$\therefore \quad \{G, *\} \cong \{\mathbb{Z}_m, +_m\}$

14 $f : \{G, *\} \to \{H, \circ\}$ where G and H are finite.

To prove: f is an isomorphism $\Leftrightarrow$ $\text{Ker}(f) = \{e_G\}$

Proof: f is an isomorphism
$\Leftrightarrow$ f is one-to-one and onto
$\Leftrightarrow$ f is one-to-one
{as the groups are finite, f is one-to-one $\Leftrightarrow$ f is onto}
$\Leftrightarrow$ $\text{Ker}(f) = \{e_G\}$ {**Theorem 22**}

15 **a** G is the only left coset of G which is a subgroup of S_5.
$G = \{e, p, p^2, p^3, p^4\}$
$= \{e, (1\,5\,3\,4\,2), (1\,3\,2\,5\,4), (1\,4\,5\,2\,3), (1\,2\,4\,3\,5)\}$

b G has $\dfrac{5!}{5} = 4! = 24$ distinct left cosets in S_5.

c $q = (1\,3\,2)$
$\therefore \quad q^{-1} = (2\,3\,1)$
$r = (1\,4\,5)$
$\therefore \quad q^{-1}r = (2\,3\,1)(1\,4\,5) = (1\,4\,5\,2\,3)$

d Using **a**, $q^{-1}r \in G$
$\therefore \quad qG = rG$
$\therefore \quad qG, rG$ are **not** disjoint, they are equal.

16 **a** $\text{Ker}(f) = \{a \in G \mid f(a) = e_H\}$ is clearly a subset of G.
$e_G \in \text{Ker}(f)$ {given theorem}
$\therefore \quad \text{Ker}(f)$ is non-empty.
Suppose a, b are two elements in $\text{Ker}(f)$, so $f(a) = f(b) = e_H$.
Consider

$$\begin{aligned} f(a * b^{-1}) &= f(a) \circ f(b^{-1}) && \{f \text{ is a homomorphism}\} \\ &= e_H \circ (f(b))^{-1} && \{\text{given theorem}\} \\ &= e_H \circ (e_H)^{-1} \\ &= e_H \circ e_H \\ &= e_H \end{aligned}$$

$\therefore \quad a * b^{-1} \in \text{Ker}(f)$

Since $\text{Ker}(f)$ is a non-empty subset of G, and since $a * b^{-1} \in \text{Ker}(f)$ for all $a, b \in \text{Ker}(f)$, by the subgroup test, $\text{Ker}(f) < G$.

b $R(f) = \{f(a) \mid a \in G\}$ is clearly a subset of H.

$f(e_G) = e_H$ {given theorem}

$\therefore\ e_H \in R(f)$

$\therefore\ R(f)$ is non-empty.

Suppose h_1, h_2 are two elements in $R(f)$,

so $h_1 = f(a_1)$

and $h_2 = f(a_2)$ for some elements $a_1, a_2 \in G$

Consider $h_1 \circ {h_2}^{-1}$

$= f(a_1) \circ f(a_2)^{-1}$

$= f(a_1) \circ f({a_2}^{-1})$ {given theorem}

$= f(a_1 * {a_2}^{-1})$

$= f(a)$ for some element $a = a_1 * {a_2}^{-1} \in G$ since $a_1, a_2 \in G$ and G is a group

$\therefore\ h_1 \circ {h_2}^{-1} \in R(f)$

Since $R(f)$ is a non-empty subset of H, and since $h_1 \circ {h_2}^{-1} \in R(f)$ for all $h_1, h_2 \in R(f)$, by the subgroup test, $R(f) < H$.

REVIEW SET D

1 **a**

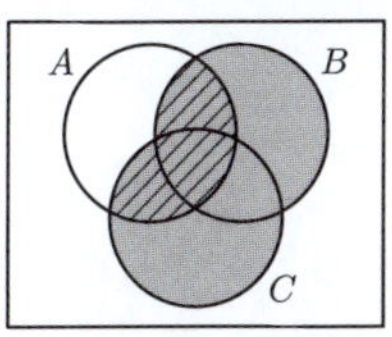

$B \cup C$ [shaded]

$A \cap (B \cup C)$ [hatched]

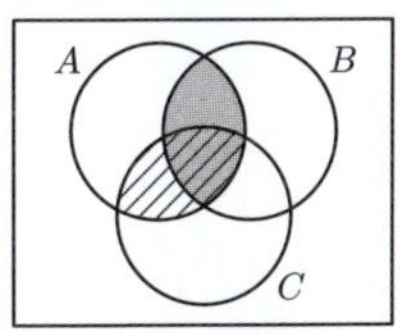

$A \cap B$ [shaded]

$A \cap C$ [hatched]

$(A \cap B) \cup (A \cap C)$ is all shaded parts

b

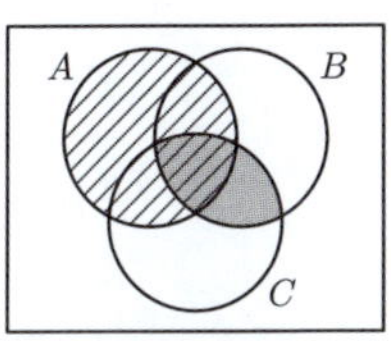

$B \cap C$ [shaded]

A [hatched]

$A \cup (B \cap C)$ is all shaded parts

$\therefore$

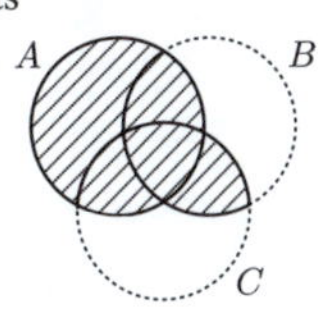

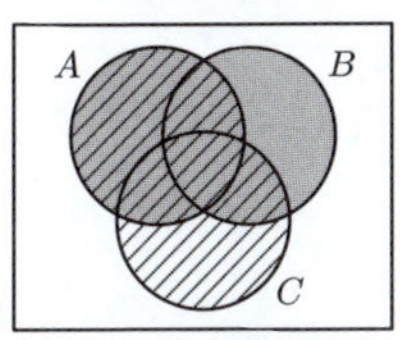

$A \cup B$ [shaded]

$A \cup C$ [hatched]

$(A \cup B) \cap (A \cup C)$ consists of all double shaded regions

$\therefore$

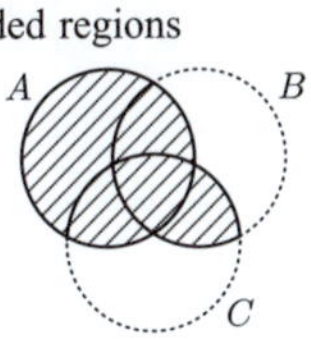

2 $A = \{1, 2, 3\}$

a $P(A) = \{\varnothing, \{1\}, \{2\}, \{3\}, \{1, 2\}, \{2, 3\}, \{1, 3\}, \{1, 2, 3\}\}$

b **i** $P(A)$ under $\cap$.

The identity would be A as for any $S \in P(A)$, $S \cap A = A \cap S = S$.

However, for example, $\{1\}$ does not have an inverse in $P(A)$ as there is no $S \in P(A)$ such that $\{1\} \cap S = A$.

$\therefore\ P(A)$ does not form a group under $\cap$.

ii $P(A)$ under $\cup$.

The identity would be $\varnothing$ as for any $S \in P(A)$, $S \cup \varnothing = \varnothing \cup S = S$.

However, for example, $\{1\}$ does not have an inverse in $P(A)$ as there is no S in $P(A)$ such that $\{1\} \cup A = \varnothing$.

$\therefore\ P(A)$ does not form a group under $\cup$.

3 $(a, b)R(x, y) \Leftrightarrow x^2 + y^2 = a^2 + b^2$ for $(a, b), (x, y) \in \mathbb{R} \times \mathbb{R}$

a $(x, y)R(x, y)$ since $x^2 + y^2 = x^2 + y^2$

$\therefore$ R is reflexive .... (1)

If $(x, y)R(a, b)$ then $x^2 + y^2 = a^2 + b^2$

$\therefore\ a^2 + b^2 = x^2 + y^2$

$\therefore\ (a, b)R(x, y)$

$\therefore$ R is symmetric .... (2)

If $(x, y)R(a, b)$ and $(a, b)R(c, d)$

then $x^2 + y^2 = a^2 + b^2$ and $a^2 + b^2 = c^2 + d^2$

$\Rightarrow x^2 + y^2 = c^2 + d^2$

$\Rightarrow (x, y)R(c, d)$

$\therefore$ R is transitive .... (3)

From (1), (2), and (3), R is an equivalence relation.

b One equivalence class is $\{(0, 0)\}$.

For each $r \in \mathbb{R}^+$, there is an equivalence class $\{(x, y) \mid x^2 + y^2 = r^2\}$, $r^2 = a^2 + b^2$ which is the set of all points on a circle centre $(0, 0)$, radius r.

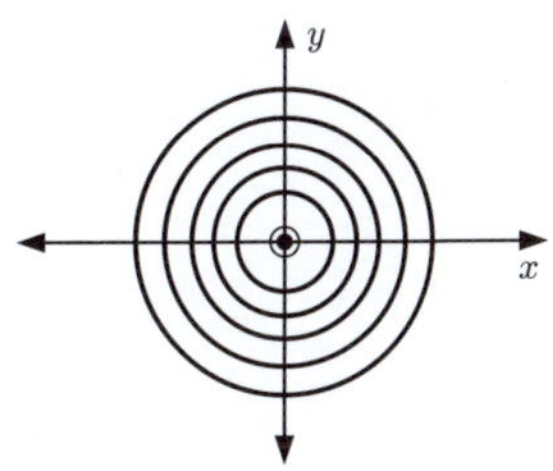

$\{(0, 0)\}$ and five circles are shown.

4 $(a, b)R(x, y) \Leftrightarrow y = b$, $(a, b), (x, y) \in \mathbb{Z} \times \mathbb{Z}$.

a $(x, y)R(x, y)$ since $y = y$

$\therefore$ R is reflexive .... (1)

If $(x, y)R(a, b) \Rightarrow b = y$

$\Rightarrow y = b$

$\Rightarrow (a, b)R(x, y)$

$\therefore$ R is symmetric .... (2)

If $(x, y)R(a, b)$ and $(a, b)R(c, d)$

then $b = y$ and $d = b \Rightarrow d = y$

$\Rightarrow (x, y)R(c, d)$

$\therefore$ R is transitive .... (3)

From (1), (2), and (3), R is an equivalence relation.

b Each point (a, b) belong to an equivalence class consisting of all points with integer coordinates lying in a horizontal line passing through (a, b).

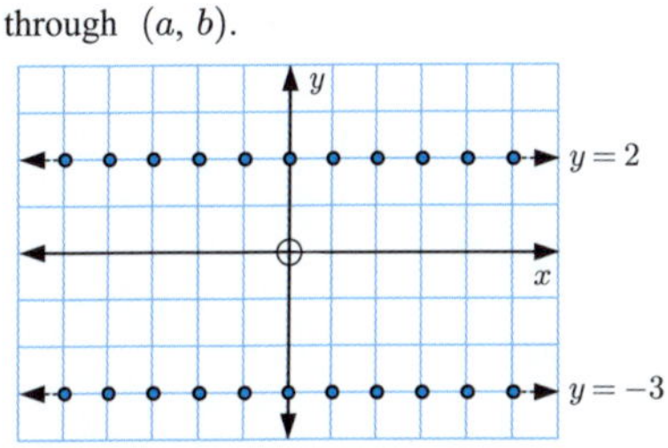

Two such classes are shown.

5 $f_1(x) = x, \quad f_2(x) = \dfrac{1}{1-x}, \quad f_3(x) = \dfrac{x-1}{x},$

$f_4(x) = \dfrac{1}{x}, \quad f_5(x) = 1 - x, \quad f_6(x) = \dfrac{x}{x-1}$

$(f_1 \circ f_1)(x) = f_1(f_1(x)) = f_1(x)$

$(f_1 \circ f_2)(x) = f_1(f_2(x)) = f_2(x)$

$(f_1 \circ f_3)(x) = f_1(f_3(x)) = f_3(x)$

$(f_1 \circ f_4)(x) = f_1(f_4(x)) = f_4(x)$

$(f_1 \circ f_5)(x) = f_1(f_5(x)) = f_5(x)$

$(f_1 \circ f_6)(x) = f_1(f_6(x)) = f_6(x)$

$(f_2 \circ f_1)(x) = f_2(f_1(x)) = f_2(x)$

$$\begin{aligned}(f_2 \circ f_2)(x) &= f_2(f_2(x)) \\ &= f_2\left(\frac{1}{1-x}\right) \\ &= \frac{1}{1-\frac{1}{1-x}} \\ &= \frac{1-x}{1-x-1} \\ &= \frac{x-1}{x} \\ &= f_3(x)\end{aligned}$$

$$\begin{aligned}(f_2 \circ f_3)(x) &= f_2(f_3(x)) \\ &= f_2\left(\frac{x-1}{x}\right) \\ &= \frac{1}{1-\frac{x-1}{x}} \\ &= \frac{x}{x-(x-1)} \\ &= x \\ &= f_1(x)\end{aligned}$$

$$\begin{aligned}(f_2 \circ f_4)(x) &= f_2(f_4(x)) \\ &= f_2\left(\frac{1}{x}\right) \\ &= \frac{1}{1-\frac{1}{x}} \\ &= \frac{x}{x-1} \\ &= f_6(x)\end{aligned}$$

$$\begin{aligned}(f_2 \circ f_5)(x) &= f_2(f_5(x)) \\ &= f_2(1-x) \\ &= \frac{1}{1-(1-x)} \\ &= \frac{1}{x} \\ &= f_4(x)\end{aligned}$$

$$\begin{aligned}(f_2 \circ f_6)(x) &= f_2(f_6(x)) \\ &= f_2\left(\frac{x}{x-1}\right) \\ &= \frac{1}{1-\frac{x}{x-1}} \\ &= \frac{x-1}{x-1-x} \\ &= \frac{x-1}{-1} \\ &= 1-x \\ &= f_5(x)\end{aligned}$$

$(f_3 \circ f_1)(x) = f_3(f_1(x)) = f_3(x)$

$$\begin{aligned}(f_3 \circ f_2)(x) &= f_3(f_2(x)) \\ &= f_3\left(\frac{1}{1-x}\right) \\ &= \frac{\frac{1}{1-x}-1}{\frac{1}{1-x}} \\ &= \frac{1-(1-x)}{1} \\ &= x \\ &= f_1(x)\end{aligned}$$

$$\begin{aligned}(f_3 \circ f_3)(x) &= f_3(f_3(x)) \\ &= f_3\left(\frac{x-1}{x}\right) \\ &= \frac{\frac{x-1}{x}-1}{\frac{x-1}{x}} \\ &= \frac{x-1-x}{x-1} \\ &= \frac{1}{1-x} \\ &= f_2(x)\end{aligned}$$

$$\begin{aligned}(f_3 \circ f_4)(x) &= f_3(f_4(x)) \\ &= f_3\left(\frac{1}{x}\right) \\ &= \frac{\frac{1}{x}-1}{\frac{1}{x}} \\ &= \frac{1-x}{1} \\ &= 1-x \\ &= f_5(x)\end{aligned}$$

$$\begin{aligned}(f_3 \circ f_5)(x) &= f_3(f_5(x)) \\ &= f_3(1-x) \\ &= \frac{1-x-1}{1-x} \\ &= \frac{-x}{1-x} \\ &= \frac{x}{x-1} \\ &= f_6(x)\end{aligned}$$

$$\begin{aligned}(f_3 \circ f_6)(x) &= f_3(f_6(x)) \\ &= f_3\left(\frac{x}{x-1}\right) \\ &= \frac{\frac{x}{x-1}-1}{\frac{x}{x-1}} \\ &= \frac{x-(x-1)}{x} \\ &= \frac{1}{x} \\ &= f_4(x)\end{aligned}$$

$(f_4 \circ f_1)(x) = f_4(f_1(x)) = f_4(x)$

$$\begin{aligned}(f_4 \circ f_2)(x) &= f_4(f_2(x)) \\ &= f_4\left(\frac{1}{1-x}\right) \\ &= \frac{1}{\frac{1}{1-x}} \\ &= 1-x \\ &= f_5(x)\end{aligned}$$

$$\begin{aligned}(f_4 \circ f_3)(x) &= f_4(f_3(x)) \\ &= f_4\left(\frac{x-1}{x}\right) \\ &= \frac{x}{x-1} \\ &= f_6(x)\end{aligned}$$

$$\begin{aligned}(f_4 \circ f_4)(x) &= f_4(f_4(x)) \\ &= f_4\left(\frac{1}{x}\right) \\ &= \frac{1}{\frac{1}{x}} \\ &= x \\ &= f_1(x)\end{aligned}$$

$$\begin{aligned}(f_4 \circ f_5)(x) &= f_4(f_5(x)) \\ &= f_4(1-x) \\ &= \frac{1}{1-x} \\ &= f_2(x)\end{aligned}$$

$$\begin{aligned}(f_4 \circ f_6)(x) &= f_4(f_6(x)) \\ &= f_4\left(\frac{x}{x-1}\right) \\ &= \frac{1}{\frac{x}{x-1}} \\ &= \frac{x-1}{x} \\ &= f_3(x)\end{aligned}$$

$(f_5 \circ f_1)(x) = f_5(f_1(x)) = f_5(x)$

$$\begin{aligned}(f_5 \circ f_2)(x) &= f_5(f_2(x)) \\ &= f_5\left(\frac{1}{1-x}\right) \\ &= 1-\frac{1}{1-x} \\ &= \frac{1-x-1}{1-x} \\ &= \frac{-x}{1-x} \\ &= \frac{x}{x-1} \\ &= f_6(x)\end{aligned}$$

$$\begin{aligned}(f_5 \circ f_3)(x) &= f_5(f_3(x)) \\ &= f_5\left(\frac{x-1}{x}\right) \\ &= 1-\left(\frac{x-1}{x}\right) \\ &= \frac{x-x+1}{x} \\ &= \frac{1}{x} \\ &= f_4(x)\end{aligned}$$

$(f_5 \circ f_4)(x)$
$= f_5(f_4(x))$
$= f_5\left(\frac{1}{x}\right)$
$= 1 - \frac{1}{x}$
$= \frac{x-1}{x}$
$= f_3(x)$

$(f_5 \circ f_5)(x)$
$= f_5(f_5(x))$
$= f_5(1-x)$
$= 1-(1-x)$
$= x$
$= f_1(x)$

$(f_5 \circ f_6)(x)$
$= f_5(f_6(x))$
$= f_5\left(\frac{x}{x-1}\right)$
$= 1 - \frac{x}{x-1}$
$= \frac{x-1-x}{x-1}$
$= \frac{-1}{x-1}$
$= \frac{1}{1-x}$
$= f_2(x)$

$(f_6 \circ f_1)(x)$
$= f_6(f_1(x))$
$= f_6(x)$

$(f_6 \circ f_2)(x)$
$= f_6(f_2(x))$
$= f_6\left(\frac{1}{1-x}\right)$
$= \frac{\frac{1}{1-x}}{\frac{1}{1-x} - 1}$
$= \frac{1}{1-(1-x)}$
$= \frac{1}{x}$
$= f_4(x)$

$(f_6 \circ f_3)(x)$
$= f_6(f_3(x))$
$= f_6\left(\frac{x-1}{x}\right)$
$= \frac{\frac{x-1}{x}}{\frac{x-1}{x} - 1}$
$= \frac{x-1}{x-1-x}$
$= 1-x$
$= f_5(x)$

$(f_6 \circ f_4)(x)$
$= f_6(f_4(x))$
$= f_6\left(\frac{1}{x}\right)$
$= \frac{\frac{1}{x}}{\frac{1}{x} - 1}$
$= \frac{1}{1-x}$
$= f_2(x)$

$(f_6 \circ f_5)(x)$
$= f_6(f_5(x))$
$= f_6(1-x)$
$= \frac{1-x}{1-x-1}$
$= \frac{1-x}{-x}$
$= \frac{x-1}{x}$
$= f_3(x)$

$(f_6 \circ f_6)(x)$
$= f_6(f_6(x))$
$= f_6\left(\frac{x}{x-1}\right)$
$= \frac{\frac{x}{x-1}}{\frac{x}{x-1} - 1}$
$= \frac{x}{x-(x-1)}$
$= x$
$= f_1(x)$

Thus the Cayley table is:

$\circ$	f_1	f_2	f_3	f_4	f_5	f_6
f_1	f_1	f_2	f_3	f_4	f_5	f_6
f_2	f_2	f_3	f_1	f_6	f_4	f_5
f_3	f_3	f_1	f_2	f_5	f_6	f_4
f_4	f_4	f_5	f_6	f_1	f_2	f_3
f_5	f_5	f_6	f_4	f_3	f_1	f_2
f_6	f_6	f_4	f_5	f_2	f_3	f_1

Closure: The elements of the table consist of all elements of S.
$\therefore$ S is closed under $\circ$.

Associative: $\circ$ is associative on S
{as function composition is associative}

Identity: The identity function is f_1 as
$f_1 \circ f_i = f_i \circ f_1 = f_i$ for $i = 1, 2, 3, 4, 5, 6$.

Inverse: From the table:
$f_1^{-1} = f_1$
$f_2^{-1} = f_3$, $f_3^{-1} = f_2$ as $f_2 \circ f_3 = f_3 \circ f_2 = f_1$
$f_4^{-1} = f_4$, $f_5^{-1} = f_5$, $f_6^{-1} = f_6$ as $f_i \circ f_i = f_1$ for $i = 4, 5, 6$
$\therefore$ every member of S has a unique inverse in S.

Hence $\{S, \circ\}$ is a group.

6 **a** $a * b = \frac{a+b}{a^2}$

No, for example,

$(1*2)*1$ whereas $1*(2*1) = 1 * \frac{3}{4}$
$= \frac{3}{1} * 1$ $\qquad = \frac{1\frac{3}{4}}{1}$
$= 3*1$
$= \frac{4}{9}$ $\qquad = \frac{7}{4}$

So, in general, $(a*b)*c \neq a*(b*c)$.

b $a * b = 2^{a+b}$

No, for example $(0*1)*2$ whereas $0*(1*2)$
$= 2^1 * 2$ $\qquad = 0 * 2^3$
$= 2*2$ $\qquad = 0*8$
$= 2^{2+2}$ $\qquad = 2^8$
$= 2^4$

So, in general, $(a*b)*c \neq a*(b*c)$.

c $a * b = a + b - 3ab$

$(a*b)*c = (a+b-3ab)*c$
$= a+b-3ab+c-3(a+b-3ab)c$
$= a+b-3ab+c-3ac-3bc+9abc$
$= a+b+c-3ab-3ac-3bc+9abc$

and $a*(b*c) = a*(b+c-3bc)$
$= a+b+c-3bc-3a(b+c-3bc)$
$= a+b+c-3bc-3ab-3ac+9abc$
$= (a*b)*c$, for all $a, b, c \in \mathbb{R}$

So, $*$ is associative on $\mathbb{R}$.

7 **a** $p = \begin{pmatrix} 1 & 2 & 3 & 4 & 5 & 6 \\ 2 & 6 & 1 & 4 & 3 & 5 \end{pmatrix}$
$= (1\ 2\ 6\ 5\ 3)$

$\therefore$ $p^{-1} = (3\ 5\ 6\ 2\ 1) = (1\ 3\ 5\ 6\ 2)$

b $p = (1\ 2\ 6\ 5\ 3)$ has order 5
$\therefore$ $p^5 = e$
$\therefore$ $p^6 = p = (1\ 2\ 6\ 5\ 3)$

8 Let $S = \{a, b, c, d, e\}$

a Consider $\mathbb{Z}_5 = \{0, 1, 2, 3, 4\}$ under $+_5$, addition modulo 5.

The Cayley table is:

$+_5$	0	1	2	3	4
0	0	1	2	3	4
1	1	2	3	4	0
2	2	3	4	0	1
3	3	4	0	1	2
4	4	0	1	2	3

Closure: From the table, for all $a, b \in \mathbb{Z}_5$, $a +_5 b \in \mathbb{Z}_5$
$\therefore$ $\mathbb{Z}_5$ is closed under $+_5$.

Associative: Associativity follows from the associative property of $+$ in $\mathbb{Z}$.

Identity: The identity is 0 in $\mathbb{Z}_5$, as $0 +_5 a = a +_5 0 = a$ for all $a \in \mathbb{Z}_5$.

Inverse: $0^{-1} = 0,\ 1^{-1} = 4,\ 2^{-1} = 3,\ 3^{-1} = 2$, and $4^{-1} = 1$

$\therefore$ each element in $\mathbb{Z}_5$ has a unique inverse.

$\therefore$ $\mathbb{Z}_5$ is a group under $+_5$.

The given table has the same structure as $\mathbb{Z}_5$ under $+_5$

$0 \mapsto a$
$1 \mapsto b$
$2 \mapsto c$
$3 \mapsto d$
$4 \mapsto e$

$*$	a	b	c	d	e
a	a	b	c	d	e
b	b	c	d	e	a
c	c	d	e	a	b
d	d	e	a	b	c
e	e	a	b	c	d

We have an isomorphism between $\{\mathbb{Z}_5, +_5\}$ which we know is a group, and $\{S, *\}$.

$\therefore$ $\{S, *\}$ is a group.

b $*$ is not associative on S.

For example, $(b * c) * d = d * d = b$ and $b * (c * d) = b * e = c$

and $b \neq c$

$\therefore$ S is not a group on $*$.

9 **a** $a * b = a \times_6 (a + b)$

$*$	0	1	2	3	4	5
0	0	0	0	0	0	0
1	1	2	3	4	5	0
2	4	0	2	4	0	2
3	3	0	3	0	3	0
4	4	2	0	4	2	0
5	1	0	5	4	3	2

b No, as for example, in row 1 all of 0, 1, 2, 3, 4, 5 are not represented.

c No, as for example, there is no identity.

10 **a** $(a, b, c) * (x, y, z)$

$= (a + x \pmod m,\ b + y \pmod m,\ c + z - xb \pmod m)$

Closure: $a + x \pmod m \in A$, $b + y \pmod m \in A$, and $c + z - xb \pmod m \in A$

$\therefore$ $(a, b, c) * (x, y, z) \in G$

$\therefore$ G is closed under $*$.

{We will now write $(p \pmod m,\ q \pmod m,\ r \pmod m)$ as (p, q, r).}

Associative: $[(a, b, c) * (x, y, z)] * (e, f, g)$

$= (a + x,\ b + y,\ c + z - xb) * (e, f, g)$

$= (a + x + e,\ b + y + f,\ c + z - xb + g - e(b + y))$

$= (a + x + e,\ b + y + f,\ c + z + g - xb - eb - ey)$

and

$(a, b, c) * [(x, y, z) * (e, f, g)]$

$= (a, b, c) * (x + e,\ y + f,\ z + g - ey)$

$= (a + x + e,\ b + y + f,\ c + z + g - ey - (x + e)b)$

$= (a + x + e,\ b + y + f,\ c + z + g - xb - ey - eb)$

$= [(a, b, c) * (x, y, z)] * (e, f, g)$

$\therefore$ $*$ is associative on G.

Identity: $(0, 0, 0)$ is the identity as

$(a, b, c) * (0, 0, 0)$

$= (a + 0,\ b + 0,\ c + 0 - 0(b))$

$= (a, b, c)$

Likewise $(0, 0, 0) * (a, b, c) = (a, b, c)$.

Inverse: Suppose $(a, b, c) * (x, y, z) = (0, 0, 0)$

$\therefore$ $(a + x,\ b + y,\ c + z - xb) = (0, 0, 0)$

So, $a + x \pmod m = 0$

$\therefore$ $x = m - a$ $\{x \in A\}$

$b + y \pmod m = 0$

$\therefore$ $y = m - b$ $\{y \in A\}$

$c + z - xb \pmod m = 0$

$\therefore$ $z = xb - c \pmod m$

$= (m - a)b - c \pmod m$

$\therefore$ $(a, b, c)^{-1}$

$= (m - a,\ m - b,\ (m - a)b - c)$

Check:

$(m - a,\ m - b,\ (m - a)b - c) * (a, b, c)$

$= (m - a + a,\ m - b + b,\ mb - ab - c + c - a(m - b))$

$= (m,\ m,\ mb - ab - am + ab)$

$= (m,\ m,\ m(b - a))$

$= (0, 0, 0)$

b $(a, b, c) * (x, y, z) = (a + x,\ b + y,\ c + z - xb)$

$(x, y, z) * (a, b, c) = (x + a,\ y + b,\ z + c - ay)$

These are equal $\Leftrightarrow bx = ay$

$\Leftrightarrow \dfrac{x}{y} = \dfrac{a}{b}$ which is not true in general.

$\therefore$ $\{G, *\}$ is not Abelian.

c a can take m values $\{0, 1, 2, 3, ..., m - 1\}$

b likewise has m values

c likewise has m values

$\therefore$ the order of G, $|G| = m^3$.

11 As G is **associative** on $*$ and $*$ is **closed** on G it remains to be proved that: (1) G has an identity, and (2) every element of G has an inverse in G.

Given: $x * a = b$ and $a * y = b$ have unique solutions in G.

(1) *Identity*: Suppose $b * a = c * a = d$ say, $d \in G$.

Since $x * a = d$ has a unique solution

$x = b = c$

$\therefore$ $b * a = c * a \Rightarrow b = c$

This is right cancellation.

Similarly,

$a * b = a * c \Rightarrow b = c$ (left cancellation).

Now suppose

$a * x = a$

$\therefore$ $a * x * a = a * a$ {right $\times$ by a}

$\therefore$ $a * x * a = a * a * x$ $\{a * x = a\}$

$\therefore$ $x * a = a * x$ {left cancellation}

$\therefore$ $x * a = a * x = a$

and x is a unique solution to $y * a = a$.

Thus for each $a \in G$, there is a unique $x \in G$ such that $a * x = x * a = a$.

Now suppose for $a, b \in G$

$a * x = x * a = a$ **and** $b * x' = x' * b = b$

and also suppose z is the unique solution to $a * z = b$

$\therefore \quad (x*a)*z = x'*b \quad \{a = x*a,\ b = x'*b\}$

$\therefore \quad x*a*z = x'*b$

$\therefore \quad x*b = x'*b \quad \{a*z = b\}$

$\therefore \quad x = x' \quad$ {right cancellation}

Thus the element $x \in G$ such that $a*x = x*a = a$ is the same for all $a \in G$.

$\therefore$ G has identity $x = e$.

(2) *Inverse*: For $a \in G$, let x be the unique solution to $a*x = e$.

$\therefore \quad a*x*a = e*a = a \quad$ {right $\times$ by a}

$\therefore \quad a*x*a*e = a*e \quad$ {e the identity}

$\therefore \quad x*a*e = e \quad$ {left cancellation}

$\therefore \quad x*a = e \quad$ {identity}

Thus $a*x = x*a = e$, and so each element in G has a unique inverse.

As the group axioms are satisfied, $\{G, *\}$ is a group.

12 Let $\dfrac{2a+1}{2b+1}$ and $\dfrac{2c+1}{2d+1}$ be in $\mathbb{Q}\setminus\{0\}$.

$$\text{Then } \frac{2a+1}{2b+1} \times \left(\frac{2c+1}{2d+1}\right)^{-1} = \frac{2a+1}{2b+1} \times \frac{2d+1}{2c+1} = \frac{4ad+2a+2d+1}{4bc+2b+2c+1} = \frac{2(2ad+a+d)+1}{2(2bc+b+c)+1}$$

where $2ad+a+d,\ 2bc+b+c \in \mathbb{Z}$

{as $\mathbb{Z}$ is closed under $\times$ and $+$}

$\therefore$ $\dfrac{2a+1}{2b+1} \times \left(\dfrac{2c+1}{2d+1}\right)^{-1}$ is a member of

$S = \{$rationals of the form $\dfrac{2a+1}{2b+1}\}$, which is a non-empty subset of $\mathbb{Q}\setminus\{0\}$.

$\therefore$ by the subgroup test, S is a subgroup of $\{\mathbb{Q}\setminus\{0\}, \times\}$.

13 **a** **i**

$\times_{20}$	1	9	11	19
1	1	9	11	19
9	9	1	19	11
11	11	19	1	9
19	19	11	9	1

ii

$\times_{20}$	1	3	7	9
1	1	3	7	9
3	3	9	1	7
7	7	1	9	3
9	9	7	3	1

iii

$\times_{20}$	1	9	13	17
1	1	9	13	17
9	9	1	17	13
13	13	17	9	1
17	17	13	1	9

b **i** $S = \{1, 9, 11, 19\}$ is **closed** under $\times_{20}$ as the Cayley table contains only elements of S.

$\times_{20}$ is **associative** on S as $\times$ is associative on $\mathbb{Z}^+$.

The **identity** is 1 as

$a \times_{20} 1 = 1 \times_{20} a = a$ for all $a \in S$.

Each element of S has a unique **inverse**;

$1^{-1} = 1,\ 9^{-1} = 9,\ 11^{-1} = 11,\ 19^{-1} = 19$.

$\therefore$ $\{S, \times_{20}\}$ is a group.

ii $S = \{1, 3, 7, 9\}$ is **closed** under $\times_{20}$ as the Cayley table contains only elements of S.

$\times_{20}$ is **associative** on S as $\times$ is associative on $\mathbb{Z}^+$.

The **identity** is 1 as

$a \times_{20} = 1 \times_{20} a = a$ for all $a \in S$.

Each element of S has a unique **inverse**;

$1^{-1} = 1,\ 3^{-1} = 7,\ 7^{-1} = 3,\ 9^{-1} = 9$.

$\therefore$ $\{S, \times_{20}\}$ is a group.

iii $S = \{1, 9, 13, 17\}$ is **closed** under $\times_{20}$ as the Cayley table contains only elements of S.

$\times_{20}$ is **associative** on S as $\times$ is associative on $\mathbb{Z}^+$.

The **identity** is 1 as

$a \times_{20} 1 = 1 \times_{20} a = a$ for all $a \in S$.

Each element has a unique **inverse**;

$1^{-1} = 1,\ 9^{-1} = 9,\ 13^{-1} = 17,\ 17^{-1} = 13$.

$\therefore$ $\{S, \times_{20}\}$ is a group.

c The groups in **ii** and **iii** are cyclic groups of order 4 and are therefore isometric.

The group in **i** is not cyclic since no element has order 4.

{This group is isomorphic to the Klein 4-group.}

14 **a** **i** Since 0 is even, $0 \in G$

$\therefore$ $0G = G$

$\therefore$ $0G = \{2\mathbb{Z}, +\} = \{$even integers$\}$

$1G = 1\{2\mathbb{Z}, +\}$

$= \{...., 1+(-4), 1+(-2), 1+0, 1+2, 1+4, ...\}$

$= \{...., -3, -1, 1, 3, 5, ...\}$

$= \{$odd integers$\}$

ii $\{\mathbb{Z}, +\}$ is the disjoint union of all the left cosets of $\{2\mathbb{Z}, +\}$. Since $\mathbb{Z} = \{$even integers$\} \cup \{$odd integers$\}$, the two left cosets found in **a i** are all the (distinct) left cosets of $\{2\mathbb{Z}, +\}$ in $\{\mathbb{Z}, +\}$.

iii Only the left coset which is the subgroup itself, is a subgroup of $\{\mathbb{Z}, +\}$. Hence $1G$ cannot be a subgroup.

$\therefore$ the set of odd integers cannot be a subgroup of $\{\mathbb{Z}, +\}$.

Alternatively, the identity $0 \notin \{$odd integers$\}$, so it cannot be a subgroup.

b **i** $0G = G$

$= \{...., -3n, -2n, -n, 0, n, 2n, 3n, ...\}$

$= [0]$

$1G = \{...., 1+(-3n), 1+(-2n), 1+(-n), 1+0, 1+n, 1+2n, ...\}$

$= \{...., 1-3n, 1-2n, 1-n, 1, n+1, 2n+1, ...\}$

$= [1]$

Similarly,

$2G = [2],\ 3G = [3],\ ....,\ (n-1)G = [n-1]$

$\therefore$ we obtain the n residue classes modulo n.

ii The cosets found in **b i** are pairwise disjoint, and a careful look shows that their union gives all the integers.

For example:

$0 \in 0G,\ 1 \in 1G,\ ....,\ n-1 \in (n-1)G$

$n \in 0G,\ n+1 \in 1G,\ ....,$ and so on

So, there are n distinct left cosets of G in $\{\mathbb{Z}, +\}$.

iii The n residue classes modulo n are precisely the n left cosets of G listed in **b i**.

Since the distinct left cosets of G partition the group $\{\mathbb{Z}, +\}$, it follows that the n residue classes modulo n partition $\mathbb{Z}$.

INDEX

Abelian group *58, 86, 98*
associative *15, 46*
bijection *37*
binary operation *44*
cancellation laws *56*
cardinal number *9*
cardinality *9*
Cartesian product *20*
Cayley table *53*
closed *44*
codomain *34*
commutative *47*
complement *15*
complex number *9*
composite function *40*
composition of permutations *68, 73*
congruence *30*
coset *103*
cycle *73*
cycle notation *73*
cyclic group *86, 99*
De Morgan's Laws *16*
difference between sets *17*
dihedral group *79*
direct proof *115*
disjoint set *14*
distributive *15, 48*
domain *21, 34*
empty relation *24*
empty set *10*
equal sets *10, 12*
equivalence *12, 120*
equivalence class *25*
equivalence relation *24*
finite group *62*
finite set *9*
function *34*
generator *86*
group *55*
groups of order n *106*
homomorphism *92*
horizontal line test *36*
identity *49, 69*
infinite group *62*
infinite set *9*
injection 35
integer *9*
intersection of sets *13*
inverse *50*
inverse function *41*
inverse permutation *69, 74*
irrational number *9*
isomorphic *96*
isomorphism *92, 96*
kernel *92*
Klein 4-group *67, 83*
Lagrange's theorem *105*
Latin square *53, 57*
left coset *103*
left identity *61*
left inverse *61*
natural number *9*
one-to-one *35*
onto *37*
order *61, 71, 74, 88, 105*
ordered pair *20*
partition *14*
permutation *68*
power set *11*
proof by contradiction *116*
proper subset *11*
range *21, 34, 92*
real number *9*
reflexive relation *21*
relation *21*
residue class *29*
right coset *103*
set *9*
subgroup *80, 94, 99*
subgroup test *82*
subset *11*
surjection *37*
symmetric difference *18*
symmetric group *70*
symmetric relation *22*
symmetries of figures *77*
transitive relation *23*
trivial subset *11*
union of sets *13*
universal set *10*
Venn diagram *11*
vertical line test *35*
well-defined *9*